Systems Maintainability

Systems Maintainability

Analysis, engineering and management

Jezdimir Knezevic

*Research Centre for Management of Industrial
Reliability, Cost and Effectiveness,
School of Engineering,
University of Exeter, UK*

**Published by Chapman & Hall, 2–6 Boundary Row,
London SE1 8HN, UK**

Chapman & Hall, 2–6 Boundary Row, London SE1 8HN, UK

Chapman & Hall GmbH, Pappelallee 3, 69469 Weinheim, Germany

Chapman & Hall USA, 115 Fifth Avenue, New York NY 10003, USA

Chapman & Hall Japan, ITP-Japan, Kyowa Building, 3F, 2-2-1
Hirakawacho, Chiyoda-ku, Tokyo 102, Japan

Chapman & Hall Australia, 102 Dodds Street, South Melbourne, Victoria
3205, Australia

Chapman & Hall India, R. Seshadri, 32 Second Main Road, CIT East,
Madras 600 035, India

First edition 1997

© 1997 J. Knezevic

Typeset in 10/12 pt Palatino by Cambrian Typesetters, Frimley, Surrey

Printed in Great Britain by St Edmundsbury Press, Bury St Edmunds

ISBN 0 412 80270 8

A catalogue record for this book is available from the British Library

To Alan Mulally, Jack Hessburg and Clive Irving, individuals who I have never met, but who considerably influenced this book.

Contents

Contents

Preface

Maintenance managers want a clean gate –
their report card in line maintenance is
based on having a clean gate and not
having pigeons roosting on the airplane's
fin.

(J. Hessburg, Chief Mechanic New Airplanes,
Boeing, in *Aviation Equipment Maintenance*,
May 1994)

The main objectives of this book are to address systems maintainability, through analytical, engineering and management procedures, tools and techniques. This has been achieved through five main parts of the text, namely:

(1) Concept of Maintainability
(2) Maintenance Analysis
(3) Maintainability Analysis
(4) Maintainability Engineering
(5) Maintainability Management.

In order to make the presentation easier, specific meanings are attached to certain words throughout this book. Thus **item** is used as a generic term for a product, module, subsystem, component and similar, when analysed as a single entity, and **system** is used as a generic term for products which are analysed as a set of many consisting items. This means that the word item will be used for a car in the same way as for an engine, a carburettor and a distributor cap, when treated as a single object. On the other hand, the word system will be used for a car when it is treated as a set of consisting objects such as the engine, transmission, body, brakes, etc.

As the main benefit of maintainability studies is the improvement of maintenance processes, which has tremendous impact on availability of system, safety, operational effectiveness and cost of ownership, the importance of maintainability is analysed in Chapter 1.

The relationship between maintainability, systems effectiveness and life cycle process has been addressed in Chapter 2.

Chapters 3, 4, 5 and 6 analyse specific aspects of maintenance tasks, like classification, duration, cost and frequency, as the main building blocks of maintenance processes.

Corresponding issues for the maintenance process have been covered in Chapters 7, 8, 9 and 10.

The quantitative analysis of maintainability characteristics of items and systems have been fully defined in Chapters 11 and 12.

The relationship between engineering design function and maintainability engineering function has been addressed in Chapter 13, with the objective of promotion of an integrated approach to the complexity of the design process. This is extremely important because, at present, members of the maintainability team, quite understandably, see themselves as the most important members of the design team, whereas people outside that department have never heard of them, or if they are aware of their existence they are not sure what they do and why they are needed. It is my hope that everybody who reads this book will clearly realize the error in both views, as well as the necessity for the existence of both groups and the need for close cooperation between them. Examples of the TGV train, Boeing 777, Airbus 330 and similar are only used to illustrate the potential for possible achievements when maintainability and maintenance personnel are recognized and respected members of the design team.

Maintainability tools, techniques and standards used in engineering design for maintainability allocation are analysed in Chapter 14.

Anthropometric evaluation for maintainability is dealt with in Chapter 15, whereas the impact of a new technology on testability, as an important part of maintainability analysis, is analysed in Chapter 16.

A brief overview of condition monitoring techniques is presented in Chapter 17.

Chapter 18 covers a method for the prediction of maintainability measures based on sequential, simultaneous and complex timing of execution of maintenance activities using corresponding measures of the consisting activities.

In Chapter 19, prediction of maintainability statistics is covered by detailed analysis of mathematical renewal theory, as the main tool for the prediction of the maintenance demands for unscheduled maintenance tasks.

Maintainability management issues are dealt with in Chapter 20.

The maintenance model used to assess the optimal maintenance procedure for the whole system, based on the ongoing study by the Research Centre MIRCE of Exeter University, is presented in Chapter 21.

Methods for the selection of the most suitable probability distribution for the modelling of the maintenance task duration are presented in Chapter 22.

Finally, Chapter 23 covers the methods for the analysis of existing empirical data obtained from maintainability demonstration tests.

In order to make this work beneficial to practising engineers and managers, large numbers of 'real-life examples' have been used in the text. It is necessary to stress that the products and manufacturers mentioned in this book have been selected for no other reason than the availability of the information in the literature or simply due to my familiarity with that particular system. Under no circumstances is the use of practical examples intended to promote or demote any of them, nor their producers, or users.

As a mechanical engineer who has spent over 15 years dealing with reliability, maintainability and supportability, theoretically and practically, I would be extremely happy if my engineering and analytical experience, as summarized in this book, can be of value to all maintainability practitioners and students alike.

J. Knezevic
Bickleigh, September 1996

Acknowledgements

My unreserved thanks go to all undergraduate and postgraduate students, together with practitioners from industry, who have through the years shaped my research direction by providing me with the 'feedback' necessary for its continuous improvement relative to systems maintainability, analysis, engineering and management issues.

I would like to extend my sincere thanks to: R. Giffin (Lockheed Martin Federal Systems, Manassas), S. Hunt (Aspire Consulting Ltd, UK), J. Kneepkens (NATO Maintenance and Supply Agency, NAMSA, Luxembourg), R. Knotts (AVRO International, UK), K. Miller (British Aerospace, Military Aircraft Division, Warton, UK), C. Nicholas (Research Centre MIRCE, University of Exeter), A. Sols (Isdefe, Spain), the late P. Sparks (Logistics Support Consultants, UK), Dr D. Verma (Loral Defense Systems, USA) and I. Watson (British Aerospace, Farnborough, UK) for their contribution to the final preparation of the manuscript.

Also, I extend my appreciation to Thelma Filbee, secretary of the Research Centre for Management of Industrial Reliability, Cost and Effectiveness (MIRCE), University of Exeter, for typing and printing many versions of this book, which also included a few 'working Saturdays'.

For her tolerance, I thank Lynn who successfully maintained my soul, body and study room during the creation of this book.

I am deeply grateful to Mr Eugene Malnik, who as a devoted member of the Society of Logistics Engineers, SOLE, frequently helped me and my students with real-life examples based on his long-standing and successful work in the Maintainability Division, Boeing, Seattle, USA. As a tribute to his professionalism I wish to include a part of his letter to my undergraduate student P. Hodson, in which he has highlighted the Boeing approach to systems maintainability and maintenance of commercial aircraft. I feel that this letter could be successfully used as the preface of any book related to maintainability, and that makes me even happier, thus:

The latest commercial aircraft produced by Boeing Company, Boeing 777, has been designed for a useful life of 20 years. Boeing recommends and the authorities of the FAA and JAA decide what maintenance is required to keep the airplane airworthy while in service. This involves defining what minimum scheduled and unscheduled maintenance must be performed in order to continue flying. Scheduled maintenance is performed at certain intervals that are tied to number of flight hours, number of cycles (such as turn-on/off, take-offs and landings), etc. It consists primarily of inspections followed by maintenance, corrosion prevention, etc. Unscheduled maintenance is performed after a failure occurs. Depending upon the criticality of the failure, maintenance is accomplished either before the airplane is returned to revenue service or within a specified interval.

When total cost is considered over the life cycle, it is evident that the operating and support costs of the airplane will eventually exceed the initial acquisition cost. In order for Boeing to make the airplane attractive to the airlines, the engineers must include maintenance cost savings in the design. This was done by increasing the reliability and maintainability. Increased reliability means fewer failures to fix. Increased maintainability means shorter maintenance times.

The figure of merit chosen to measure reduction of the follow-on costs was schedule reliability. In other words, how often will the airplane, or fleet of airplanes, meet the scheduled take-off time? The target for initial delivery is 97.8% with improvement to 98.8% after fleet maturity. In order for the airplane to meet such a high number it must be inherently reliable. Double and triple redundancy is used in critical areas, allowing deferral of maintenance to an overnight time while the back-up system or systems keep the plane flying until that time.

Maintenance must be able to be completed during the scheduled downtimes, whether it be during a 45 minute turnaround between flights or during an overnight. This implies having good means of identification and isolation of failures, as well as good access to the equipment. An innovative computer aided human model was used to prove good maintenance access without the use of expensive mock-ups. Fault identification and isolation is enhanced with the use of extensive built in testing with fault messages displayed on the computer screens available to the mechanics. Great care was used to ensure that maintenance messages are prioritized, understandable, do not give extraneous information, and are accurate. Accompanying fault isolation and maintenance manuals complement this information.

Reliability requirements were passed along to equipment manufacturers by specifying mean time between failures (*MTBF*) and target mean time between unscheduled removals (*MTBUR*). The latter was estimated to be between 0.8 and 0.9 of *MTBF*, but could be verified only by service experience. It was recognized that unscheduled removals also counted the times that equipment is wrongfully removed because of the haste that a gate mechanic expends in trying to clear a fault during a 45 minute turnaround. The tendency is to replace the first suspected unit or groups of units in order to eliminate the obvious faults from the process. Thus the maintenance messages must give the right information that avoids removing good items. Specifying both *MTBF* and *MTBUR* means both inherent reliability and field reliability could be controlled.

For fault tolerant systems or items, the reliability index was mean time between maintenance alerts (*MTBMA*). Maintenance alerts are the maintenance messages that are documented on equipment internal failure that did not immediately affect function.

Boeing also documented 'lessons learned' data to record service history and feedback from other airplanes in order to avoid the same mistakes in the design of the new airplane. The airline representatives stayed in touch by attending design reviews and other meetings of concurrent engineering teams. From time to time their field mechanics visited Boeing to provide their inputs. The result was a working together relationship that benefited both sides and will result in increased reliability and maintainability.

PART ONE

Concept of Maintainability

CHAPTER 1

Introduction

Everything that the human race has done and thought is concerned with the satisfaction of felt needs and the assuagement of pain.

(Einstein, 1991)

In daily life needs for food, transport, entertainment, cure from illness, shelter and protection from natural forces, hot water, fast and accurate calculation, communication, bridging of rivers and canyons, and many other things are clearly manifested.

Human-made solutions to these needs include: bread, cars, aircraft, ships, hotels, radios, TV sets, medicines, houses, hotels, cookers, kettles, slide rules, calculators, computers, radars, telephones, pencils, bridges, tunnels, motorways, and so forth.

Although the felt needs cover a very large spectrum of solutions, the word **system** is commonly used as a generic name for all of them. The most commonly used systems in daily life are:

- *Aeronautical and aerospace*: which are created to satisfy the need for air transport, through supersonic and subsonic aircraft (jet, vertical take-off), spacecraft, missiles, rockets, remotely piloted vehicles, spacelabs and similar.
- *Agricultural*: which satisfy the need for production, processing, handling and storage of food, and related products such as tractors, combines, barns, silos, granaries, processing buildings, freezers and many others.
- *Structural*: which are created to satisfy needs for large office buildings, manufacturing plants, sporting arenas, housing complexes and so on.
- *Chemical process and processing*: which facilitate production of chemicals such as plastics, paints, synthetics, alkalis, dyes, polymers, insecticides, fungicides, oil, fuel and many other comparable outputs.
- *Civil engineering*: like highways, bridges, tunnels, dams, canals, waterways, sanitary and sewage treatment, disposal, water and gas networks, airports, railway stations, hotels, shopping centres and many others, which are created as a result of specific human needs.
- *Electrical and electronic*: which are created in response to the need for creation, transfer and utilization of energy. This includes electrical

power systems, control systems, computer systems, communication systems, electronic systems (radar, navigation, fire control, missile guidance, signal processing equipment, etc.), electro-optical devices and instrumentation appliances, as well as small electrical/electronic components (transistors, semiconductors, switches, etc.).

- *Mechanical*: the main tasks of which are to convert energy into useful mechanical forms. This covers both power-generating machines and machines that transform or consume this power in order to perform a particular function. Typical examples of mechanical systems are: engines, turbines, motors, control mechanisms, transportation systems (cars, bicycles, trains, space vehicles, etc.), refrigeration and air-conditioning systems, propulsion systems (steam, gas, nuclear) and cryogenic systems.
- *Metallurgical, mining and materials*: which are created as a response to the need for dealing with various forms and applications of metals, alloys and materials in general. These are related to the initial location and evaluation of various materials from earth, the accomplishment of land reclamation after the mining and extraction functions have been completed and conversion of commodities (ores) into basic metals or comparable alloys. Some of the tasks accomplished by these systems are related to changing the chemical or physical characteristics of metals (extrusion, reforming, hardening and similar).
- *Nuclear*: which are created to deal with all aspects of fission and fusion reactions (initiation, control of reactive materials, storage, disposal, decontamination) in order to provide power generation and medical applications.
- *Ocean, marine and nautical*: which are manifested through the existence of ships, submarines, hydrofoils, underwater sea laboratories, sea towers and similar marine structures.
- *Petroleum*: which basically deal with the need for exploration, location, development and recovery of petroleum resources through tasks like drilling, separation, processing, transportation and storage of crude oil, gases and related products.

In summary, it could be said that the only common characteristic among all the above-mentioned systems is their ability to satisfy a felt need by performing a specific function.

1.1 CONCEPT OF A SYSTEM

A system is a set of mutually related items, brought together in order to perform a specific function.

A system is a set of interrelated items which always has some characteristic or behaviour pattern that cannot be exhibited by any of its subsets. A system is more than the sum of its item parts. However, the items of a system may themselves be systems, and every system may be part of a larger system in a hierarchy.

Systems consist of items, attributes and relationships, which are described as follows.

- *Items*: the operating parts of a system consisting of input, process and output.
- *Attributes*: the properties of the items of a system.
- *Relationships*: the links between items and attributes.

The systems viewpoint looks at a system from the top down rather than from the bottom up. Attention is first directed to the system as a black box that interacts with its environment. Secondly, attention is focused on how the smaller black boxes (subsystems) combine to achieve the system objective. The lowest level of concern is then with individual items. Addressing systems, subsystems and items in a hierarchy forces consideration of all necessary relationships. Items and attributes are important, but only in order to achieve the ultimate purpose of the whole system through the functional relationships linking them.

In any particular situation it is important to define the system under consideration by specifying its limits or boundaries. Everything that remains outside the boundaries of the system is considered to be the environment. However, no system is completely isolated from its environment. Material, energy and/or information must often pass through the boundaries as input to the system. In reverse, material, energy and/or information that passes from the system to the environment is called output. Thus, generally speaking, the throughput is that which enters the system in one form and leaves the system in another form.

The total system, at whatever level in the hierarchy, consists of all items, attributes and relationships needed to accomplish a function. Each system has a purpose for which all system items, attributes and relationships have been organized. Constraints placed on the system limit its operation and define the boundary within which it is intended to operate. Similarly, the system places boundaries and constraints on its subsystems.

At any level of hierarchy of any complex industrial system there are inputs and outputs. The outputs of one subsystem of an item are the inputs to another. A system receives inputs from the environment and makes outputs to the environment. It is through inputs and outputs that all the items interact and communicate. Inputs can be physical entities, such as materials and products, electrical impulses, mechanical forces or information.

1.2 CONCEPT OF FUNCTIONABILITY

Functionality is the most important characteristic of any human-made system and it is related to its inherent ability to perform a specific function. For example, a kettle is a human-made system which satisfies a need for heating water to boiling point.

A given system is not only expected to peform a specified function but it is also expected to meet specified requirements, commonly known as performance and attributes. Thus a kettle which needs, say, 45 minutes to heat a litre of water to boiling point does not exhibit a satisfactory performance, for certain users. The expression satisfactory performance is a common description for the requirements which the system has to satisfy while performing the specified function. Most often performance requirements are related to size, weight, volume, shape, capacity, flow rate, speed, acceleration, and many other physical and operational characteristics. At the same time the kettle should possess satisfactory attributes such as attractiveness, ease of operation, cleaning and handling, and so forth. Most of these are quantifiable numerically, but still there are some which are only describable in a qualitative form. In either case they are usually well defined and demanded.

It is also necessary to specify the operating conditions under which the system is expected to be used. In the case of a kettle, the operating conditions are primarily related to the mains supply voltage, vibrations, humidity and similar factors.

Therefore, for the 'satisfaction of felt needs and assuagement of pain' the aspects of functionality, performance, attributes and operating conditions have to be brought together in order to obtain a full picture about the system which can satisfy a felt need. This has been done by Knezevic (1993), where the concept of functionability has been intro- duced as an embracing mechanism of all three aspects, as shown in Figure 1.1. Thus, functionability is defined (Knezevic, 1993) as:

> the inherent ability of an item/system to perform a required function with specified performance and attributes, when it is utilised as specified.

In the above definition the word 'inherent' is used to stress that all decisions related to the functionability of a system are made at the design phase. For instance, a functionable motor vehicle is one which performs a transportation function, satisfying specific performance such as speed, fuel and oil consumption, acceleration, load (number of passengers and luggage), riding comfort, and many other features when it is used under specified operating conditions (type of road surface, terrain configuration, outside temperature, fuel grade, etc.). Of course, in reality the list is much longer and more exhaustive, because the real 'felt needs' are more defined and tuned to the higher level of precision.

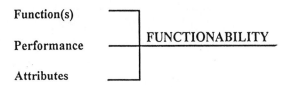

Figure 1.1 Concept of functionability.

From the above definition it is clear that there is a significant difference between functionality and functionability of a system considered. The former is related purely to the function performed whereas the latter takes into consideration the level of performance obtained. For example, most worn-out engines consume a larger quantity of oil than when they were introduced into service. Thus, despite the fact that the motor vehicles with worn-out engines still perform the transportation function, the level of performance achieved (fuel and oil consumption, maximum speed, and similar) is not at the level of the new system.

1.3 CONCEPT OF A FUNCTIONABILITY PROFILE

In spite of the fact that a system is functionable at the beginning of its operational life, every user is fully aware that irrespective of how perfect the design of a system may be, or the technology of its production or the materials from which it is made, during its operation certain irreversible changes will occur. These changes are the result of processes such as corrosion, abrasion, accumulation of deformations, distortion, overheating, fatigue, diffusion of one material into another, and similar. Often these processes superimpose on each other; they interact with each other and cause a change in the system, as a result of which its output characteristics will change. The deviation of these characteristics from the specification values is considered a failure.

The failure of the system can therefore be defined as an event whose occurrence results in either the loss of ability to perform the required functions or the loss of ability to satisfy the specified requirements. Regardless of the reason for its occurrence, a failure will cause the transition of a system from the satisfactory state to a new, unsatisfactory state, known as the state of failure.

Thus, from the point of view of ability to satisfy 'felt needs' according to the specifications established, all human-created systems could be in one of the two possible states:

- State of Functioning, SoFu.
- State of Failure, SoFa.

For human creations like rockets, satellites, batteries, light bulbs, resistors, fuses, chips, bricks and similar, a transition to the state of

Figure 1.2   Functionability profile of non-restorable engineering system.

failure means retirement. Their functional pattern can be represented as shown in Figure 1.2. Engineering systems of this type are known as non-restorable simply because it is impossible to restore their ability to perform a function once transition to the state of failure has taken place.

Conversely, there are a large number of systems whose functionability can be restored, and they are known as restorable systems. Thus, when one says that a specific system is restorable it is understood that after it has failed its ability to perform a specified function, it can be restored. Consequently, the term restorability will be used to describe the ability of the system to be restored after its failure.

In order to restore the ability for performing a function of a system it is necessary to perform specified tasks known as maintenance tasks. The most common maintenance tasks are cleaning, adjustment, lubrication, painting, calibration, replacement, repair, refurbishment, renewal, and so on; very often it is necessary to perform more than one task in order to restore the functionability of a system. Apart from the maintenance tasks caused by the failure during operation, a system may require some additional tasks to be performed, in order to retain it in a state of functioning. Generally speaking these tasks are less complex than those needed for the restoration of functionability, and are typified by cleaning, adjustment, checking and inspection.

From the point of view of functionability, during the operational life a restorable system fluctuates between SoFu and SoFa until its retirement, as shown in Figure 1.3. The established pattern is termed the functionability profile because it maps states of the system during its utilization process (Knezevic, 1993). Usually calendar time is used as the unit of operational time against which the profile is plotted.

It is extremely important for the user to have information about the functionability, cost, safety and other characteristics of the system under consideration at the beginning of its operational life. However, it is equally, or even more, important to have information about the characteristics which define the pattern of its functionability profile, as the main reason for the acquisition of any system is the satisfactory performance of its expected function. Simply, the system is useful when, and only when, it performs the required function. A commercial aircraft makes money only when it is in the sky flying the ticket-paid passengers between two destinations. The situation is the same with

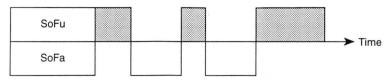

Figure 1.3 Functionability profile of restorable engineering system.

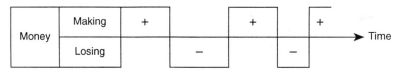

Figure 1.4 Contributions of a system in different states.

cars, kettles, computers, motorways, bridges, etc. This statement could be illustrated graphically, as shown in Figure 1.4, where plus (+) signs stand for:

- a money-making phase of all income-generating systems, like commercial aircraft, trains, production lines, taxi cabs, ice cream makers and similar;
- a positive contribution phase of all public and non-profit making systems, like motorways, hospitals, weapons, traffic lights, and similar, which do not generate income in a direct sense but certainly provide satisfactory feelings for their users.

Thus, one of the main concerns of the users of any system is the pattern of its functionability profile, with a specific emphasis on the proportion of the time during which the system under consideration will be available for the fulfilment of functionability. Clearly, the following two factors are chiefly responsible for its specific shape:

(a) *Inherent characteristics* of a system like reliability, maintainability and supportability, which directly determine the frequency of the occurrence of failures, the complexity of the restoration tasks and the ease of the support of the tasks required.
(b) *Utilization characteristics* of a system, which are driven by the users' operational scenario, maintenance policy/concept and the logistics function, the objective of which is to manage the provision of the resources needed for the successful completion of all operation and maintenance tasks (Blanchard, 1991).

Consequently, the proportion of the time during which the system under consideration is functionable depends on the interaction between the inherent characteristics of a system from the design, and the utilization function given by the users' specific requirements and actions.

Example 1.1

A quick look at the logbook of the first Boeing 747 owned by Pan Am, registration number N747PA, clearly illustrates the interaction between the operation and maintenance processes during the 22 years of service. Thus, this particular aircraft has:

- been airborne 80 000 hours;
- flown 37 000 000 miles (*c.* 60 000 000 km)
- carried 4 000 000 passengers;
- made 40 000 take-offs and landings;
- consumed more than 271 000 000 gallons (*c.* 1 220 000 000 litres) of fuel.

These are some statistics related to the SoFu, driven by Pan Am's business plan.

In order to meet the above given operational scenario, among many other resources consumed, this aircraft has:

- gone through 2100 tyres;
- used 350 brake systems;
- been fitted with more than 125 engines;
- had the passenger compartment and lavatories replaced four times;
- had structural inspections for metal fatigue and corrosion which have needed more than 9800 individual X-ray frames of film;
- had the metal skin on its superstructure, wings and belly replaced five times.

Replacement of the above-mentioned items and others, coupled with all other maintenance tasks performed during the 22 years of operation, accumulated 806 000 maintenance hours.

1.4   THE IMPORTANCE OF AVAILABILITY

Clearly, from the point of view of the user, the shape of the functionability profile is one of the most important characteristics of the system, even more than the proportion of the time during which the system under consideration is available for the satisfaction of needs. Thus, every time when the system is not functionable, the needs have to be met in some other way. A commercial aircraft makes money only when it is flying, i.e. transporting passengers between two destinations. The situation is the same with cars, kettles, computers, etc.

Clearly the shape of the functionality profile depends on the characteristics of the design, such as reliability, maintainability and supportability, and the operational and maintenance policy adopted by the user supported by the logistics function which is related to the provisioning of the operational and maintenance resources needed. The lack of availability of the system could last for a long time due to the shortage of spare items, adequate facilities, trained personnel, special tools and equipment, etc. Consequently, the proportion of the time during which the system under consideration is available for the operation depend on the characteristics inherent from the design and the characteristics related to the performance of the logistics function.

Designers, especially those working on aerospace and military products, have been under huge pressure from the users, in the last 30 years, to provide some information regarding the expected shape of the functionability profile, together with a recommended list of type and quantity of resources needed for its achievement.

Availability is a characteristic which quantitatively summarizes the functionability profile of an item. It is an extremely important and useful measure in cases where the user has to make decisions regarding the acquisition of one item among several competing possibilities. For example, which item should the user select where item *A* has more favourable reliability measures, *B* is superior regarding the maintainability and *C* could be supported the best? Clearly, deciding is a very difficult task because the information given is related to different characteristics. Thus, in order to make an objective decision regarding the acquisition of the new item it is necessary to use information which encompasses all related characteristics. Thus, availability is a measure which provides a fuller picture about the functionability profile.

Further illustration of the recognition of the importance of the maintenance process to successful airline operation is the fact that the new 777, which was introduced into service in June 1995, is the first Boeing model to have a Chief Mechanic, whereas there has always been a Chief Pilot on every Boeing model (*Airliner*, January–March, 1995).

Example 1.2

The *Daily Mail*, in the UK, reported on 13 December 1990, in the article entitled 'Warships wasting years stuck in dock' that a British Parliamentary investigation revealed that: 'a frigate or destroyer spends eight years of its average 22-year "life" under maintenance, and only half of the remaining 14 years would be spent at sea'.

In practice, this means that the frigate/destroyer of this type under this specific operational scenario is available to the Navy for only around 60% of the time when needed.

1.5   CONCEPT OF MAINTAINABILITY

Although it is extremely important for the operators/users to know the functionability, durability and reliability characteristics of the system at the beginning of its operational life, it is equally, or even more important for them to have information regarding issues like:

- Which maintenance tasks should be performed?
- When should the maintenance tasks be performed?
- How difficult is it to perform a maintenance task?
- How safe is it to perform a maintenance task?
- How many people are required to perform a maintenance task?
- How much is the restoration going to cost?
- How long is the system going to be in a state of failure?
- What equipment is required?
- What worker skills are needed to perform the prescribed activities?

In most cases the answers to these questions provided by designers/manufacturers are very basic and limited. For example, in the case of motor vehicles the answers cover no more than the list of maintenance activities which should be performed during regular service every 6000 miles (9600 km) or so. All the above questions remain unanswered, and the users are left to find the answers by themselves. The reason for this is the fact that, up to now, the main purpose and concern of designers has been the achievement of functionability, whereas the ease of maintaining functionability by the users has been almost ignored. Traditionally it was the 'problem' of the maintenance personnel, not the designers.

However, today the situation is gradually changing, thanks to aerospace and military customers who recognized the importance of information of this type and who made it a characteristic equally desirable as performance, reliability, availability and similar.

As no scientific disciplines were able to help designers and producers to provide an answer to the above questions, the need arose to form a new discipline. Maintainability theory was created, defined (Knezevic, 1993) as:

> a scientific discipline which studies complexity, factors and resources related to the tasks needed to be performed by the user in order to maintain the functionability of a system/product, and works out methods for their quantification, assessment, prediction and improvement.

Maintainability theory is rapidly growing in importance because of its considerable contribution towards the reduction of maintenance cost of a system during its utilization.

At the same time, in order to be used in daily practice, maintainability as a characteristic of human-made items/systems has to be defined. In technical literature several definitions for maintainability can be found. For example, the US Department of Defense's MIL-STD-721C (1966) defines maintainability as a characteristic of design and installation which is expressed as the probability that an item will be retained in or restored to a specified condition within a given period of elapsed time, when maintenance is performed in accordance with prescribed procedures and resources.

However, in this book the definition proposed by Knezevic (1993) is used:

> Maintainability is the inherent characteristic of an item/system related to its ability to be maintained in functionable state when the required maintenance task or tasks are performed as specified.

It is necessary to stress that maintainability theory provides a very powerful tool for engineers and managers to quantify and assess the ability of their systems to be maintained in SoFu during operational life.

1.6 MAINTAINABILITY IMPACT ON AVAILABILITY

The majority of users state that they need the equipment availability as badly as they need safety, because they cannot tolerate having equipment out of operation. There are several ways that designers can control that. One is to build items/systems that are extremely reliable and, consequently, costly. The second is to provide a system that, when it fails, is easy to restore. However, if everything is made highly reliable and everything is easy to repair, the producer has got a very efficient system which no one can afford to buy. Consequently, the question is how much a utility of the system is needed, and how much one is prepared to pay for it? For example, how important is it for the train operator to move the train from the platform, when 1000 fare-paying passengers expect to leave the gate at 6.25 a.m.? Clearly, the passengers are not interested in what the problem is, or whether it is a designer's error, or a manufacturer's, maintainer's, operator's or somebody else's problem. They are only interested in leaving at 6.25 a.m. in order to arrive at their chosen destination at 7.30 a.m. Thus, if any problem develops, it needs to be rectified as soon as possible.

Consequently, maintainability is one of the main factors in achieving a high level of operational availability, which in turn increases users' or customers' satisfaction.

Example 1.3

The main objective of this example is to illustrate the impact of maintainability on operational availability of motor vehicles. The

Table 1.1 Service intervals, durations and replacement times

Model	Major service		Replacement time in hours						
	Interval (miles)[a]	Duration (hr)	Clutch	Exhaust	Rear damper	Headlamp	Windscreen	Front bumper	Alternator
Montego 1.6	12 000	2.6	3.9	1.2	1.5	0.4	2.5	1.0	0.6
Peugeot 205	12 000	1.8	3.7	1.0	1.4	0.5	2.0	0.3	0.5
Astra GTE	9 000	1.4	1.2	0.9	0.6	0.6	0.2	0.8	0.6
Jetta 1.8	10 000	2.0	2.9	0.9	0.5	0.4	0.7	0.4	0.5
Toyota Carina	10 000	2.0	3.9	1.3	0.8	0.4	2.9	0.8	0.7
Lada 1500	6 000	3.6	3.2	1.8	0.9	0.2	0.5	0.5	0.7
Cavalier	9 000	1.3	1.2	0.9	0.6	0.7	1.3	0.6	0.5
Golf 1.6	10 000	2.0	3.3	0.9	0.6	0.7	1.3	0.6	0.5
Sierra 1.6	12 000	2.4	2.0	0.6	0.4	0.4	2.1	0.4	0.4
Nissan Micra GL	6 000	2.8	3.3	0.7	1.6	0.2	1.8	0.6	0.6
Renault 5 TL	12 000	3.6	4.4	1.3	0.4	0.4	0.4	0.4	0.4
Alfa 33 1.5	12 000	3.0	4.4	0.5	0.4	0.2	1.8	0.4	0.3

[a]1 mile = 1.6 km.

inherent maintainability characteristics for several motor vehicles are given in Table 1.1. They clearly indicate the impact of the design decisions on the maintenance resources, frequency, and ultimately operational availability.

Based on the data in Table 1.1 for the specific operational scenario, where it was assumed that during a three-year period the total mileage covered by each type of motor vehicle was 75 000 miles, the total hours spent by the users on maintaining functionability are given in Table 1.2, together with the operational availability achieved.

Table 1.2   Impact of maintainability on availability

Model	No. of services	MTIM^p (hr)	MTIM^c (hr)	MTIM (hr)	Availability
Montego 1.6	6	15.6	17.2	32.8	0.9927
Peugeot 205	6	10.8	14.1	24.9	0.9945
Astra GTE	8	11.2	14.1	25.3	0.9955
Jetta 1.8	7	14.0	11.1	25.1	0.9944
Toyota Carina	7	14.0	17.0	31.0	0.9931
Lada 1500	12	43.2	18.3	61.5	0.9863
Cavalier	8	9.1	8.9	18.0	0.9960
Golf 1.6	7	14.0	15.9	29.9	0.9934
Sierra 1.6	6	14.4	9.9	24.3	0.9946
Nissan Micra GL	12	33.6	13.8	47.4	0.9895
Renault 5 TL	6	21.6	17.0	38.6	0.9914
Alfa 33 1.5	6	18.0	15.7	33.7	0.9925

Another area to be considered under maintainability is trouble-shooting the various modules within the allowed time. For the airlines, this is usually only about one hour at the gate prior to its departure to the next destination, whereas for a racing car or weapon system this is usually a few minutes. An easily manageable device is needed for the diagnosis of all different modules in order to determine their state and identify the failed one within them. Practice shows that false removals cost about the same as an actual failure when the component under investigation is removed and replaced. Reducing this would be a big cost saver. Devices of such capabilities have been developed in the aerospace industry, as a result of maintainability studies and research. For example, the design of a Boeing 777 includes an 'on-board maintenance system' with the objective to assist the airlines with a more cost-effective and time-responsive device to avoid expensive gate delays and flight cancellations (Proctor, *Aviation Week & Space Technology*, February 1995). For similar purposes the Flight Control Division of the

Wright Laboratory in the USA has developed a fault detection/isolation system for F-16 aircraft, which allows maintainers, novice as well as expert, to find failed components.

With the older fleets, both in the military and commercial sector, there is a great need for easier detection of corrosion. When these aircraft were built, they were designed for a certain life cycle, not for the extended service imposed on them. As the number of flying hours on an aircraft increase, the chance of corrosion and structural fatigue increases. One of the objectives of maintainability is the development of the system for detection and identification of failures before they make aircraft safety critical.

One of the common perceptions is that maintainability is simply the ability to reach a component to change it. However, that is only a small aspect. Maintainability is actually just one dimension of system design and a system's maintenance management policy. For example, it could be required from the designer that only three screws are acceptable on a certain partition panel in order to get speedy access inside. However, this request has to be placed into larger context and it becomes a trade-off. If the item behind that panel only needs to be checked once in every five to six years, or say 50 000 miles (80 000 km), it does not make sense to concentrate much intellectual effort and spend project money on quick access. Thus, a lot of fasteners and connectors could be tolerated and the item may not be quickly accessible, but all of that has to be traded off against the cost and operational effectiveness of the system.

Additionally, decision makers have to be aware of the environment in which maintainers operate. It is much easier to maintain an item on the bench, than at the airport gate, in a war theatre, amongst busy morning traffic, or in any other result-oriented and schedule-driven environment. Thus, the trade-off process has to take into account the operational environment and the significance of the consequences if the task is not completed satisfactorily, when the trade-off is made. According to Hessburg, the Chief Mechanic of new airplanes from Boeing: 'Maintenance managers want a clean gate, their report card in line maintenance based on having a clean gate and not having pigeons roosting on the airplane's fin. So it is necessary to try to influence the design that way, and say, "here's what mechanics have to do at the gate".' (*Aviation Equipment Maintenance*, May 1994).

The majority of users are currently showing concern over the competitive advantage that maintainability and maintenance can provide to a company. To illustrate the economic importance of maintenance, a recent study of engineering maintenance practices show that:

- United States airlines spend 9 billion dollars, or approximately 11% of their operating cost, on maintenance.
- The military sector has even higher concern for the maintenance cost

which accounts for about 30% of the life cycle cost of a weapons system. In 1987/88 the Royal Air Force spent around 1.9 billion pounds on the maintenance of aircraft and equipment.

• British manufacturing industry, according to a 1988 report produced by the Department of Trade and Industry, spends 3.7% of annual sales value each year on maintaining direct production systems. Translating the above percentage to the sum of money spent in UK industry on maintenance it amounts to 8.0 billion pounds in a year.

1.7   MAINTAINABILITY IMPACT ON SAFETY

Finally, performance of any maintenance tasks is related to an associated risk, both in terms of the non-correctly performed specific maintenance task, and the consequences of performing the task on the other item of the system, i.e. possibly of inducing a failure on the system while doing maintenance.

Example 1.4

An Airbus A320, owned by Excalibur Airways, performed an un-commanded roll to the right due to loss of spoiler control just after take-off from Gatwick Airport in London, UK, in August 1993.

A report released in February 1994 by the Air Accidents Investigation Branch (AAIB) stated 'the emergency rose, not from any mechanical malfunction, but from a complex chain of human errors by the maintenance crew and by both pilots.' Apparently, during the flap change, maintenance did not comply with the maintenance manual. The spoilers were placed in maintenance mode and the collars and flags were not fitted. Also, the reinstatement and functional check of the spoilers after flap installation were not carried out.

In addition, the pilots failed to notice during the independent functional check of the flight controls that spoilers two through five on the right wing did not respond to the right roll commands.

The AAIB made 14 safety recommendations to the Civil Aviation Authority including formally reminding technicians of their respons-ibility to ensure all work is carried out in compliance with the maintenance manual and no work otherwise is to be certified. It also recommended that Airbus amend the A320 maintenance manuals concerning flap removal, and that the flap refitting and spoiler deactivation chapters include specific warnings to reinstate and function the spoilers after deactivation.

(Source: *Aviation Equipment Maintenance*, April 1995, p. 8.)

Example 1.5

In the article entitled 'Hangar error' published in January 1992, in the journal *Aerospace*, the following three maintenance-related accidents were exposed:

(a) At Chicago airport a DC-10 rolled onto its back after take-off and crashed into a caravan park, killing all 272 on board and two people on the ground. The cause of the accident was an engine separation due to fatigue in a cracked engine mounting which resulted from an improper forklifting short cut. After the accident, other DC-10 maintenance engineers said they had used the same short cut. One had actually heard a sharp cracking noise from the structure but had not dared to report it.
(b) The total engine failure of a TriStar was caused by the incorrect insertion of oil chip-detectors, with O-ring seals missing. There had been 12 previous similar occurrences in the same airline, seven leading to unscheduled landings. This was a classic case of boredom and complacency in the hangar.
(c) The total engine failure of a Boeing 767 was caused by misreading a dipstick in gallons instead of litres. Many other cases could be added to the list of maintenance-related accidents which had very nearly happened before but were not reported. It is very difficult to admit to the boss that the litres had been misread for gallons!

Example 1.6

Analysis of major civil aviation accidents resulting from non-satisfactory completion of maintenance tasks shows that between 1981 and 1985 there were 19 maintenance-related failures which in total claimed 923 lives. The biggest accident took place on 12 August 1985, when a JAL-owned Boeing 747 decompressed due to fatigue because of an improperly repaired bulkhead, killing 520 people.

The same analysis shows that between 1986 and 1990 there were 27 maintenance-related failures claiming 190 lives. The most tragic of them was the crash of a United-owned DC-10, in 1989, when the fatigue of a fan disc of the second engine caused complete hydraulic and flight control failure, and loss of 111 lives.

(Source: *Aerospace*, January 1992.)

1.8   UNDESIRABLE MAINTAINABILITY PRACTICES

Several real-life examples are cited here, in order to illustrate some of the undesirable maintainability decisions made in the past, which have caused considerable problems to the users.

Example 1.7

This example concerns the engine starting system on Hunter aircraft. As the rapid starting of the heavy and large Avon 200 engine was a dominant operational characteristic, the designers concentrated on a small turbine starter powered by an iso-propyl-nitrate. Its high inertia forced the turbine to work at the peak of its performance. In the case of overspeed it could have damaged the engine, which would certainly have been catastrophic in the air. Consequently, the design was reviewed and a relay system introduced in order to shut down the start cycle in case the starter turbine had not disengaged by 1600 rpm. This was a good design decision, especially from the safety point of view, but very little consideration had been made regarding the reliability and maintainability issues. Hence, due to the very high failure rate of the redesigned system, the aircraft availability was drastically reduced, especially due to the fact that the relay system could not be changed on site, unless the mechanic of the day had '3 meter-long arms'. Consequently, the engine had to come out. Unfortunately, to achieve this the back of the aircraft had to be removed. To achieve this activity the engine and flying control connectors had to be disconnected. The final result in the field was: 40 manhours to change a relay, of which approximately 5 minutes was spent actually changing the relay itself. On top of that, every time the squadron went on detachment, the maintainers had to take along a full set of bulky support equipment to satisfy the inevitable need to change a few relays.
(Source: Air Commodore O. Truelove, RAF, presentation at Exeter University, UK, 1989.)

Example 1.8

The engine change on the Harrier GR3. In order to perform this task the wing of the aircraft must be removed and in order to achieve this it is necessary to disconnect a variety of control systems. The total task requires 24 hours of elapse time involving an assortment of heavy and bulky support equipment.
(Source: Air Commodore O. Truelove, RAF, presentation at Exeter University, UK, 1989.)

Example 1.9

The Times, on 11 February 1995, reported the following story:
A Renault 25 TX, with nearly 75 000 miles on the clock, had been almost completely trouble-free during its life. The alarm bells rang only mildly when the heater stopped working and the temperature gauge refused to move, but then after 10 minutes driving sprang straight into the red. The technician at the Renault garage sounded mournfully like a

doctor diagnosing a long, painful and exotic illness. 'The heating matrix has gone, about the worst thing that could have happened. Most unusual. Jolly bad luck.' The heating matrix is an oblong metal construction 30 cm by 15 cm by 5 cm, shaped like a small radiator, the main function of which is to provide warm air to heat the car. They are supposed never to go wrong, so manufacturers snuggle them deep in the car where they can remain unmolested until the vehicle is scrapped. However, when they do fail, trouble and cost follow.

The price of the heating matrix itself was £57.50. The total cost of the replacement, however, was £553.30, including value added tax. This is because it took 10.5 hours to get the old one out and put the new one in. The mechanics had to dismantle virtually the whole dashboard, remove most of its innards and, key-hole-surgery-style, negotiate the matrix out through the glove compartment. The work took a couple of days and the user had no use of the vehicle while the major surgery progressed.

Renault's head office in Britain confirmed that 10.5 hours was the correct amount of labour time needed to replace the matrix on that particular model. However, Renault pointed out that on their latest model, the Laguna, as a result of design change, the same item could be replaced within 1.5 hours. The matrix is now accessible through the engine rather than the glove compartment.

1.9 DESIRABLE MAINTAINABILITY PRACTICES

Certainly there are many more desirable maintainability practices, where efforts have been made at the design stage with the objective of making positive contributions towards the ease, accuracy and safety of maintaining the functionability of the system, by the user, during the utilization phase.

Example 1.10

During the course of most Formula 1 races, cars make at least one mid-race stop at their pits in order to change tyres. The outcome of this maintenance task can occasionally mean the difference between first and second place. Consequently, in order to reduce the time spent in the pit to the minimum, the wheels of the F1 cars are designed in such a way that only one central wheelnut provides a sufficient force for attachment to the hub. Typical replacement times for all four wheels are given in Table 2.3.

The task requires 15 mechanics, three to remove and replace each wheel, two on quick-lift jacks, and the chief mechanic who holds a board

Table 1.3 Tyre change times, 1993 British Grand Prix

Team	Driver	Time (s)
McLaren	A. Senna	5.11
Benetton	M. Schumacher	5.50
Ligier	M. Brundle	6.75
Williams	D. Hill	7.61
Williams	A. Prost	8.02
Lotus	A. Zanardi	9.21

in front of the car with signs 'Brakes on/Go'. These may be joined by another mechanic to steady the car.

The situation is similar with other modules of a racing car, as illustrated by the list below which shows average replacement times in minutes for other items.

Engine	60
Gearbox	30
Four shock absorbers	12
Pedal box, seat and harness	10

Example 1.11

A complete operational turnaround for a SAAB-Grippen fighter aircraft, in the Swedish Air Force, including refuelling, reloading the gun, mounting six air-to-air missiles and making an inspection, can be performed with the minimum equipment in less than ten minutes by five conscripts under the supervision of one technician. No tools are required to open and close the service panels, which are at a comfortable working height. All lights, indicators and switches needed during the turnaround are in the same area of the aircraft, together with the connections for fuel and communication with the pilot.

(Source: *The Grippen Logistics Concept*, Publication 950601, Saab Military Aircraft, 1995.)

Example 1.12

GE Aerospace transportable solid state radar FPS-17 provides a system availability of 0.996. This is achieved through huge reliable solid-state components, continuous automatic performance monitoring and fault isolation, and a mean time to repair of less than 30 minutes.

The situation is very similar with a Tactical Solid-State Radar AN/TPS-59. The reduction of maintenance costs (depot repair) is achieved through design improvements such as those listed below:

(a) All printed wire boards are plug-in.
(b) All integrated circuits are plug-in.

(c) Of all electrical connections 90% are implemented by plug-in connectors or screws and lugs.
(d) All printed wire boards employ solder masking to prevent solder shorts, and silk screening to prevent component misplacement.
(e) Continuous on-line automatic performance monitoring, off-line fault location, and permit maintenance by medium-skill-level personnel during operations.

> (Source: Manager – Marketing, GE Aerospace, Radar Systems Department, Syracuse, NY, USA.)

1.10 CLOSING REMARKS

The main purpose of existence of any human-made item/system is the provision of utility by performing a required function with expected performance and attributes. Hence, once the functionability is provided, the main concern of the user is to achieve the highest possible availability and safety for the least possible investment in resources.

The performance of any maintenance task is related to associated costs, both in terms of the cost of maintenance resources and the cost of the consequences of not having the system available for operation. Therefore, maintenance departments are one of the major cost centres, costing industry billions of dollars or pounds each year, and as such they have become a critical factor in the profitability equation of many companies. Thus, as maintenance actions are becoming increasingly costly, maintainability engineering is gaining recognition day by day.

It is clear, from the brief analysis of the role and the importance of maintainability given above, that it represents one of the main drivers in achieving users' goals regarding availability, reliability, cost of ownership, reputation and similar objectives. For example, in the journal *Aviation Week & Space Technology*, of 22 January 1996, it was reported that by the year 2000 the USA Air Force will begin looking at upgrades of a heavy airlifter aircraft C-5A. The comment was that although the structure of the aircraft is considered to be good, 'the reliability and maintainability leaves a lot to be desired'. It is inevitable that in the future considerations and comments like this will significantly increase and that the impact of these considerations on the final selection of the systems will be far greater.

Thus, the analysis of concepts, tools, techniques and models available to the maintainability specialists for the prediction, assessment and improvement of their decisions related to the ease, accuracy, safety and economy of performing tasks related to maintaining systems in a functionable state during their utilization, which directly influence the length of time which a system will spend in SoFa, are the main concerns of this book.

CHAPTER 2

Maintainability and systems effectiveness

Let's make sure that when we buy a tank, or a ship, or a plane, that we know – going in – exactly what it's going to take down the road to operate it, and keep it fixed for as long as we have it. Furthermore, since 60 to 80 percent of our total system cost is in the operating and maintaining category, let's do what we can, during the design phase, to minimize those costs by designing a product that's easier to operate or repair. But above all, let's make sure we know what we're going to need, when we're going to need it, and then go about an orderly process of obtaining all these things at the right time, in the right place, and in the right amounts to keep our product operating.

(Senior Officer of USA Department of Defense)

The analysis of the reasons for the creation of systems which exhibit non-desirable maintainability characteristics, some examples of which were given in Chapter 1, have shown inadequate early planning and the lack of a total integrated approach to system design process from the outset. In essence, it has been found that maintainability, as one of many design disciplines (like producibility, inspectability, testability, reliability, supportability, manability, transportability and similar), has a tremendous impact on the ultimate effectiveness of a system (Blanchard, 1991).

Further, it has become increasingly evident that these and other important issues, such as environment, social impact and similar, are not adequately addressed in the design process. If considered at all, they are introduced 'after the fact', which can be costly experience should changes be required at this stage.

In order to achieve set goals it is necessary to integrate specialities like maintainability, producibility, reliability, supportability and others into the overall design process. The main efforts are concentrated towards creation of analytical tools which assist designers in foreseeing the consequences of decisions made to the investment of the resources needed for the production and utilization process of the future system at a very early stage of the design when changes are possible at almost no extra cost.

Thus, the main objective of this chapter is to set up the scene for the life cycle engineering analysis whose main objective is to bring competitive systems into being in a way which minimizes their deficiencies and life cycle cost.

This approach is applicable to both small- and large-scale systems. Further, it is applicable to many different categories (aircraft, ship, kettle, bridge, motorway, etc.). Although the nature of the requirements may vary from one application to the next, the procedure is essentially the same.

2.1   LIFE CYCLE PROCESS

The cycle begins at the moment when the idea of a new system is born and finishes at the moment when the system is safely disposed. Thus, the main processes through which any human-created system goes are:

- specification
- design
- production
- utilization
- retirement

each of which are briefly analysed below.

2.1.1   The specification process

The specification process is a set of tasks performed in order to identify the needs and requirements for the new system and transform them into a technically meaningful definition.

In the first phase of the life cycle of a system, the needs and requirements which the future system should satisfy have to be clearly specified. The main reason for a need for a new system could be:

(a) a new need for a new function to be performed; or
(b) deficiency of a presently used system due to:
  - functional deficiencies,
  - inadequate performance,
  - inadequate attributes,
  - extremely high maintenance cost,
  - inadequate support resources (spares, facilities, personnel, etc.),
  - low demand from the market,
  - low profit provided to the company.

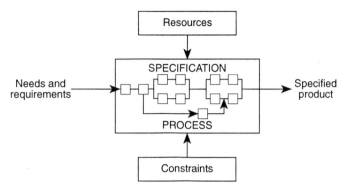

Figure 2.1   The specification process.

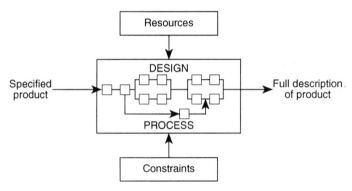

Figure 2.2   The design process.

The input characteristics of the specification process are the needs for the new system, and output characteristics are the fully described functionability of the future system (Figure 2.1).

2.1.2   The design process

The design process is a set of tasks performed in order to transform the specification for a new system into its full technical definition (Figure 2.2).

The main tasks performed during the design process are:

- management
- planning
- research
- engineering design
- documentation development

- design of software
- building a prototype
- test and evaluation.

Thus, the main objective of the design process is to determine and define all items of which a future system consists and to define their attributes as well as their relationships in order for the system to meet a needed function according to specified requirements.

By taking into account a full specification at a very early stage of the design process, known as the conceptual stage, the possible alternatives of the future system could be examined in a broad form in order to outline the most favourable solution. At this stage no details of the system are addressed. Thus, the outcome of the conceptual stage is a general concept and plan for the future system, details of which should be examined more closely further down the line.

Therefore, the conceptual design has to be converted into a detailed set of drawings and specifications related to every single item of which the system is to be built. As this process progresses, experiments, tests and analyses are performed in order to select the best solutions among all possible alternatives. At the later stage of design, prototypes or models are usually built as a result of the design refinements made.

Consequently, the design is 'frozen', and the fully defined system in qualitative and quantitative form moves into the next process of its life cycle, known as the production and/or construction process.

2.1.3   The production/construction process

The production/construction process is a set of tasks performed in order to transform the full technical definition of the new system into its physical existence (Figure 2.3).

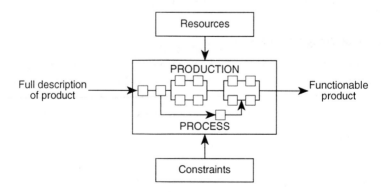

Figure 2.3   The production process.

The main tasks performed during this process are:

- management
- operation analysis
- manufacturing
- assembling/construction
- quality assurance
- provision of initial support
- testing
- delivery.

During the production/construction process the system is physically created in accordance with the design definition. At the same time various tasks are performed in order to ensure that all set-up requirements and specifications are met.

At the end of this process a system physically exists which fully satisfies all needs and requirements and is ready to be utilized.

2.1.4 The utilization process

The utilization process is a set of tasks performed with objectives to utilize the inherent functionability of a new system in order to satisfy an identified need (Figure 2.4).

The main tasks performed during this process of a system's life are:

- management
- distribution
- operation
- maintenance
- support
- modification.

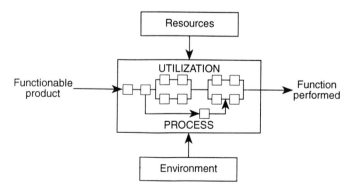

Figure 2.4 The utilization process.

In spite of the fact that all human-made systems are able to perform a specified function in a specified manner, it is only possible through the operation process which could be defined as:

> the set of tasks performed in order to utilize the inherent ability of the system to satisfy needed function with required performance and attributes.

Thus, at the beginning of the utilization process the system under consideration exists and it is functionable. Consequently, every system of the same kind will have the same functionability. For example, every single Ford Sierra 1.6 LX ever produced will have equal ability to perform a transportation function with equal performance (maximum speed, acceleration, fuel consumption and similar) and attributes (road holding, visibility, ventilation and so forth) when used as recommended.

After acquisition the system is in a state of functioning and each individual user starts using it in order to satisfy his or her individual needs or business demands. The behaviour of the system during its operational life is fully defined by its functionability profile in the manner discussed in Chapter 1 (see Figure 1.3).

2.1.5   The retirement process

The retirement process is a set of tasks performed in order to phase a system out of inventory, at the end of its useful life, together with the recycling/disposal of the items of which it consists (Figure 2.5).

The main tasks performed at this last process are:

- management
- phase-out
- disposal
- documentation.

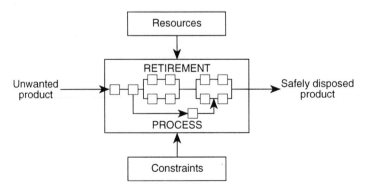

Figure 2.5   The retirement process.

Withdrawal from the operation is the last process in the life cycle of the system. This means the actual end of its operational life due to loss of functionability or to any other reason like obsolete design, increased maintenance cost, low productivity, danger for user or general public, lack of spare items, or destruction due to disaster or accident.

2.2 LIFE CYCLE COST

The execution of each task of each life cycle process analysed above requires some resources for its successful execution.

Typical resources needed for the successful completion of every specification, design, production/construction, utilization and retirement task, could be grouped into the following categories:

- *Personnel*: adequately educated and trained, with the required experience and attitude.
- *Material*: a generic name which includes all natural and artificial materials needed for the accomplishment of each task.
- *Equipment and tools*: including all tools, machinery and equipment needed for the execution of the diversity of tasks during the life cycle process.
- *Facilities*: which are needed for completion of each task, such as physical plant, real estate, portable buildings, inspection pits, dry docks, wind tunnels, housing, maintenance shops, research and development laboratories, repair and overhaul facilities.
- *Technical data*: needed for design, manufacture, utilization and disposal of the system. Typical examples are maintenance instructions, inspection and calibration procedures, overhaul procedures, modification instructions, facilities information and drawings and specifications that are necessary in the performance of system design and production functions. Such data not only cover the system but test and support equipment, production machinery, transportation and handling equipment, training equipment and facilities.
- *Maintenance computer resources (MCR)*: which refers to all computer equipment and accessories, software, programme tapes/disks, databases and so on, necessary in the performance of life cycle functions.
- *Energy*: needed for the execution of each task within the life cycle process.

Clearly, the life cycle cost of a system consists of the costs related to the monetary value of resources used for the completion of each task needed to be executed during each life cycle process (specification, design, production, use and retirement).

In summary, it could be said that besides functionability and availability, which have been briefly discussed in Chapter 1, the third main characteristic of any system is its life cycle cost.

Example 2.1

Assume that you are in the process of defining the maintenance concept early in the conceptual design of System 'XX-25', and that you wish to evaluate the feasibility of either designing the system for two levels of maintenance support, or for three levels of maintenance support (more details in Chapter 8).

In the design for two levels (organizational and supplier maintenance), the system packaging scheme includes six items, operating in series, and integrated into the overall system housing. In the event that corrective maintenance is required, faults are isolated to a specific item through a built-in test capability; the applicable item is removed and replaced with a spare; and the faulty item is sent back to the supplier for repair.

In the design for three levels (organizational, intermediate and supplier maintenance), the system packaging scheme includes three items operating in series within Unit 'A', and three items operating in series within Unit 'B'. The units are integrated into the overall system housing. In the event that corrective maintenance is required, faults are isolated to the unit level through a built-in test capability; the applicable unit is removed and replaced with a spare; and the faulty unit is returned to the intermediate maintenance shop for repair. Unit repair includes isolation to a faulty item using external test equipment; the item is removed and replaced with a spare; and the malfunctioned item is returned to the supplier for repair. The two configurations are illustrated in Figures 2.6 and 2.7.

The objective of this exercise is to determine the most suitable level of maintenance support (repair) based on the information available. Thus,

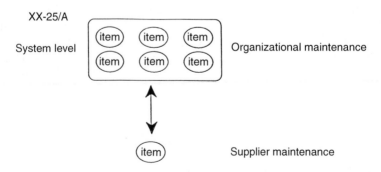

Figure 2.6 Configuration XX-25/A.

XX-25/B

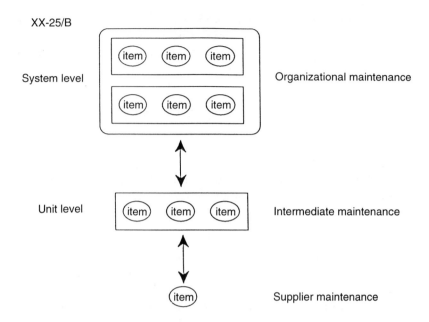

System level   Organizational maintenance

Unit level   Intermediate maintenance

Supplier maintenance

Figure 2.7   Configuration XX-25/B.

regarding the two feasible options XX-25/A and XX-25/B, the following data have been collected:

(a) System 'XX-25' is expected to operate 2000 hours per year by a single user, and the estimated system acquisition cost (design and production) for XX-25/A is £250 000 and for XX-25/B is £175 000. The difference in cost is primarily attributed to the more extensive built-in test capability required for XX-25/A.

(b) Assume that all items are equivalent in terms of reliability, and that the average item failure rate is 0.001 failures per hour. Also, assume that all repair times are equivalent.

(c) The average personnel cost per maintenance task (MT) is £100 at the organizational level, £200 at the intermediate level and £300 at the supplier level.

(d) For XX-25/A, three spare items are required for organizational maintenance support, and the cost per item is £20 000. For XX-25/B, two spare units are required for organizational maintenance support and two spare items are required for the intermediate level of maintenance support. The cost of a spare unit is £50 000, and the cost of a spare item is £15 000. These figures include the costs of provisioning and inventory maintenance.

(e) The cost of external test equipment to support unit-level repair is

£75 000, and item-level repair is £50 000. These figures include the costs of operating and maintaining the support equipment.
(f) The cost of facility utilization at the intermediate level of maintenance is £75/MT, and for the supplier level of maintenance is £30/MT.
(g) Transportation costs associated with unit-level maintenance are £100/MT, and for item-level maintenance are £75/MT. Assume that the cost of maintenance data is £25/MT for XX-25/A and £40/MT for XX-25/B.

Solution
Based on the data available, the expected number of maintenance tasks (*NMT*) during one year of operation could be determined according to the following expression:

$$NMT \ (2000 \ hr) = 0.001 \times 2000 = 2$$

As all items are identical and connected in series, from the reliability point of view, the expected number of maintenance tasks for both configurations will be 12.

Making use of the data available, summaries of the results of the level of repair analysis are presented in Table 2.1.

Table 2.1   Proposed solutions

Cost elements	XX-25/A	XX-25/B
Acquisition	250 000	175 000
Personnel	4 800	7 200
Spare parts	60 000	130 000
Test equipment	50 000	125 000
Facilities	360	1 260
Transportation	900	2 100
Data	300	480
Total	366 360	441 040

Based on the total cost obtained for both alternatives, the configuration XX-25/A is preferred.

2.3   SYSTEMS TECHNICAL EFFECTIVENESS

In Chapter 1 (Figure 1.1) it was demonstrated that the functionability of any system consists of the:

● function(s) it performs,
● performance with which it performs the function,
● attributes which it possesses while performing the function.

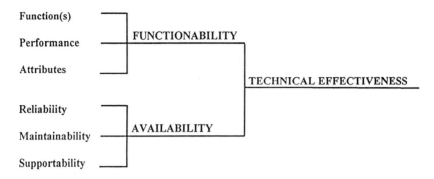

Figure 2.8 Systems technical effectiveness.

Additionally, inherent characteristics of the system like reliability, maintainability and supportability have a major potential impact on the functionability profile of the system, summarized in the unique characteristic known as availability.

Consequently, functionability, reliability, maintainability and supportability brought together form an overall measure of the 'goodness' of the system known as the technical effectiveness, which represents the inherent capability of the system, as shown in Figure 2.8.

2.4   SYSTEMS OPERATIONAL EFFECTIVENESS

Once introduced into the utilization process, each system is exposed to three different processes, namely:

(a) *Operation*: which consists of tasks needed to be performed in order for the system to satisfy required need.
(b) *Maintenance*: which consists of tasks needed to be performed in order to maintain the system in functionable state during its life.
(c) *Logistics*: which consists of tasks needed to be performed in order to provide all necessary resources for the operation and maintenance of the system.

Clearly, each of the above-mentioned processes are planned and managed in accordance with the business plans of the system's owners.

The operational effectiveness is the joint product of the technical effectiveness and the operational, maintenance and logistics policies of the owner, as shown in Figure 2.9.

2.5   CLOSING REMARKS

In summary it could be said that the biggest opportunities to make an impact on maintainability characteristics are at the design stage.

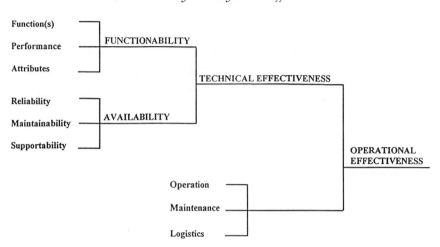

Figure 2.9 Systems operational effectiveness.

Consequently, the biggest challenge for the maintainability engineers is to assess the impact of the design on the maintenance process at the early stage of design, when changes and modifications are possible at almost no extra cost and time. Thus, the battle for maintenance is won or lost at design stage, through maintainability analysis, engineering and management.

2.6   CASE STUDY: MAINTAINABILITY EFFECTIVENESS OF AIRBUS A330

This case study is based on the article entitled 'Maintainable A330' by J.M. Ramsden, published in *Aerospace*, October 1992.

While airliners which need only fuel and oil remain a dream, the A330 has been designed with airline maintenance engineers 'in the loop' to advance the art of reliability – a commodity much demanded by today's airlines.

The maintenance architecture of the A330/A340 has been designed in partnership with the airlines through an Industry Steering Committee (ISC). This Airbus committee consists of about 30 members and they have been meeting fequently (usually for 2 or 3 days) since November 1989. ISC members are the senior maintenance engineers of 14 airlines, namely: Air Inter, All Nippon, Austrian, British Airways, Continental, Iberia, LTU, Lufthansa, Northwest, Sabena, TAP, Thai, TWA and UTA. Also represented on the ISC, in addition to Airbus and its members and the engine manufacturers, are the airworthiness authorities, including the European JAA and the FAA Maintenance Review Board, which attend every meeting.

Maintenance Working Groups (MWGs) have been formed for specific tasks. These typically comprise seven or eight airline maintenance specialists plus representatives of the manufacturers and airworthiness authorities and one advisory ISC member.

Up to 1992, 67 meetings of the A330/A340 MWGs had been held. This represents about 2700 person-days of design–maintenance liaison and 3100 person-days including senior engineering liaison at ISC meetings. In addition to these are mock-up reviews. Called 'maintainability visualizations and demonstrations', they are usually attended by up to ten airline maintenance specialists. Meetings have been held at Garrett Raunheim (APU), BAe Filton (wing), Aerospatiale St Nazaire (centre fuselage), Casa Getafe (tailplane), PW East Hartford (PW4168), Rohr Chula Vista (GE CF6-80E1), R-R Derby (Trent) and elsewhere.

The airline specialists have raised more than 1000 comment sheets, all of which have been answered by the manufacturers' design and maintenance offices. Design changes have included pipe and conduit clamping and routing to improve access and to minimize vibration; grease nipples relocated; electrical breakdown points repositioned; cable protection improved; and connections 'Murphy proofed'. Changes resulting from the mock-up reviews have been few because the ISCs and MWGs got hangar opinion at the early design stage.

Mock-ups are still essential despite the accurate modelling power of three-dimensional Catia computer design. Catia indicates whether something will foul, but a hands-on check still pays. For example, pipe-bending software is in the programme, but it does not always reproduce the 'bendback' which can obstruct access. Nor will Catia subjectively judge whether one dark night someone might use a wire bundle as a hand hold, or duct as a step. An engine manufacturer who dresses engines vertically will make vertical mock-ups, which change the maintenance engineer's view of, say, thrust reverser cable adjustment. However, Catia still saves a lot of time.

The biggest impact on operational effectiveness is the time during turnaround. The airline maintenance teams working with Airbus have produced mathematical models of the minimum workforce needed to turn the A330 round.

The A330 airframe structure has been designed for inspectability and an economic repair life of 60 000 flying hours or approximately 40 000 flights, without major maintenance tasks or corrosion. Airbus has always been particular about corrosion protection, even though it is not an airworthiness certification item and 'doesn't sell'. Until the early 1980s few airlines bothered about corrosion protection in view aircraft. After expensive restructuring of some US aircraft, and alarming corrosion events including fuselage decompressions and wing spar failures, the airlines came to have higher expectations.

In the early 1980s the International Air Transport Association set out its expectations in Technical Document 2637 ('Corrosion Control and Protection'). Airbus regards these provisions as minimum, according to structures designer Jean-Jacques Cuny, and corrosion protection as important as crack tolerance and detection.

Inspectability is a fundamental design feature. Fuselage lapjoints below the window line are not bonded, as they were on A300B2/B4s. Disbonding is hard to detect (as Aloha's 737 inspectors discovered) and the original two rivet rows increased stress level and the risk of microcracking and corrosion of the bonding lines. Other ageing factors were imperfect surface preparation and edge rimming.

The design was improved, notably by fastening fuselage lapjoints with three rivet rows, and by milling rather than glueing doublers ('A320 standard'). The early aircraft were modified with the help of Airbus, which also introduced a fleet inspection programme. No serious disbondings were found but, as Cuny says, 'It is too much to ask the airlines to guarantee the health of a bondline by NDT inspection'. In the A330 and A340 the entire bilge area is riveted and milled, with no bonding at all.

The best manufacturing quality control will sometimes let small flaws through, and in-service loads will usually find them. So the Airbus policy is to back up corrosion protection (shotpeening and coatings) with low stress levels, and to design for external fatigue-crack inspectability.

The A330/A340 bottom wing skins, for example, are designed to be inspected only from the outside. The inspector does not have to crawl around fuel tanks wearing breathing apparatus. The same design philosophy is applied to that historically critical jet structure, the pressure bulkhead. In the A330/A340 it is designed so that cracks will become visible first from the outside, without the need to remove cabin insulation. The 'trick' is to exploit the difference in mechanical properties between titanium and aluminium to make the crack appear on the outside before it gets to a significant length.

The pressure bulkhead is also designed so that even if a crack goes undetected and it propagates to a critical length, only one panel will fail. Control lines and pipes are segregated to survive such a failure and, perhaps, even an onboard explosion. The fuselage structure, including the bulkhead, can tolerate a 3 foot square (0.9 m^2) hole. The key is to design for inspectability from the outside wherever internal inspection is difficult or unattractive.

Corrosion protection in the A330 begins with selection of the improved 7000 series of aluminium–zinc alloys, combining high strength with optimum corrosion resistance and fracture toughness. High strength aluminium alloys such as 7079 and 7178 are not used, nor are magnesium alloys. Titanium alloys, noted for their corrosion resistance, are used for certain primary structures including pylons, flap tracks, wing spar failsafe straps, wing tank manhole covers and APU firewalls and access doors.

Drainage has been a major consideration since the start of design. Holes and gaps are provided throughout the A330/A340 structure to ensure natural drainage to collection points. All holes and gaps are made before anti-corrosion treatment. Care has been taken to design out structural 'receptacles' in which fluids can collect. Insulation blankets are kept away from the skin so that they do not impede drainage. Fluids accumulating in the bilge area are drained overboard through automatic pressure drain valves. Special attention is paid to the protection of lavatory areas – for example, floors and walls are integrated to form a sealed element.

Large proportions of the outer fuselage skin are aluminium clad, and all important wing machined components are vacublasted or shotpeened to improve corrosion and fatigue resistance. All fasteners are wet-installed with polysulphide sealant, and all structural joints are assembled with sealant, overcoated in areas prone to hydraulic fluid. Aluminium alloys at the detail stage are anodized and primed, or primed and polyurethane topcoated. All interfaces with titanium are wet blasted and primed. Microbiological attack by a chromate leaching primer is used for fuel tanks protection.

The complete exterior surface of the aircraft is painted, except for titanium and stainless steel parts.

BITE (built-in-test-equipment) has grown into the A330's Centralized Maintenance System (CMS). The old-generation BITE black boxes and lights, which had to be decoded inside avionics bays, are incorporated in 'onboard maintenance'. CMS tells the pilot what is wrong in flight, and on the ground enables the maintenance engineer to access exactly what the pilot has been dealing with, rather than troubleshooting a 'clonking noise' or a wrong trail or a random failure. It gives the maintenance engineer a complete history of events and aircraft status, and the ability to scan other systems for related problems.

The Centralized Maintenance System is operated from the cockpit, displaying failure messages in plain English, interrogating BITE, and initiating system tests. It can also be linked with Acars, the airborne communications and reporting system. This radios CMS data ahead so that base is ready with the right tools and spares, perhaps saving a major minimum equipment list (MEL) delay. Air Inter is already using Acars in this way. The Centralized Maintenance System does not require a fault to be 'techlogged' by the crew – the system will inform the maintenance engineers of a problem and help to diagnose it. The A330 'tells' the engineer or pilot what is wrong with it. With CMS the ground crew can check complete aircraft maintenance status on every turn-around, though, as one Airbus engineer says, 'training will be required to avoid too much data'.

Airbus is certificating three makes of wheels and carbon brakes: Messier/Goodrich, ABS (Loral) and Bendix. Two sorts of tyre are being certificated: Michelin radials and Goodyear X-plies. Bridgestone (Japan) X-plies and radials, and Dunlop (France) X-plies, may also be certificated. Tpis, the cockpit tyre pressure indication system, is standard. Primarily a safety aid, allowing pilots to check their tyres before takeoff and landing, Tpis is also a valuable maintenance aid.

'Everywhere in the airline industry today the pressure is on to reduce maintenance and engineering costs', says a senior Airbus official. 'Airlines can't do without pilots and fuel, but licensed engineers are at a premium and aircraft are now regarded as machines for carrying passengers, not for employing engineers. There are no sensational maintainability breakthroughs, just constant incremental improvements. We have to increase 75 hour inspections to 150 or even 350 hours. The word "annual" has to mean 15 months. Perhaps when we achieve that we will deserve 15 month annual salaries.'

PART TWO

Maintenance Analysis

CHAPTER 3

Maintenance tasks

Even the most junior mechanic can keep an airplane in the hangar
if something isn't right.

(American Airline)

All users desire their systems to be able to stay in a state of functioning
(SoFu) as long as possible for obvious reasons. In order to achieve that it
is necessary to 'assist' the system in maintaining its functionability
during operation, by performing appropriate tasks. As the main
building block of the maintenance process is a maintenance task, the
main objective of this chapter is to analyse its main characteristics,
categories and features.

3.1   CONCEPT OF A MAINTENANCE TASK

A maintenance task is a set of activities which need to be
performed, in specified manner, by the user in order to maintain
the functionability of the item/system.

(Knezevic, 1995)

Thus, a need for execution of a specific task on an item/system, the
functionability of which has to be maintained by a user, represents the
input into the maintenance task, and the successful task completion is
the output of the maintenance task, as shown in Figure 3.1.

It is necessary to stress that the number of activities, their sequence,
and the type and quantity of resource required mainly depends on the
decisions taken during the design phase of the item/system. In a sense,
the order of magnitude of the length of the elapsed time required for the
restoration of functionability (5 minutes, 5 hours, or 2 days) could only be
taken at a very early stage of the design process, through decisions related
to the complexity of the maintenance task, accessibility of the items,
safety of the restoration, testability and physical location of the item, as

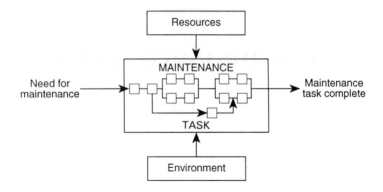

Figure 3.1 Maintenance task.

well as the decisions related to the requirements for the maintenance support resources (facilities, spares, tools, trained personnel and similar).

It is necessary to stress that some resources are needed to facilitate the successful completion of the maintenance task. As the main task of these resources is to facilitate the maintenance process they will be called **maintenance resources**, MR. The resources needed for the successful completion of every maintenance task could be grouped into the following categories.

- *Maintenance Supply Support, MSS*: A generic name which includes all spares, repair items, consumables, special supplies and related inventories needed to support the maintenance process.
- *Maintenance Test and Support Equipment, MTE*: Includes all tools, special condition monitoring equipment, diagnostic and check-out equipment, metrology and calibration equipment, maintenance stands and servicing and handling equipment required to support maintenance tasks associated with the item/system.
- *Maintenance Personnel, MP*: Required for the installation, check-out, handling and sustaining maintenance of the item/system and its associated test and support equipment. Formal training for maintenance personnel required for each maintenance task should be considered.
- *Maintenance Facilities, MFC*: Refers to all special facilities needed for completion of maintenance tasks. Physical plant, real estate, portable buildings, inspection pits, dry docks, housing, maintenance shops, calibration laboratories and special repair and overhaul facilities must be considered related to each maintenance task.
- *Maintenance Technical Data, MTD*: Necessary for check-out procedures, maintenance instructions, inspection and calibration procedures,

overhaul procedures, modification instructions, facilities information, drawings and specifications that are necessary in the performance of system maintenance functions. Such data not only cover the system, but also test and support equipment, transportation and handling equipment, training equipment and facilities.

- *Maintenance Computer Resources, MCR*: Refers to all computer equipment and accessories, software, programme tapes/disks, databases and so on, necessary in the performance of maintenance functions. This includes both condition monitoring and diagnostics.

On the other hand, it is important to remember that each task is performed in a specific work environment which could make significant impact on the safety, accuracy and ease of task completion. The main environmental factors could be grouped as follows.

- Space impediment (which reflects the obstructions imposed on maintenance personnel during the task execution which requires them to operate in awkward positions).
- Climatic conditions (rain/snow, solar radiation, humidity, temperature, etc., which could make significant impact on the safety, accuracy and ease of task completion).
- Platform on which the maintenance task is performed (on-board a ship/ submarine, space vehicle and similar).

Example 3.1

In order to illustrate the above concept, a very simple maintenance task will be used. It is related to changing a wheel on a small passenger car. Thus, the objective of this task is to restore functionability of a faulty tyre by replacing the wheel and tyre assembly with a functionable one. The list of specified activities which have to be performed in a sequence is shown in Table 3.1.

Table 3.1   List of maintenance activities

Order no.	Description of activity
1	Remove spare wheel from car boot
2	Take off wheel trim
3	Loosen all four bolts on existing wheel
4	Position and secure the jack
5	Raise car
6	Remove bolts and take off the wheel
7	Replace the wheel and tighten bolts by hand
8	Lower jack
9	Tighten all four bolts
10	Install the wheel trim
11	Place the old wheel and jack in boot

Maintenance tasks, like this one, are specified in the user manual which is delivered to the user together with the car, at the beginning of the operation of the system. Also, all maintenance resources needed for the successful completion of the tasks considered likely to be performed by the user have been provided by the manufacturer of the car to the user as a part of the overall package. The list of resources needed for the task analysed is given in Table 3.2.

Table 3.2 List of required maintenance resources

Resources category	Specific resource
Personnel	Existing (driver, no training required)
Supply support	Spare wheel
Equipment	Mechanical jack
Tools	Screwdriver, spanner 19 mm
Facilities	Existing
Data	Tyre's pressure
Technical information	User's manual
Computer resources	N/a

3.2 MAINTENANCE TASK CLASSIFICATION

According to the objective of performing a maintenance task, all of them could be classified into three categories:

(a) corrective maintenance tasks;
(b) preventive maintenance tasks;
(c) conditional maintenance tasks.

Each maintenance task is briefly examined below.

3.3 CORRECTIVE MAINTENANCE TASKS

Corrective maintenance tasks, CRTs, are the tasks which are performed with the intention of restoring the functionability of the item or system, after the loss of the function or performance. A typical corrective maintenance task consists of the following activities:

- failure detection
- failure location
- disassembly
- restoration/replacement

- assembly
- test/check
- verification.

Graphical representation of a CRT maintenance task is given in Figure 3.2. The duration of the task is denoted as DMT^c, which represents the elapsed time needed for the successful completion of the corrective maintenance task.

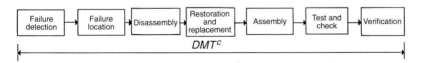

Figure 3.2 Graphical representation of a typical corrective maintenance task.

3.4 PREVENTIVE MAINTENANCE TASKS

A preventive maintenance task, PRT, is a task which is performed in order to reduce the probability of failure of the item/system or to maximize the operational benefit. A typical preventive maintenance task consists of the following maintenance activities:

- disassembly
- restoration/replacement
- assembly
- test/check
- verification.

Graphical representation of a PRT maintenance task is given in Figure 3.3. The duration of the task is denoted as DMT^P, which represents the elapsed time needed for the successful completion of the preventive maintenance task.

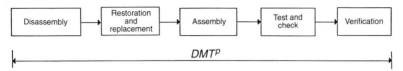

Figure 3.3 Graphical representation of a typical preventive maintenance task.

Maintenance tasks of this type are performed before the transition to the SoFa occurs with a main objective of reducing:

- maintenance cost
- probability of failure.

Table 3.3 Types and frequency of preventive tasks used in
the aircraft industry

Type	Frequency (flying hours)	Duration (elapsed days)
Major check	18 000	25–30
Inter check	4500	9
Service check	800–3000	1–2
Ramp check	125–500	0.3

Most common preventive maintenance tasks are replacements,
renewal, overhaul and similar. It is necessary to stress that these tasks
are performed, at fixed intervals such as every 3000 operating hours,
every 10 000 miles or 500 landings, regardless of the real condition of the
items/systems.

Typical examples of preventive maintenance tasks used by a majority
of operators of a large commercial aircraft are given in Table 3.3.

3.5 CONDITIONAL MAINTENANCE TASKS

Traditionally, corrective maintenance and preventive maintenance tasks
have been favourite among maintenance managers. However, during
the last 20 years, the disadvantages of these approaches have been
recognized by many industrial organizations. Therefore, the need for
the provision of safety and reduction of the maintenance cost have led to
an increasing interest in development of alternative maintenance tasks.
Consequently, the approach which seems to be the most attractive for
minimizing the limitations of existing maintenance tasks is the conditional maintenance task, COT. This maintenance task recognizes that a
change in condition and/or performance is the principle reason for
carrying out maintenance, and execution of preventive maintenance
tasks should be based on the actual condition of the item/system. Thus,
through monitoring some parameter(s) it should be possible to identify
the most suitable instant of time at which preventive maintenance tasks
should take place.

Consequently, a conditional maintenance task represents a maintenance task which is performed to gain insight into the condition of the
item/system or discover hidden failure, in order to determine the further
course of action regarding the maintenance of the functionability of the
item/system, from the point of view of the user.

The conditional maintenance task is based on condition monitoring
activities which are performed in order to determine the physical state of
an item/system. Therefore, the aim of condition monitoring, whatever

form it takes, is to monitor those parameters which provide information about the changes in condition and/or performance of an item/system. The philosophy of condition monitoring is therefore the assessment of the current condition of an item/system by the use of techniques which can range from human sensing to sophisticated instrumentation, in order to determine the need for performing a preventive maintenance task.

A typical conditional maintenance task consists of the following maintenance activities:

- condition assessment
- condition interpretation
- decision making.

Graphical representation of a COT maintenance task is given in Figure 3.4. The duration of the task is denoted as DMT^m, which represents the elapsed time needed for the successful completion of the conditional maintenance task.

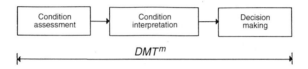

Figure 3.4   Graphical representation of a typical conditional maintenance task.

Therefore, the conditional maintenance task recognizes that a change in condition and/or performance is the principle reason for carrying out maintenance, and execution of preventive maintenance tasks should be based on the actual condition of the item/system. Thus, the condition assessment through condition monitoring of selected parameter(s) enables the user to identify the most suitable instant of time at which preventive maintenance tasks should take place. Thus, the preventive maintenance tasks are not performed as long as the condition of the item/system is acceptable.

3.5.1   Condition monitoring parameters

In order to assess the condition of the item/system, in engineering practice, there are two distinguishable types of conditional parameters used (Knezevic, 1993).

(a)   Relevant Condition Indicator, RCI

This is a monitorable parameter which indicates the condition of the item/system, at the instant of checking. Typical examples of RCIs are:

- the level of pressure, power steering fluid, vibration, noise, oil, water, brake fluid, coolant, windscreen washer liquid;
- engine idle speed;
- alternator/water pump belt tension;
- clutch pedal travel, handbrake lever travel;
- wheel geometry.

According to the RCI, the condition of the item/system is satisfactory as long as it maintains a value beyond its critical level, RCI_{cr}. When this level is reached the required maintenance task must be performed, because the failure will occur as soon as the parameter reaches its limit value, RCI_{lim} as illustrated by Figure 3.5. It is necessary to stress that the RCI could have an identical value at different instances of operating time.

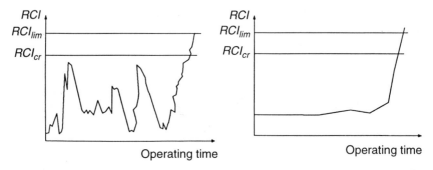

Figure 3.5 Change of *RCI* during operational time.

(b)   Relevant Condition Predictor, RCP

This is a monitorable parameter which describes the condition of the item at every instant of operating time. Usually this parameter is directly related to the shape, geometry, weight and other characteristics which describe the condition of the item under consideration. Typical examples of RCP are thickness of wall pipes, brake pads/shoe, brake disc, driven plate of the clutch, crack length, depth of tyre tread, diameter of a cylinder and similar. Generally speaking, the condition of the item/system is satisfactory as long as the RCP maintains a value beyond its critical level, $RCPcr$. At this point the required preventive maintenance task must be performed, because the failure will occur as soon as the parameter reaches its limit value, RCP_{lim}. It is necessary to stress that the RCP cannot have identical values at two or more instances of time, as illustrated by Figure 3.6. This means that the RCP is continuously increasing/decreasing with operating time.

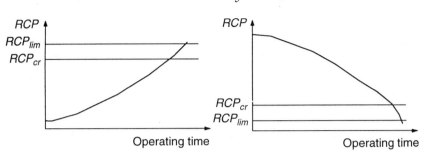

Figure 3.6 Change of *RCP* during operational time.

3.5.2 Type of conditional maintenance task

According to the way in which the obtained information about the condition of the item/system is used to assess the condition of the item/system, two different conditional maintenance tasks can be distinguished (Knezevic, 1995):

(a) *Inspection*: a specific conditional maintenance task the result of which is a statement about the condition of the item, i.e. the condition is satisfactory or it is unsatisfactory, which is determined according to the RCI.

(b) *Examination*: another type of conditional maintenance task the result of which is a numerical description of the condition of the item at that moment through RCP.

3.6 CASE STUDY

In order to illustrate the impact of maintainability on maintenance tasks, the comparison between scheduled maintenance tasks, at 6000 and 12 000 miles (9600 and 19 200 km), for the Austin Mini and Ford Cortina are summarized in Tables 3.4 and 3.5.

Table 3.4 Scheduled maintenance tasks at 6000 miles (9600 km)

Activities	Austin Mini			Ford Cortina		
	MDMT	SL	NMPS	MDMT	SL	NMPS
Air filter change	0.1	U	1	0.1	U	1
Oil filter change	0.4	U	1	0.2	U	1
Points gap check	0.3	S	1	0.1	S	1
Plug change	0.2	U	1	0.2	U	1
Engine oil change	0.3	U	1	0.2	U	1

Table 3.4 *Continued*

Activities	Austin Mini			Ford Cortina		
	MDMT	SL	NMPS	MDMT	SL	NMPS
Check gearbox oil	n/a	–	–	0.1	U	1
Check rear axle oil	n/a	–	–	0.1	U	1
Check brake pads	0.3	U	1	0.1	U	1
Check rear shoes	0.4	S	1	0.2	S	1
Grease front swivels	0.4	S	1	n/a	–	1
Check fan belt	0.1	U	1	0.1	U	1
Check lights	0.1	U	1	0.1	U	1
Check levels	0.1	U	1	0.1	U	1
Total time	2.7			1.6		

MDMT = mean duration of maintenance task (hours); *SL* = skill level (S, skilled; U, unskilled); *NMPS* = number of maintenance personnel.

Table 3.5 12 000 miles scheduled maintenance tasks

Activities	Austin Mini			Ford Cortina		
	MDMT	SL	NMPS	MDMT	SL	NMPS
Change carburettor	0.5	S	1	0.1	S	1
Tune carburettor	0.5	S	1	0.1	S	1
Set tappets	0.6	S	2	0.2	S	1
Remove and replace rocker cover	0.3	S	1	0.2	S	1
Grease rear shaft	0.2	U	1	n/a	–	–
Adjust clutch freeplay	0.2	S	1	0.2	S	1
Adjust rear brakes	0.4	S	1	Auto	n/a	n/a
Change thermostat winter/summer operation	0.5	S	1	0.3	S	1
Total time	3.2			1.1		

MDMT = mean duration of maintenance task (hours); *SL* = skill level (S, skilled; U, unskilled); *NMPS* = number of maintenance personnel.

One of the most frequently performed corrective maintenance tasks on internal combustion engines is the headgasket change. In order to illustrate the meaning of the *MDMT*, a comparison of the duration of this task in the Austin Mini and Ford Cortina is summarized in Table 3.6.

Table 3.6 *MDMT* for the headgasket change

Maintenance activities	Austin Mini			Ford Cortina		
	MDMA	SL	NMPS	MDMA	SL	NMPS
Remove carburettors	0.5	S	1	n/a	–	–
Detach exhaust pipe	0.5	S	1	0.2	U	1
Remove manifold	0.4	S	1	0.3	U	1
Remove rocker assembly	0.2	S	1	0.2	S	1
Drain radiator and remove top hose	0.3	S	1	0.3	S	1
Remove head bolts/ nuts	0.5	S	1	0.2	S	1
Lift head	0.2	S	1	0.2	S	1
Clean head/block faces and remove gasket	0.8	S	1	0.2	S	1
Replace gasket and head	0.1	S	1	0.1	S	1
Install head nuts/ bolts	0.3	S	1	0.1	S	1
Replace rocker assembly	0.3	S	1	0.2	S	1
Adjust tappets	0.4	S	2	0.3	S	1
Connect top hose	0.1	S	1	0.1	S	1
Replace manifold	0.3	S	1	0.2	S	1
Replace exhaust pipe	0.6	S	1	0.3	S	1
Replace carburettors	0.4	S	1	n/a	–	–
Total time	5.9			2.9		

MDMA = mean duration of maintenance activity; SL = skill level (S, skilled; U, unskilled);
NMPS = number of maintenance personnel.

Clearly, the differences between the mean duration of this mainten-ance task are inherited from the design. As the result of the engine design of the Austin Mini, removal of carburettors is required, which adds almost an extra hour to the duration of this task.

CHAPTER 4

Duration of maintenance tasks

A major concern of the user of repairable systems is the shape of the functionability profile, as shown in Figure 1.3, but of even greater concern is the proportion of the elapsed time during which the product under consideration will be available for the performance of the function (Figure 4.1). Thus, it is very important to provide the answer to the following question: What is the nature of the duration of maintenance task (*DMT*)?

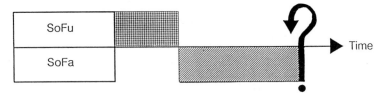

Figure 4.1 Duration of maintenance task.

In other words, is the *DMT* constant for each execution of the maintenance task considered, or does it differ from trial to trial?

In order to illustrate this point, in Table 4.1 elapsed time needed for the replacement of the wheel, according to the procedure shown in Table 3.1, by a group of second-year students from the School of Engineering of Exeter University in the UK, are given. Ten students performed this task individually on the same car following a list of specified activities which were to be performed in sequence. The tools required for execution of this task were laid out beside the wheel to be changed.

Table 4.1 The duration of replacement of the wheel

Student	1	2	3	4	5	6	7	8	9	10
Elapsed time (s)	230	259	442	286	397	365	332	279	321	351

In setting the task, an attempt was made to minimize the effect of various external factors by:

(a) Performing the task in a garage in order to achieve stable environmental conditions.
(b) Using only engineering students. An attempt was also made to select a group with a similar mental approach, minimizing personal factors.

However, elapsed time differences indicate varied levels of skill, motivation, experience and physical ability.

Generally speaking, if the duration of elapsed time of several trials of a specific maintenance task is analysed, it will be seen that one of them will be completed at the instant denoted by b_1, another at instant b_2 and the n will be executed by instance b_n (Figure 4.2).

The illustration only confirms what everyone familiar with the maintenance of engineering items already knows: the execution of each trial of a specific maintenance task will be completed after a different duration of elapsed time. Thus, the duration of elapsed time needed for the completion of each maintenance task is a specific characteristic of each trial.

The question which naturally arises here is why are different durations of an elapsed time needed for the execution of identical maintenance tasks?

In order to provide the answer to this question it is necessary to

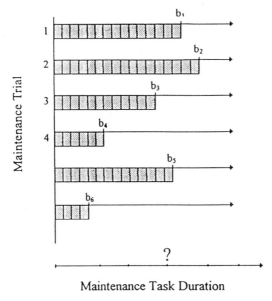

Figure 4.2 Maintenance pattern of several trials of a specific maintenance task.

analyse all factors which are responsible. The following three groups are the most influential.

- *Personal factors*: which represent the influence of the skill, motivation, experience, attitude, physical ability, eyesight, self-discipline, training, responsibility and other similar characteristics of the personnel involved.
- *Conditional factors*: which represent the influence of the operating environment and the consequences of the failure to the physical condition, geometry and shape of the item under restoration.
- *Environmental factors*: which represent the influence of factors such as temperature, humidity, noise, lighting, vibration, time of the day, time of the year, wind, noise and similar on the maintenance personnel during the restoration.

Thus, the different durations of elapsed time for the execution of each individual trial of the maintenance task are the result of the influence of the above-mentioned factors.

Consequently, the nature of the parameter *DMT* for the maintenance task also depends on the variability of these parameters. The relationship between the influential factors and parameter *DMT* could be expressed by the following equation:

$$DMT = f \text{ (personal, conditional and environmental factors)} \quad (4.1)$$

Analysing the above expression it could be said that as a result of the large number of influential parameters in each group on the one hand and their variability on the other hand, it is impossible to find a rule which would deterministically describe this very complex relation denoted by '*f*'. The only way forward with maintainability analysis is to call upon probability theory, which offers a 'tool' for the probabilistic description of the relationship defined by the above expression.

Thus, in spite of the fact that each maintenance task consists of the specified activities which are performed in the specified sequence, the elapsed time needed for the execution of all of them might differ from trial to trial.

In conclusion it could be said that it is impossible to give a deterministic answer regarding the instant of operating time when the transition from the SoFa to the SoFu will occur for any individual trial of the maintenance task under consideration. It is only possible to assign a certain probability that it will happen at a certain instant of maintenance time or that a certain percentage of trials will or will not be completed by the specific instant of elapsed time.

Since it was clearly demonstrated earlier that the maintenance task considered represents a flow of maintenance activities, the timing of which can only be described in probabilistic terms, let us establish the relationship between the concept of the probability system and *DMT*.

The duration of the maintenance task could be considered as a random experiment and its successful completion as an elementary event which represents its outcome. The function which assigns a corresponding numerical value t_i to every elementary event b_i from sample space S is a random variable; in this particular case it will be called duration of maintenance task, *DMT*, as shown in Figure 4.3.

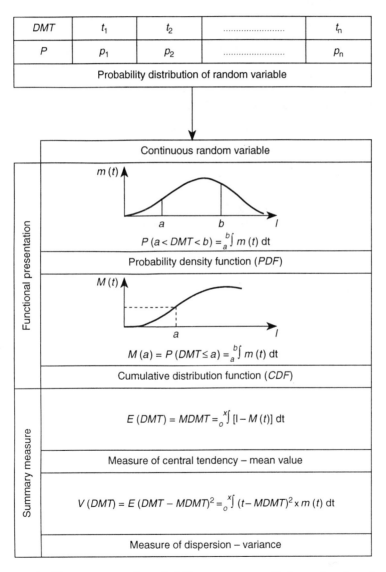

Figure 4.3 The concept of probability system applied to the duration of maintenance task.

Thus, the probability that random variable *DMT* will take on value t_i is $p_i = P(DMT = t_i)$. The numerical values taken on by random variables and the probability of their occurrence define a probability distribution which can be expressed by different indicators. This establishes the complete analogy between the probability system defined in probability theory and the ability of a system to be maintained in SoFu (Knezevic, 1993).

It should be noted at this point that, although time is the true variable, it may often be more convenient to use some other readily available variables which represent time of use: days, hours, minutes, etc.

The most frequently used characteristics of *DMT* are:

• maintainability function;
• mean duration of maintenance task;
• variance of duration of maintenance task.

A brief definition and description of these characteristics follows.

4.1 MAINTAINABILITY FUNCTION

This function, denoted as $M(t)$, represents the probability that the maintenance task considered will be successfully completed before or at the specified moment of maintenance elapsed time t, thus:

$M(t) = P$(maintenance task will be completed before or at elapsed time t)

$$= P(DMT \leq t)$$

$$= \int_0^t m(t)dt \tag{4.2}$$

where $m(t)$ is the probability density function of *DMT*.

The maintainability function for a hypothetical maintenance task is shown in Figure 4.4.

4.2 MEAN DURATION OF A MAINTENANCE TASK

This maintainability measure, denoted as $E(DMT)$, represents the expectation of the random variable *DMT* can be used for calculation of this characteristic of the restoration process, thus:

$$E(DMT) = MDMT = \int_0^\infty t \times m(t)dt \tag{4.3}$$

The above characteristic could also be expressed in the following way:

$$E(DMT) = \int_0^\infty [1 - M(t)]dt \tag{4.4}$$

which represents the area below the function which is complementary to the maintainability function.

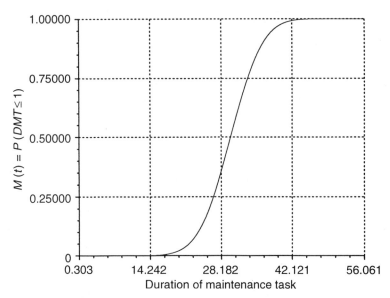

Figure 4.4 Maintainability function for hypothetical task.

4.3 VARIANCE OF DURATION OF A MAINTENANCE TASK

This maintainability measure, denoted as $V(DMT)$, represents the expectation of the square deviation about the mean, and it is defined by the following expression:

$$V(DMT) = E(DMT - MDMT)^2 = \int_0^\infty (t - MDMT)^2 \times m(t)dt \quad (4.5)$$

The above characteristic represents the degree of the dispersion of individual values from the mean. Clearly, the larger dispersions are from the mean, the more the individual values differ from each other, and the more apparent the spread, within the distribution, becomes.

It is worth mentioning that the positive square root of the variance for a distribution considered is known as the standard deviation, $SD(DMT)$, and it is determined by the following equation:

$$SD(DMT) = \sqrt{V(DMT)} \quad (4.6)$$

Thus, the standard deviation is a measure which shows how closely the values of random variables are concentrated around the mean.

4.4 WELL-KNOWN THEORETICAL DISTRIBUTIONS

In probability theory there are several rules which define the functional relationship between the possible values of random variable X and their probabilities, $P(X)$. Rules which have been developed by mathematicians will be analysed here. As they are purely theoretical, i.e. they do not exist in reality, they are called theoretical probability distributions. Instead of analysing the ways in which these rules have been derived, the analysis in this chapter concentrates on their properties.

It is necessary to emphasize that all theoretical distributions represent the family of distributions defined by a common rule through unspecified constants known as parameters of distribution. The particular member of the family is defined by fixing numerical values for the parameters which define the distribution. The probability distributions most frequently used in maintainability engineering are examined in this chapter.

In the theory of probability there are several different rules for distribution functions each of which are derived in a purely theoretical way. The most frequently used rules for distribution functions in maintenance engineering practice are exponential, normal, lognormal and Weibull.

Each of the above-mentioned rules define a family of distribution functions. Each member of the family is defined with a few parameters which in their own way control the distribution. All parameters can be classified in the following three categories:

(a) Scale parameter, A_m, which defines the location of the distribution on the horizontal scale.
(b) Shape parameter, B_m, which controls the shape of the distribution curves.
(c) Source parameter, C_m, which defines the origin or the minimum value which a random variable can have.

Thus, individual members of a specific family of probability distributions are defined by fixing numerical values for the above parameters.

4.4.1 The exponential distribution

This type of probability distribution is fully defined by a single parameter which governs the location of the distribution, as well as its shape. Thus, according to the notation introduced above, $A_m = B_m$. The mathematical expression for the probability density function, in the case of the exponential distribution, is defined by the following rule:

$$m(t) = \frac{1}{A_m} \exp - \left(\frac{t}{A_m} \right), \, t > 0 \qquad (4.7)$$

Making use of the general expression for the maintainability function, in the case of the exponential probability distribution it will have the following form:

$$M(t) = P(DMT \leq t) = 1 - \exp - \left(\frac{t}{A_m} \right) \quad (4.8)$$

In Figure 4.5 several curves of cumulative distribution functions are shown for different values of A_m.

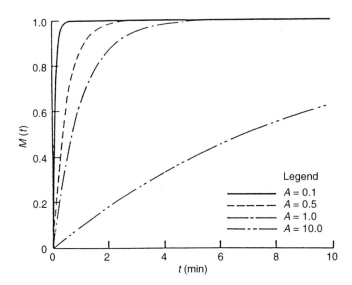

Figure 4.5 $M(t)$ for the exponential distribution.

It can be shown that the mean and variance of the exponential distribution are:

$$E(DMT) = MDMT = A_m \quad (4.9)$$

$$V(DMT) = A_m^2 \quad (4.10)$$

The standard deviation in the case of the exponential distribution rule has a numerical value identical to the mean and the scale parameter, $SD(DMT) = E(DMT) = A_m$.

Table 4.2 summarizes the exponential distribution characteristics for *DMT*.

Table 4.2 Exponential distribution characteristics for *DMT*

$m(t)$	$\frac{1}{A_m} \exp\left(-\left(\frac{t}{A_m} \right) \right)$
$M(t)$	$1 - \exp\left(-\left(\frac{t}{A_m} \right) \right)$
$E(DMT)$	A_m
$V(DMT)$	A_m^2

Example 4.1

On average it takes 10 days to restore a specific machine. Find the chance that less than 5 days will be enough to successfully complete the restoration of this type of machine.

Solution
Based on the available data, we are dealing with the probability distribution which is governed by the exponential rule where the scale parameter A_m is 10 days. Accordingly, the probability density function is

$$m(t) = \frac{1}{10} \exp - \left(\frac{t}{10} \right)$$

and the required chance is $P(DMT \leq 5)$. The required probability is determined by the value of $M(t)$ where $t = 5$, thus:

$$P(DMT \leq 5) = M(5) = \int_0^5 \frac{1}{10} \exp - \left(\frac{t}{10} \right) dt = 1 - \exp - \frac{5}{10} = 1 - 0.61 = 0.39$$

Example 4.2

Suppose that the length of the failure diagnosis is an exponential random variable with parameter $A_m = 3$ minutes. Find the probability that the failure finding task will be (a) more than 3 minutes, or (b) between 3 and 6 minutes.

Solution
Letting *DMT* denote the length of the time needed for the successful diagnosis of failure, the desired probabilities are:

(a) $P(DMT > 3) = \int_3^\infty 1/3 \exp - \left(\frac{t}{3} \right) dt = \exp - 1 = 0.368$

(b) $P(3 < DMT < 6) = \int_3^6 1/3 \exp - \left(\frac{t}{3} \right) dt = \exp - 1 - \exp - 2 = 0.233$

4.4.2   Normal distribution

This is the most frequently used and most extensively covered theoretical distribution in the literature. The normal distribution is continuous for all values of *DMT* between $-\infty$ and $+\infty$. It has a characteristic symmetrical shape, which indicates that the mean, the median and the mode have the same numerical value. The mathematical rule for its probability density function is as follows:

$$m(t) = \frac{1}{\sqrt{2\pi}B_m} \exp -\frac{1}{2} \left(\frac{t - A_m}{B_m} \right)^2 \qquad (4.11)$$

where A_m is a scale parameter and B_m is a shape parameter.

This rule pairs a probability density $m(t)$ with each and every possible value of *DMT*.

As a deviation of t from the scale parameter A_m enters as a squared quantity, two different t values, showing the same absolute deviation from A_m, will have the same probability density according to this rule. This dictates the symmetry of the normal distribution. Parameter A_m can be any finite number, while B_m can be any positive finite number.

The maintainability function for the normal distribution is:

$$M(t) = \int_{-\infty}^{t} \frac{1}{\sqrt{2\pi}B_m} \exp -\frac{1}{2} \left(\frac{t - A_m}{B_m} \right)^2 \, dt \qquad (4.12)$$

In Figures 4.6 and 4.7 several maintainability functions are given for the normal distribution, corresponding to different values of A_m and B_m.

As the integral in equation 4.12 cannot be evaluated in a closed form, statisticians have constructed the table of probabilities which complies to

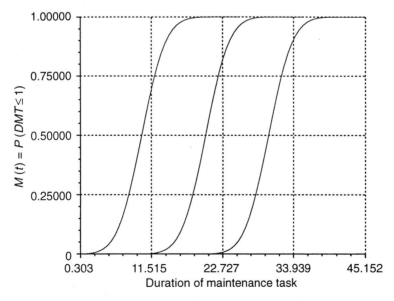

Figure 4.6 $M(t)$ for the normal distribution with different values of scale A_m.

Duration of maintenance tasks

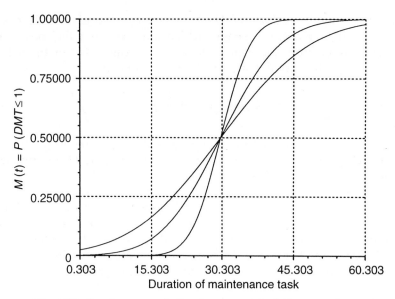

Figure 4.7 $M(t)$ for the normal distribution with different values of shape parameter B_m.

the normal rule for the standardized random variable, Z. This is a theoretical random variable with parameters $A_m = 0$ and $B_m = 1$. The relationship between standardized random variable Z and random variable DMT is established by the following expression:

$$z = \frac{t - A_m}{B_m} \tag{4.13}$$

Making use of the above expression equation 4.11 becomes simpler:

$$m(z) = \frac{1}{\sqrt{2\pi}} \exp \left(-\frac{1}{2}z^2 \right) \tag{4.14}$$

The standardized form of the distribution makes it possible to use one table only for the determination of *PDF* for any normal distribution, regardless of its particular parameters (see Appendix Table 1).

The relationship between $m(t)$ and $m(z)$ is:

$$m(t) = \frac{m(z)}{B_m} \tag{4.15}$$

By substituting $\dfrac{t - A_m}{B_m}$ with z equation 4.12 becomes:

$$M(t) = \int_{-\infty}^{z} \frac{1}{\sqrt{2\pi}} \exp -\frac{1}{2}z^2 \, dz = \Phi \left(\frac{t - A_m}{B_m} \right) \tag{4.16}$$

where Φ is the standard normal distribution function defined by:

$$\Phi(z) = \int_{-\infty}^{t} \frac{1}{\sqrt{2\pi}} \exp\left(-\frac{1}{2}z^2\right) dt \qquad (4.17)$$

The corresponding standard normal probability density function is:

$$m(z) = \frac{1}{\sqrt{2\pi}} \exp\left(\frac{-z^2}{2}\right) \qquad (4.18)$$

Most tables of the normal distribution give the cumulative probabilities for various standardized values. That is, for a given z value the table provides the cumulative probability up to, and including, that standardized value in a normal distribution.

The expectation of a random variable, in the case of the normal distribution, is equal to the scale parameter A_m, thus:

$$E(DMT) = A_m \qquad (4.19)$$

whereas the variance is:

$$V(DMT) = B_m^2 \qquad (4.20)$$

Table 4.3 summarizes the normal distribution characteristics for DMT.

Table 4.3 Normal distribution characteristics for *DMT*

$m(t)$	$\dfrac{1}{\sqrt{2\pi}B_m} \exp\left[-\dfrac{1}{2}\left(\dfrac{t - A_m}{B_m}\right)^2\right]$
$M(t)$	$\displaystyle\int_{-\infty}^{t} \dfrac{1}{\sqrt{2\pi}B_m} \exp\left[-\dfrac{1}{2}\left(\dfrac{t - A_m}{B_m}\right)^2\right] dt$
$E(DMT)$	A_m
$V(DMT)$	B_m^2

Example 4.3

A catering department installed 2000 coffee machines, each of which requires an annual service with average duration of 200 minutes with a standard deviation of 40 minutes.

(a) How many machines might be expected to be serviced in the first 100 minutes of maintenance?
(b) What is the probability of a coffee machine being serviced between 180 and 240 minutes of maintenance?

Solution

The duration of service time of this particular machine can be represented with a random variable, DST, whose probability distribution is defined as $N(200,40)$.

(a) $P(DST \leq 100) = M(100) = ?$

According to equation 4.13 $z = (100 - 200)/40 = -2.5$. From Appendix Table 1 the required probability is $\Phi(-2.5) = 0.00621$. Thus, the expected number of coffee machines to be serviced within the first 100 minutes is $200 \times 0.00621 = 12.4 = 12$.

(b) $P(180 \leq DST \leq 240) = M(240) - M(180) = ?$

The required probability can be determined by making use of equation 4.16, thus: $M(240) = \Phi(1.0) = 0.84135$ and $M(180) = \Phi(-0.5) = 0.30848$. Therefore, the probability of a coffee machine being successfully serviced in the specified interval of maintenance time is $P(180 \leq DST \leq 240) = 0.53287$.

4.4.3 Lognormal distribution

In some respects, this probability distribution can be considered as a special case of the normal distribution because of the derivation of its probability function. If a random variable $Y = \ln DMT$ is normally distributed $N(A_Y, B_Y)$, then the random variable DMT follows the lognormal distribution.

Thus, the probability density function for a random variable DMT is defined as:

$$m_{DMT}(t) = \frac{1}{xB_Y\sqrt{2\pi}} \exp\left[-\frac{1}{2}\left(\frac{\ln t - A_y}{B_y} \right)^2 \right] \geq 0 \qquad (4.21)$$

The relationship between parameters A_{DMT} and A_Y is defined by:

$$A_{DMT} = \exp\left(A_Y + \frac{1}{2}B_Y^2 \right) \qquad (4.22)$$

and the relationship between B_{DMT} and B_Y is:

$$B_{DMT} = \sqrt{A_{DMT}^2(\exp(B_Y^2) - 1)} \qquad (4.23)$$

The maintainability function, $M(t)$, for the lognormal distribution is defined by the following expression:

$$M_{DMT}(t) = (DMT \leq t) = \int_0^t \frac{1}{tB_Y\sqrt{2\pi}} \exp\left[\frac{1}{2} - \left(\frac{\ln t - A_Y}{B_Y} \right)^2 \right] dt \qquad (4.24)$$

As the integral cannot be evaluated in close form the same procedure is applied as in the case of normal distribution. Then, making use of the standardized random variable, equation 4.24 transforms into:

$$M_{DMT}(t) = P(DMT \leq t) = \Phi\left(\frac{\ln t - A_Y}{B_Y}\right) \quad (4.25)$$

where Φ is the standard normal distribution function whose numerical values can be found in Appendix Table 1, for:

$$z = \frac{\ln t - A_Y}{B_Y} \quad (4.26)$$

Several cumulative distributive functions are shown in Figures 4.8 and 4.9.

The measures of central tendency in the case of lognormal distributions are defined by: (a) location parameters (equations 4.27 to 4.29):

Mean: $\quad MDMT = E(DMT) = \exp\left(A_Y + \dfrac{1}{2} B_Y^2\right)$ (4.27)

Median: $MdDMT = \exp(A_Y)$ (4.28)

Mode: $\quad MoDMT = \exp(A_Y - B_Y^2)$ (4.29)

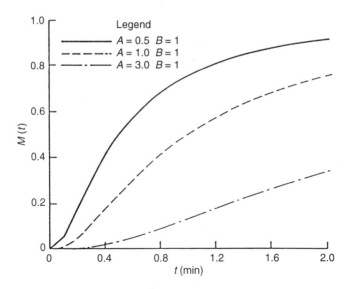

Figure 4.8 $M(t)$ for lognormal distribution with constant shaped parameters.

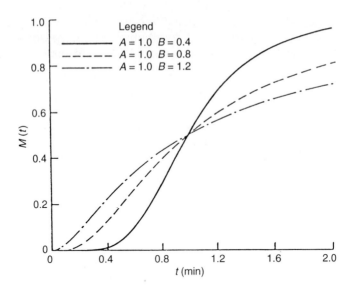

Figure 4.9 $M(t)$ for lognormal distribution with constant scaled parameters.

and (b) the deviation parameter (the variance):

$$V(DMT) = \exp(2A_Y + B_Y)^2 \, [\exp(B_Y^2 - 1)] \qquad (4.30)$$

Table 4.4 summarizes the lognormal distribution characteristics for *DMT*.

Table 4.4 Lognormal distribution characteristics for *DMT*

$m(t)$	$\dfrac{1}{tB_Y\sqrt{2\pi}}\exp\left[-\dfrac{1}{2}\left(\dfrac{\ln t - A_Y}{B_Y}\right)^2\right]$
$M(t)$	$\Phi\left(\dfrac{\ln t - A_Y}{B_Y}\right)$
$E(DMT)$	$\exp(A_Y + \dfrac{1}{2}B_Y^2)$
$V(DMT)$	$\exp(2A_Y + B_Y)^2[\exp(B_Y^2 - 1)]$

So far the range of random variables considered has been between 0 and + infinity. The lognormal distribution can be successfully applied in the cases where the random variable has a range $C_m \leqslant DMT \leqslant +\infty$, where C_m can take any value greater than zero. As it has already been pointed out, the parameter C_m is called a source parameter and represents a value below which probability is equal to 0. This type of lognormal distribution is known as the **three parameter distribution**. The relevant equations for the three parameter lognormal distribution are given in Table 4.5.

Table 4.5 Three parameter lognormal distribution characteristics for *DMT*

$m(t)$	$\dfrac{1}{tB_Y\sqrt{2\pi}}\exp\left[-\dfrac{1}{2}\left(\dfrac{\ln(t-C_m)-A_Y}{B_Y}\right)^2\right]$
$M(t)$	$\Phi\left(\dfrac{\ln(t-C_m)-A_Y}{B_Y}\right)$
$E(DMT)$	$C_m + \exp(A_Y + \dfrac{1}{2}B_Y^2)$
$V(DMT)$	$C_m + \exp(2A_Y + B_Y)^2[\exp(B_Y^2 - 1)]$

Example 4.4

The refuelling of motor vehicles can be represented by a lognormal distribution with parameters $A_m = 1.75$ and $B_m = 0.57$. What percentage of the vehicles can be refuelled within 5 minutes?

Solution

$$P(\text{refuelling time of motor vehicle} \leqslant 5) = M(5) = \Phi\left(\frac{\ln(5) - 1.75}{0.57}\right) = 0.40$$

Thus, 40% of vehicles could be refuelled in less than 5 minutes.

Example 4.5

If we are interested in the distribution of the refuelling time of motor vehicles, the more realistic distribution function will be defined by three parameters because the refuelling cannot be completed in under 3 minutes. Thus, the random variable *DMT* is defined as ln(1.75, 0.57, 3). What percentage of the vehicles are expected to be refuelled within less than 5 minutes?

Solution

$$P(\text{refuelling of motor vehicle} \leqslant 5) = M(5) = \Phi\left(\frac{\ln(5 - 3) - 1.75}{0.57}\right) = 0.031$$

Thus, less than 3% of vehicles are expected to be refuelled within 5 minutes.

4.4.4 Weibull distribution

This distribution has no characteristic shape, such as the normal distribution for example, but it has a very important role in the statistical analysis of experimental data. The shape of this distribution is governed by its parameters.

The rule for the probability density function of the Weibull distribution is:

$$m(t) = \frac{B_m}{A_m - C_m} \left(\frac{t - C_m}{A_m - C_m} \right)^{B_m - 1} \exp - \left(\frac{t - C_m}{A_m - C_m} \right)^B_m \tag{4.31}$$

where $A_m > 0$, $B_m > 0$ and $C_m > 0$. As the parameter C_m, in some cases, is often set equal to zero this can be rewritten as:

$$m(t) = \frac{B_m}{A_m} \left(\frac{t}{A_m} \right)^{B_m - 1} \exp - \left(\frac{t}{A_m} \right)^B_m \tag{4.32}$$

The cumulative distribution function for the Weibull distribution is:

$$M(t) = 1 - \exp - \left(\frac{t - C_m}{A_m - C_m} \right)^B_m \tag{4.33}$$

or

$$M(t) = 1 - \exp - \left(\frac{t}{A_m} \right)^B_m \tag{4.34}$$

Figure 4.10 shows the $M(t)$ for the Weibull distribution with different shaped parameters and Figure 4.11 shows the $M(t)$ with different scaled parameters.

The expected value of the Weibull distribution is given by:

$$E(DMT) = (A_m - C_m) \times \Gamma \left(\frac{1}{B_m} + 1 \right) \tag{4.35}$$

or

$$E(DMT) = A_m \times \Gamma \left(\frac{1}{B_m} + 1 \right) \tag{4.36}$$

where Γ is the gamma function, numerical values of which can be found in Appendix Table 4. The variance of the Weibull distribution is given by:

$$V(DMT) = (A_m - C_m)^2 \left[\Gamma \left(1 + \frac{2}{B_m} \right) - \Gamma^2 \left(1 + \frac{1}{B_m} \right) \right] \tag{4.37}$$

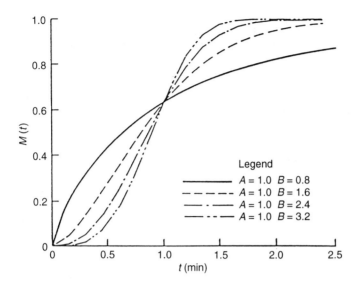

Figure 4.10 $M(t)$ for the Weibull distribution with different shaped parameters.

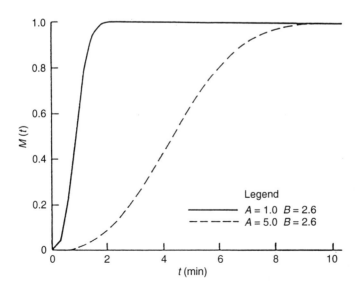

Figure 4.11 $M(t)$ for the Weibull distribution with different scaled parameters.

or

$$V(DMT) = A_m^2 \left[\Gamma\left(1 + \frac{2}{B_m}\right) - \Gamma^2\left(1 + \frac{1}{B_m}\right) \right] \qquad (4.38)$$

Numerical values for the gamma function are given in Appendix Table 4 for $0.5 \leqslant B_m \leqslant 20$ and Weibull distribution characteristics for *DMT* are summarized in Table 4.6.

Table 4.6   Weibull distribution characteristics for *DMT*

$m(t)$	$\dfrac{B_m}{A_m - C_m}\left(\dfrac{t - C_m}{A_m - C_m}\right)^{B_m-1}\exp\left[-\left(\dfrac{t - C_m}{A_m - C_m}\right)^B_m\right]$
	$\dfrac{B_m}{A_m}\left(\dfrac{t}{A_m}\right)^{B_m-1}\exp\left[-\left(\dfrac{t}{A_m}\right)^B_m\right]$
$M(t)$	$1 - \exp\left[-\left(\dfrac{t - C_m}{A_m - C_m}\right)^B_m\right]$
	$1 - \exp\left[-\left(\dfrac{t}{A_m}\right)^B_m\right]$
$E(DMT)$	$(A_m - C_m) \times \Gamma\left(\dfrac{1}{B_m} + 1\right)$
	$A_m \times \Gamma\left(\dfrac{1}{B_m} + 1\right)$
$V(DMT)$	$(A_m - C_m)^2\left[\Gamma\left(1 + \dfrac{2}{B_m}\right) - \Gamma^2\left(1 + \dfrac{1}{B_m}\right)\right]$
	$A_m^2\left[\Gamma\left(1 + \dfrac{2}{B_m}\right) - \Gamma^2\left(1 + \dfrac{1}{B_m}\right)\right]$

Example 4.6

Assume that the overhaul of a certain machine can be represented by the Weibull distribution with $B_m = 4$, $A_m = 200$ hours and $C_m = 100$ hours. Calculate the probability that the machine will not be overhauled within the first 150 hours of maintenance.

Solution
The required probability can be calculated by applying equation 4.33 thus:

$$P(DMT > 150) = 1 - P(DMT \leqslant 150) = 1 - M(150)$$

$$= \exp\left[-\left(\frac{150 - 100}{200 - 100}\right)^4 \right] = 0.939$$

Thus, there is a chance of 93.9% that the overhaul will not be successfully completed within 150 hours of maintenance.

4.5 CLOSING REMARKS

In Chapter 2 it was concluded that the maintainability could be quantitatively expressed through the duration of elapsed time during which the specified maintenance task is successfully completed. Also, it was said that as a result of the large number of influential parameters in each of three groups on the one hand and their variability on the other hand, it is impossible to find a rule which would deterministically describe this very complex relation denoted by '*f*' in equation 4.1. Thus, it is impossible to give a deterministic answer regarding the instant of operating time when the transition from the SoFa to the SoFu will occur for any individual trial of the maintenance task under consideration. It is only possible to assign a certain probability that it will happen at a certain instant of maintenance time or that a certain percentage of trials will or will not be completed by the specific instant of elapsed time.

CHAPTER 5

Cost of maintenance tasks

The main objective of this chapter is to analyse the direct maintenance cost associated with execution of each maintenance task as well as the direct cost for application of selected maintenance policy.

5.1 DIRECT COST OF MAINTENANCE TASKS

The direct cost associated with each maintenance task, CMT, is related to the cost of maintenance resources, CMR, directly used during execution of the task. Thus, it is a function of:

$$CMT = f(C_{MSS}, C_{MTE}, C_{MPS}, C_{MFC}, C_{MTD}, C_{MCR}) \qquad (5.1)$$

where C_{MSS} is the cost of maintenance supply support, C_{MTE} is the cost of maintenance test and support equipment, C_{MPS} represents the cost of maintenance personnel, C_{MFC} is the cost of maintenance facilities, C_{MTD} is the cost of maintenance technical data and C_{MCR} represents the cost of maintenance computer resources.

It is necessary to stress that the type and quantity of all maintenance resources required for successful completion of any maintenance task are inherited from the design of the item/system and they are fully addressed during the maintainability analysis of the design process.

The cost of personnel involved with a specific maintenance task is a function of the following variables:

$$C_{MPS} = f(DMT, HCP) \qquad (5.2)$$

where DMT represents a random variable related to the duration of elapsed maintenance time, and HCP represents the monetary value of the hourly cost of maintenance personnel used for the execution of a specific maintenance task.

Most frequently, in daily practice, engineers deal with the average (mean) value of direct cost of a maintenance task, which could be defined as follows:

$$MCMT = C_{MSS} + C_{MTE} + C_{MFC} + C_{MTD} + C_{MCR} + (MDMT \times HCP) \qquad (5.3)$$

under the assumption that the cost of all maintenance resources, apart from personnel, is constant.

It is necessary to underline that the cost defined by equation 5.3 could differ considerably between different types of maintenance tasks. The main reason for this is the fact that preventive maintenance tasks are performed at a predetermined instant of time, before the failure takes place, which means that maintenance resources used for the successful completion of this task are those which are absolutely necessary. However, in the case of the corrective maintenance tasks there is a high possibility of development of secondary failures caused by the occurrence of primary failure, which in turn could demand use of some additional resources. For example, the failure of a fan belt in a car engine could cause overheating of the engine which in turn could cause some further damage to the engine.

5.1.1 Direct cost of corrective maintenance tasks

The direct cost associated with each corrective maintenance task, CMT^c, is related to the cost of maintenance resources needed for the successful completion of the task, CMR^c. Thus, the general expression for the cost of each corrective maintenance task will have a form as shown below:

$$CMT^c = f(C^c_{MSS}, C^c_{MTE}, C^c_{MFC}, C^c_{MTD}, C^c_{MCR}, DMT^c, HCP^c) \quad (5.4)$$

In daily practice, engineers deal with the average (mean) value of corrective maintenance costs, which is denoted as $MCMT^c$, thus

$$MCMT^c = C^c_{MSS} + C^c_{MTE} + C^c_{MFC} + C^c_{MTD} + C^c_{MCR} + (MDMT^c \times HCP^c) \quad (5.5)$$

5.1.2 Direct cost of preventive maintenance tasks

The direct cost associated with each preventive maintenance task, CMT^p, is related to the cost of maintenance resources needed for the successful completion of the task. Thus, the general expression for the cost of each preventive maintenance task will have a form as shown below:

$$CMT^p = f(C^p_{MSS}, C^p_{MTE}, C^p_{MFC}, C^p_{MTD}, C^p_{MCR}, DMT^p, HCP^p) \quad (5.6)$$

Again, in daily practice, engineers deal with average (mean) values of preventive maintenance costs, which are denoted as $MCMT^p$.

$$MCMT^p = C^p_{MSS} + C^p_{MTE} + C^p_{MFC} + C^p_{MTD} + C^p_{MCR} + (MDMT^p \times HCP^p) \quad (5.7)$$

5.1.3 Direct cost of conditional maintenance tasks

The direct cost associated with each conditional maintenance task, CMT^m, is related to the cost of maintenance resources needed for the successful completion of the task, CMR'''. Consequently, the general expression for the cost of each conditional maintenance task will have a form as shown below:

$$CMT''' = f(C'''_{MSS}, C'''_{MTE}, C'''_{MFC}, C'''_{MTD}, C'''_{MCR}, DMT''', HCP''') \quad (5.8)$$

The average (mean) value of conditional maintenance cost, denoted as $MCMT'''$, could be obtained according to the following expression:

$$MCMT''' = C'''_{MSS} + C'''_{MTE} + C'''_{MFC} + C'''_{MTD} + C'''_{MCR} + (MDMT''' \times HCP''')$$
$$(5.9)$$

CHAPTER 6

Frequency of maintenance tasks

There are a large number of human-made systems whose functionability has to be maintained during the utilization process by the user. The process during which the ability of the system to perform a function is maintained, is known as maintenance process, and it is defined (Knezevic, 1995) as:

> a flow of maintenance tasks performed by the user in order to maintain the functionability of the system during its operational life, in accordance with the adopted maintenance policy.

One of the main characteristics of maintenance processes is the frequency of maintenance task. Thus, the main objective of this chapter is to present methodology for the determination of the frequency of corrective, preventive and conditional maintenance tasks.

6.1 FREQUENCY OF CORRECTIVE MAINTENANCE TASKS

As has already been discussed, in Chapter 3, the corrective maintenance task is performed after the occurrence of failure. Thus, frequency of demand for corrective maintenance tasks is driven by the durability characteristics of the item considered. The random variable which numerically describes this design characteristic is known as the duration of functionable life, *DFL*. It is fully described by the following parameters.

- Type of distribution, TD_f,
- the scale parameter, A_f,
- the shape parameter, B_f,
- the source parameter, C_f.

Once the above characteristics are known, the length of satisfactory operation of an item is fully defined by the probability distribution of this continuous random variable. Like any other random variable, *DFL* could be defined by one of the well-known theoretical distributions (exponential, normal, lognormal, Weibull and similar) through some of the measures dealt with below.

6.1.1 Failure function

Failure function, $F(l)$, represents the cumulative probability that the transition to a state of failure (or simply failure) will occur before or at the moment of the operational length, l, thus:

$$F(l) = P(DFL \leq l) = \int_0^l f(l)dl \tag{6.1}$$

where $f(l)$ is the probability density function of DFL. Table 6.1 shows the failure function for some well-known theoretical distributions.

Table 6.1 Failure function, $F(l)$ for well-known theoretical distributions

Distribution	Expression	Range
Exponential	$1 - \exp\left(\dfrac{t}{A_f}\right)$	$l \geq 0$
Normal	$\Phi\left[\dfrac{l - A_f}{B_f}\right]$	$-\infty \leq l \leq +\infty$
Lognormal	$\Phi\left[\dfrac{\ln(l - C_f) - A_f}{B_f}\right]$	$t \geq C_f,\ C_f \geq 0$
Weibull	$1 - \exp\left[-\left(\dfrac{(l - C_f)}{(A_f - C_f)}\right)^{B_f}\right]$	$l \geq C_f,\ C_f \geq 0$

A_f is a scale parameter, B_f is a shape parameter and C_f is a source (minimum value) parameter of the theoretical probability distributions. Φ is the standard normal distribution function, numerical values of which can be found in Appendix Table 1.

It should be noted at this point that operational length, l, could be represented in operational units, such as miles read from an odometer, hours of operation, number of cycles, number of switches, number of landings, total accumulated energy, etc.

6.1.2 Durability function

Durability function, $D(l)$, presents the probability that the duration of functionable life will be maintained up to a particular operational length, l, thus:

$$D(l) = P(DFL > l) = \int_l^\infty f(l)dl \tag{6.2}$$

Since the item, at any instant of time, can be in one of two possible states, for any value of l the equality $F(l) + D(l) = 1$ is valid. Table 6.2 shows the durability function for well-known theoretical distributions.

Table 6.2 Durability function, $D(l)$ for well-known theoretical distributions

Distribution	Expression	Range
Exponential	$\exp\left(\dfrac{l}{A_f}\right)$	$l \geqslant 0$
Normal	$\Phi\left[\dfrac{A_f - l}{B_f}\right]$	$-\infty \leqslant l \leqslant +\infty$
Lognormal	$\Phi\left[\dfrac{A_f - \ln(l - C_f)}{B_f}\right]$	$l \geqslant C_f,\, C_f \geqslant 0$
Weibull	$\exp\left[-\left(\dfrac{(l - C_f)}{(A_f - C_f)}\right)^{B_f}\right]$	$l \geqslant C_f,\, C_f \geqslant 0$

Parameters and functions as in Table 6.1.

6.1.3 Mean duration of functionable life

Mean duration of functionable life, *MDFL*, can be determined by using the expectation of the random variable, *DFL*, thus:

$$MDFL = E(DFL) = \int_0^\infty l \times f(l)\mathrm{d}l \qquad (6.3)$$

The same result can be obtained by taking the integral of the durability function, because $D(l) = 1 - F(l)$, thus:

$$MDFL = E(DFL) = \int_0^\infty D(l)\mathrm{d}l \qquad (6.4)$$

Table 6.3 shows mean duration of functionable life for well-known distributions.

It is necessary to stress that the type of distribution function and values of A_f and B_f are only mathematically significant, whereas the real

Table 6.3 Mean duration of functionable life, $MDFL = E(DFL)$ for well-known distributions

Distribution	Expression
Exponential	A_f
Normal	A_f
Lognormal	$\exp(A_f + \dfrac{1}{2}B_f^2)$
Weibull	$A_f \times \Gamma(1 + 1/B_f)$

Γ is a gamma function numerical values of which are given in Appendix Table 2.

significance comes in the physical processes occurring in the item and its interaction with environment and operator. Thus, the processes like fatigue, corrosion, abrasion, thermal deformation and similar are the main drivers of the distribution types and their parameters.

6.2   FREQUENCY OF PREVENTIVE MAINTENANCE TASKS

Preventive maintenance tasks are performed at fixed intervals, FMT^P, which are a function of the life distribution of the items considered, representing the failure mechanism which causes the failure of the item considered. There are several optimization criteria, such as:

- minimum maintenance cost
- maximum availability
- required reliability.

6.3   MINIMUM COST AS AN OPTIMIZATION CRITERION

The total direct cost associated with each maintenance task, CMT, is related to the cost of maintenance resources, CMR, used during the task execution, on one hand, and the cost of consequences, which in this case is related to the potential revenue lost for the user due to the loss of utility of the system, CLR, on the other. Thus,

$$CMT = CMR + CLR \qquad (6.5)$$

It is necessary to underline that the cost defined by the above expression could differ considerably for different maintenance tasks. The reason for this is the fact that preventive maintenance tasks are performed at a predetermined instant of time when all maintenance support resources are provided in advance, whereas in the case of the corrective maintenance task, the maintenance resources are not always available. Thus, the general expression for the cost of this type of maintenance task will be:

$$CMT^P = CMR^P + (MDMT^P + MDST^P) \times HR \qquad (6.6)$$

The mean total cost of the preventive maintenance task is a function of the cost of the corrective task performed during the interval FMT^P, and the cost of performing a preventive task, at FMT^P.

The total cost of corrective tasks in the interval FMT^P is equal to the product of the mean number of failures between two preventive replacements expressed through $MNF(FMT^P)$ and the cost of each corrective task, CMT^C.

Thus, the average cost for preventive maintenance task per unit of operation, for a specific FMT^p, denoted as $UMC^p(FMT^p)$, can be represented mathematically as follows:

$$UMC^p(FMT^p) = \left[\frac{MNF(FMT^p) \times CMT^c + CMT^p}{FMT^p} \right]_{min} \quad (6.7)$$

The above equation can be used iteratively to find the values of FMT^p for which the average unit cost will be minimum.

It is necessary to say that in the above expression $MNF(FMT^p)$ could be replaced by $F(FMT^p)$ because the difference between them within the length of time FMT^p is very small. Thus:

$$UMC^p(FMT^p) = \left[\frac{F(FMT^p) \times CMT^c + CMT^p}{FMT^p} \right]_{min} \quad (6.8)$$

where $F(FMT^p)$ is the cumulative distribution function of the DFL, $F(FMT^p) = P(DFL \leq FMT^p)$.

It is important to underline that the model used is not restricted to any particular probability distribution of the duration of the functionable length of the item involved, i.e. it is applicable to the Weibull, normal, lognormal, exponential or any other probability distribution.

Example 6.1

For brushes of an electrical motor determine the optimal maintenance frequency which will provide the minimum total cost of maintenance. The appropriate data are presented in Table 6.4. The estimated hourly rate of revenue is £10.

Table 6.4 Input data for Example 6.1

Maintenance task Failure distribution	Replacement Normal
A_f (km)	60 000
B_f (km)	18 000
$C^c_{MSS} = C^p_{MSS}$ (£)	265
$C^c_{MPS} = C^p_{MPS}$ (£)	500
$MDMT^c = MDMT^p$ (hr)	100

Solution
In the interests of script economy, only the salient details of this case are provided. Thus,

$$MCMT^c = MCMT^p = 265 + 500 + 100 \times 10 = £1765$$

$$UMC^p(FMT^p) = \left[\frac{\Phi(FMT^p - 60\ 000/1800) \times 1765 + 1765}{FMT^p} \right]_{min}$$

Results of the iteration performed for the calculation of the preventive maintenance interval for maintenance policies are presented in Table 6.5.

Table 6.5 Costs of preventive maintenance task

	Maintenance interval, FMT^P (1000 km)			
	20	*40*	*60*	*80*
$UMC^P(FMT^P)$	0.023	0.01628*	0.0221	0.0267

*Indicates optimal result.

Thus, the optimal frequencies of performing a preventive replacement of brushes in the case analysed is 40 000 kilometres.

6.4 MAXIMUM AVAILABILITY AS AN OPTIMIZATION CRITERION

Performing maintenance tasks more frequently could reduce the corrective downtime and hence improve availability. However, on the other hand, execution of any maintenance task requires time. Thus, the more frequently preventive tasks are performed the less the item/system is available for use. It is clear that a balance is required between these two conflicting tasks. Consequently, a common denominator is the interval FMT^P between maintenance tasks. Thus, by expressing availability as a function of the maintenance interval FMT^P, the maximum availability for a preventive task can be found. The expression is shown below:

$$A(FMT^P) = \left[\frac{MDFL(FMT^P)}{MDFL(FMT^P) + MDMT + [MDMT + MDST] \times F(FMT^P)} \right]_{max}$$
(6.9)

Example 6.2

The ability of an item of equipment to maintain functionability is represented by a Weibull distribution (W) with scale parameter $A_f = 1400$, and shape parameter $B_f = 2.6$. What is the maximum availability of the item and what task should be adopted given that $MDMT = 4$ hours and $MDST = 45$ hours.

Solution

The procedure used for solving this type of problem follows similar lines to that used in Example 6.1, which in practice means that several iterations are needed for equation 6.9.

The recommended task for the item in this example is preventive with a maximum availability of 98% achieved by performing maintenance tasks at $FMT^p = 460$ hours.

6.5 REQUIRED RELIABILITY AS AN OPTIMIZATION CRITERION

In a large number of engineering systems it is imperative that a very low probability of failure is maintained while the required function is performed under required conditions. The most common reason for this is safety for the user and the environment. In these cases the required reliability level, R_r, is the optimization criterion from the point of view of maintenance. Where considerations of safety are paramount, preventive maintenance is obligatory.

In determining the most beneficial schedule for preventive and conditional maintenance tasks it is necessary to know all safety and environment related requirements, which could be issued by the national, international or profession-specific standards, regulations or recommendations. For example, the safety demands, expressed through the achieved hazard rates for propulsion system, required by the CAAM Committee Initial Report on Propulsion System and APU Related Aircraft Safety Hazard 1982 through 1991, are given in Table 6.6, for some of safety critical failures.

Table 6.6 Safety achievement – hazard rate

Hazard	Rate/engine hour
High energy non-containment	3.6×10^{-8}
Uncontrolled fire	0.3×10^{-8}
Engine separation	0.2×10^{-8}
Major loss of thrust control	5.6×10^{-8}

The time for performing a preventive maintenance task, FMT^p, is determined according to the durability function, $D(t)$, based on the probability distribution of the functionable life length to failure, and it must satisfy the following equation:

$$D(FMT^p) = P(DFL > FMT^p) > R_r \qquad (6.10)$$

6.6 OPTIMAL FREQUENCY FOR A CONDITIONAL MAINTENANCE TASK

Determination of the optimal maintenance procedure in the case of conditional maintenance is based on mathematical models. They take into account the probability distribution of the characteristics which describe the condition of the item, RCI, as well as the relevant costs related to the cost of performing checking, cost of failure, cost of equipment used and similar. In most cases the optimization criterion is the required reliability or level of probability that the failure will not occur between two examinations.

In the case of the preventive maintenance task, prescribed maintenance activities take place at a predetermined time completely independent of the real condition of the system.

The time for performing these activities, FMT^m, is determined according to the durability function, $D(t)$, based on the probability distribution of the functionable life length to failure, and it must satisfy the following equations:

$$D(FMT^m) = P(DFL > FMT^m) > R_r \qquad (6.11)$$

$$P[RCI(FMT^m_{i+1}) \leqslant RCI^{cr}|RCI(FMT^m_i)] \leqslant RCP^{cr}) = R_r + \frac{D(FMT^m_{i+1})}{D(FMT^m_i)} \qquad (6.12)$$

. Thus,

$$D(FMT_{i+1}) = R_r \times D(FMT^i) \qquad (6.13)$$

Hence, it is necessary to determine FMT_{i+1} which satisfies the above expression, as a function of the expression for the durability function $D(l)$.

It is important to underline that the model used is not restricted to any particular probability distribution of the time to failure of the item involved, i.e. it is applicable to the Weibull, normal, lognormal, exponential or any other probability distribution.

Example 6.3

In order to examine the impact of the failure mechanism on the frequency of the inspections, three different mechanisms, X, Y and Z, will be examined. Each mechanism is a model with Weibull probability distribution, with specific parameters as given in Table 6.7.

In this particular example the required reliability level is 0.88.

Table 6.7 Distribution parameters for the mechanisms examined

Mechanism	Scale parameter (hr)	Shape parameter	Hazard function
X	100	0.77	Decreasing
Y	100	1.00	Constant
Z	100	1.77	Increasing

Solution

The generic expression for the durability function in the case where the time to failure could be successfully modelled by the Weibull distribution is:

$$D(i) = \exp\left[- \left(\frac{1}{A_f}\right)^{B_f} \right]$$

Consequently, making use of equation 6.13, the frequencies of performing a conditional maintenance task are determined according to the following expression:

$$FMT_i = A_f \times ((-\log(R_r)^{(\frac{1}{B_f})})$$

For the three failure mechanisms analysed, the calculated values for the inspection frequencies are given in Table 6.8.

Table 6.8 Inspection intervals, FMT_i^m (hr)

	Mechanism		
i	X	Y	Z
1	6.915	12.783	31.281
2	10.096	12.783	14.994
3	11.791	12.783	11.913
4	13.046	12.783	10.269
5	14.062	12.783	9.198
6	14.939	12.783	8.425
7	15.704	12.783	7.833
8	16.391	12.783	7.359
9	17.016	12.783	6.968
10	17.591	12.783	6.638

The results obtained clearly demonstrate that the main driver for the inspection frequency is the failure mechanism, and that in the case of decreasing hazard function the frequency also decreases, whereas in the case of increasing hazard function the frequency of inspections are decreasing, as expected.

CHAPTER 7

Maintenance cost

United States airlines spend 9 billion dollars, approximately 11 percent of their operating cost, on maintenance.

(*Aviation Week & Space Technology*, June 1995)

Execution of any maintenance task is related to associated costs, both in terms of the cost of maintenance resources and the cost of the consequences of not having the system available for operation. Therefore, maintenance departments are one of the major cost centres, costing industry billions of pounds each year, and as such they have become a critical factor in the profitability equation of many companies. Thus, as maintenance actions are becoming increasingly costly, maintainability engineering is gaining recognition day by day. Consequently, the majority of users are currently showing concern over the competitive advantage that maintenance processes can provide to a company.

Example 7.1

The Boeing 747 and the Boeing 767 are two wide-body aircraft types which have been widely used by the airline industry. The former (which entered service in 1970) is the main long-haul airliner in use worldwide, while the latter (which entered service in 1982) is a medium-haul successor to the original Boeing 707 jet airliner. Costs of four major US operators (Northwest, Pan Am, Trans World and United Airlines) in US dollars per block hour, for the financial year 1986, for both aircraft, are given in Table 7.1.

7.1 COST OF A MAINTENANCE PROCESS

For many systems/products, maintenance cost constitutes a major segment of the cost of ownership. Resources are normally expressed in monetary terms as costs. For purposes of maintenance resource

Table 7.1 Costs for the Boeing 747 (121 aircraft) and 767 (62 aircraft) based on Friend (1992)

Cost category	B747		B767	
	US$/block hr	*%*	*US$/block hr*	*%*
Crew	787	17	504	23
Fuel	2030	44	720	32
Depreciation and rental	672	14	596	27
Total maintenance	1117	24	377	17
Other	36	1	31	1
Total	4642	100	2228	100

analysis, MRA, costs can be classified into various categories. The most common cost categories of the maintenance process are briefly analysed below.

7.2 FIXED AND VARIABLE COSTS

Fixed costs have to be met irrespective of the number of maintenance tasks performed (e.g. test equipment, ground equipment, facility costs and similar). Although fixed costs are assumed to remain unchanged in response to changes in the level of activities, they may change in response to other factors, such as price changes.

Variable costs are those which are dependent upon the volume of maintenance tasks performed. These costs are usually for direct material and direct labour.

It is worth noting that many costs contain elements of both fixed and variable costs. For example, a maintenance department may have a given number of personnel at fixed salaries and offer a wide range of tasks. However, the amount of maintenance work undertaken and replacement items required on equipment may vary in relation to the output of plant and equipment. Consequently, the annual maintenance costs for plant and equipment over several years would consist of both fixed and variable elements. Precise determination of what is fixed and what is variable may not be possible or may require detailed and expensive measurement techniques and records. Fixed cost of lost utility flow is related to the cost of salaries of operators/employers, heating, insurance, taxes, facilities, electricity, telephone and similar which are incurred while the item is in a state of failure (SoFa). These costs should not be neglected, because they could be even higher than the other cost categories.

7.3 DIRECT, INDIRECT AND OVERHEAD COSTS

Direct costs are costs which can be clearly charged against each task. Direct material plus direct labour cost is generally referred to as prime cost.

Indirect costs, on the other hand, are difficult to assign to a particular activity.

Overhead (or burden) costs consist of all costs other than direct material and direct labour costs. Traditionally, maintenance has been included in overhead costs and thus has been difficult to identify. Typical overhead costs include indirect materials, indirect labour, taxes, insurance, rent, maintenance and repairs, depreciation, supervisory and administrative personnel, heat, light and fuel.

7.4 INITIAL MAINTENANCE COST

First or initial investment cost is the total investment necessary to establish a maintenance system ready for operation. Such costs are generally non-recurring during the life of the item/system. Initial investment costs for test equipment, for example, might include the acquisition cost, training, installation, transportation, initial tooling and support equipment.

7.5 OPPORTUNITY COST

Maintenance is linked with costs, but the consequences which arise from not performing maintenance are also linked to corresponding costs which could often be considerably more. As the scope of maintenance and its frequency of execution cannot be unlimited, it must be controlled and optimized according to some criteria. Estimated opportunity costs, which represents the monetary value of the consequences of the loss of revenue incurred by British Airways regarding the Boeing 747, are given in Table 7.2.

Table 7.2 Typical cost/penalties – Boeing 747

Fuel dump (50 t on turnback)	$15 000
Delay	$1000/minute
Aircraft out of service	$90 000/day
Embody fleet corrective action (60 engines at $300 000 each)	$18 million
Fleet grounded	$ millions!!
Loss of passengers/revenue (to recover a lost Club World passenger)	$105 000

Thus, the opportunity cost or cost of lost revenue, *CLR*, is directly proportional to the product of the length of the time which the system spends in the SoFa and the hourly income rate, *HR*, to the user obtained by the utilization of the product.

7.6   CASE STUDY: PRATT & WHITNEY JT8D ENGINE

The Pratt & Whitney JT8D is the most populous engine in the world, with a total of 11 000 built since its introduction to revenue service in 1964 (*Aircraft Maintenance International*, July/August 1995). This type of engine powers the Aeropatiale Caravelle, Dassault Mercure 100, McDonald Douglas DC-9 and Boeing 727 and 737. In total it powers 4232 aircraft operated by 339 airlines worldwide.

According to Pratt & Whitney, the average maintenance cost for JT8Ds, based on flights of 500 nautical air miles with typical engine performance, is US$60.83 per engine flight hour.

According to American Airlines the maintenance cost is below this cost, whereas according to Federal Express the maintenance cost is 'in the region of 80 to 140 USA dollars' per engine flight hour. Both sources agree that as there are so many different ways of measuring maintenance cost, it is difficult to pin down any specific number.

The typical cost for an engine overhaul is between US$775 000 and 800 000. This should in return give an on-wing operational period of around 12 000 flying hours.

Another way of measuring the maintenance cost is by looking at engine lease rates. This practically means that when an engine is brought in for repair and another one leased out, the lease rate should cover the cost of the maintenance of the first engine. The maintenance cost under this regime is around US$75 per engine flight hour at present.

In order to fully appreciate the importance of maintenance cost it is necessary to look at the overall picture. Thus, with 11 000 engines produced during 31 years the statistics show that Pratt & Whitney made, on average, 354 engines per year. According to the company's claim the whole JT8D family has accumulated over 450 million flight hours thus far. Consequently, on average each engine flies 2556 hours per year. Making use of the maintenance cost figure calculated above, per engine flight hour, the total annual maintenance cost, for the 9805 active engines, is around US$2.1 billion.

Maintenance concept

The main concern of the maintenance concept is the vision of the way in which the systems will be supported during their utilization phase. It defines levels of maintenance support, repair policies, maintenance environment and similar factors. The levels of maintenance could be generically categorized into the following three types (Blanchard *et al.*, 1995):

- organizational/user
- intermediate
- depot/producer.

Each of the above-listed maintenance concepts will be briefly addressed in this chapter.

8.1 ORGANIZATIONAL/USER LEVEL

This type of maintenance level comprises all maintenance tasks which are performed at the operational site (airplane, ship, vehicle, factory, home, etc.). Generally, it includes work performed by the using organization on its own equipment. Organizational-level personnel are usually involved with the operation and use of equipment, and have minimum time available for detailed system maintenance. Maintenance at this level normally is limited to periodic checks of equipment performance, visual inspections, cleaning of equipment, some servicing, external adjustments and the removal and replacement of some components. Personnel assigned to this level generally do not repair the removed components, but forward them to the intermediate level. From the maintenance standpoint, the least skilled personnel are assigned to this function.

8.2 INTERMEDIATE LEVEL

This type of maintenance level refers to the maintenance tasks performed by mobile, semimobile and/or fixed specialized organizations and installations. At this level, items concerned may be repaired by the removal and replacement of major modules, assemblies or piece parts. Scheduled maintenance requiring specialized equipment for disassembly may also be accomplished. Available maintenance personnel are usually more trained/skilled and better equipped than those at the organizational level and are responsible for performing more detailed maintenance. Mobile or semimobile units are often assigned to provide close support for dispersed operational equipment. These units may constitute vans, trucks or portable shelters containing some test and support equipment and spares. The mission is to provide on-site maintenance (beyond that accomplished by organizational-level personnel) to facilitate the return of the system to its full operational status on an expedited basis. A mobile unit may be used to support more than one operational site. A good example is the maintenance vehicle that is deployed from the airport hangar to an airplane parked at a commercial airline terminal gate and needing extended maintenance.

Fixed installations (permanent shops) are generally established to support both the organizational-level works and the mobile or semi-mobile units. Maintenance works that cannot be performed by the lower levels, due to limited personnel skills and test equipment, are performed here. High personnel skills, additional test and support equipment, more spares and better facilities often enable equipment repair to the module and piece part level. Fixed shops are usually located within specified geographical areas.

8.3   DEPOT/PRODUCER LEVEL

This constitutes the highest level of maintenance and supports the accomplishment of maintenance tasks whose complexity is beyond the capabilities available at the intermediate level. Physically, the depot may be a specialized repair facility supporting a number of systems or types of equipment in the inventory, or it may be the equipment manufacturer's plant. Depot facilities are fixed and mobility is not a problem. Complex and bulky equipment, large quantities of spares, environmental control provisions and so on, can be provided if required. The high volume potential in depot facilities fosters the use of assembly-line techniques, which, in turn, permits the use of relatively unskilled labour for a large portion of the workload with a concentration of highly skilled specialists in certain key areas such as fault diagnosis and quality control.

The depot level of maintenance includes the complete overhauling, rebuilding and calibration of equipment as well as the performance of highly complex maintenance tasks. In addition, the depot provides an inventory supply capability. The depot facilities are generally remotely located to support specific geographical area needs or designated product lines.

Example 8.1

The main objective of this example is to illustrate the process of the determination of the most suitable maintenance level, for the system under consideration, based on the minimum maintenance cost.

In order to illustrate the principles of the determination of the most suitable maintenance level, a hypothetical system, XX250, has been analysed. The system is utilized at 60 sites which are coordinated by the five operational bases. It is expected that the system, throughout the deployed period, will be operational for 452 600 hours.

For the purpose of the analysis only the main three subsystems will be addressed. Thus, System 'XXX' consists of units XX1, XX2 and XX3. When a corrective maintenance task is required, a BIT capability enables rapid checkout and fault isolation to the unit level. In the event of failure the organizational maintenance level is engaged, where the applicable unit is removed, replaced with a spare and the faulty unit is sent to the intermediate-level maintenance shop for the execution of the corrective maintenance task required. Unit repair is accomplished through fault isolation to the applicable item (in this example item XX1/1), removal of the faulty item and replacement with a spare, and checkout of the unit to verify satisfactory operation. The functionability of the faulty item can then be restored at either the intermediate or depot level of maintenance.

Based on the data provided, determine the most suitable level of maintenance. The available data associated with item XX1/1 are given below:

- The cost of item XX1/1 is £3500.
- The mean duration of functionable life, *MDFL*, of the item is 1100 hours.
- The mean duration of the corrective maintenance task, $MDMT^c$, is 4 hours.
- The maintenance task requires one technician at the hourly rate of £25 for intermediate and £40 for depot maintenance level.
- Supply support involves the following three categories of cost:
  (i) the cost of spare assemblies in the pipeline (10 spare items are required in the pipeline when maintenance is accomplished at the intermediate level and 20 spare items will be required when maintenance is accomplished at the depot level);
  (ii) the cost of material needed for repair of faulty items, which is £200 per maintenance task;
  (iii) the cost of inventory maintenance, which is 20% of the inventory value.
- When item repair is accomplished, special test and support equipment is required for diagnosis and checkout. The allocated cost per test set to cover item XX1/1 and its maintenance is £25 000.

- Transportation cost is considered to be negligible when maintenance is accomplished at the intermediate level. However, if the maintenance task is executed at the depot level it will involve a cost of £30 per round trip (at one trip per maintenance task).
- The initial formal training costs of maintenance personnel when considering the item repair option are as follows: 3 days are required for depot-level maintenance and 15 days are required for intermediate maintenance, at the cost of £500 per day.
- There is a burden rate of £30 per maintenance task at intermediate and £25 per maintenance task at depot level.

Solution

Making use of the data available, the total costs for both levels of maintenance are determined. The totals and summaries of cost categories are given in Table 8.1.

Table 8.1 Level of repair analysis for assembly XX1/1

	Intermediate level cost (£)	*Depot level cost (£)*
Acquisition cost	210 000	210 000
Personnel cost	40 700	65 120
Spare assemblies	35 000	70 000
Spare components	81 400	81 400
Inventory maintenance	23 280	30 280
Test/support equipment	125 000	25 000
Transportation	Negligible	12 210
Training	7 500	1 500
Facilities/data	12 210	10 175
Total cost	535 090	505 685

Thus, according to the total cost obtained for both alternatives, it could be concluded that the maintenance task analysed should be performed at the depot level.

8.4 COTS/NDI INTENSIVE SYSTEMS

In order to reduce development times and resources, NATO countries led by the DoD have encouraged an extensive use of 'Commercial Off-The-Shelf (COTS)' items, also known as 'Non-Developmental Items (NDI)'. This new COTS/NDI-driven approach has brought an increased dependence of the design/development team on the vendor/supplier environment. Consequently, supportability issues become more prominent due to the needs for engineering changes initiated by system support related considerations.

In the view of the above analyses it became clear that the methodologies for the determination of maintenance concepts should also reflect the impact of the new paradigm. Consequently, the well-established three level maintenance support concept (four levels in the case of the RAF, see appendix) became obsolete and inefficient.

Analyses performed as a part of this study have shown that cost-effective and affordable maintenance concepts for COTS/NDI-intensive systems should be based on two levels, namely:

- User, which is equivalent to the organizational level.
- Producer, which is equivalent to the depot level.

Thus, this newly created two-level maintenance concept, known as **user–producer**, provides the maximum efficiency of the systems support structure as well as the needed flexibility in dealing with COTS/NDI-intensive systems.

8.5   IN-HOUSE AND OUTSIDE MAINTENANCE

One of the main concerns of the maintenance management process is determination of the maintenance tasks which should be performed in-house and those which should be contracted out. Thus, the problem is the selection of the composition of the maintenance tasks performed by one or the other source.

Increase in the range of maintenance resources requires an increase in the capital maintenance cost. However, increase in the in-house capabilities reduces the need to use outside contractors. In this case a balance is required between costs associated with using in-house resources and the costs of using outside resources. A difficult costing problem arises since not only does the cost charged by the outside resources have to be considered, but also the cost associated with loss of control of maintenance work by management. For example, by using outside resources there is the possibility of greater downtime occurring and the consequent cost of lost revenue associated with it.

Thus, the selection of the more favourable alternative at any particular time depends upon:

- the nature of the maintenance task required;
- the maintenance resources available in-house;
- the workload committed within the organization;
- the costs associated with the various alternatives.

It should be stressed that these alternatives are not mutually exclusive since maintenance work can be performed by engaging both in-house and outside resources.

Example 8.2

Boeing 747 is by far the most successful wide-body aircraft ever built. In Table 8.2 a total breakdown of all models/variations of this aircraft is given together with the total number produced of each and their current status.

Table 8.2   Existing fleet census, Boeing 747

Model	Owned	Leased in	Total
B747SP	21	16	37
B747SR	14	0	14
B747-100/100F/100B/EUD/SCD	91	42	133
B747-200B/EUD/SF/Combi/C/F	213	145	358
B747-300/Combi/SR	50	27	77
B747-400/Combi/F	211	115	326
Total	600	345	945

Table 8.3   The top five operators of Boeing 747: in-house capabilities and outsourcing needs

Operator	Models	Quantity	% in-house
Japan Airlines	B747-100	6	100
	B747-200	31	100
	B747-300	13	100
	B747-400	31	100
British Airways	B747-100	15	90
	B747-200	16	90
	B747-400	32	100
United Airlines	SP	10	100
	B747-100	18	100
	B747-200	9	80
	B747-400	24	100
Singapore Airlines	B747-200	6	97
	B747-300	12	97
	B747-400	32	97
Air France	B747-100	6	100
	B747-200	21	100
	B747-300	4	100
	B747-400	13	100

The market analyses performed by the journal *Aircraft Maintenance International* in the June 1995 issue have shown the proportion of in-house and outside maintenance capability among the top ten B747 operators. The results of the survey for the top five operators are shown in Table 8.3.

Thus, out of 75 operators of Boeing 747s, the top five own 299 aircraft (or nearly 32%). It is necessary to notice that worldwide, at the beginning of 1996, there were 66 airframe overhaul vendors capable of dealing with B747s. Their distribution per continent is given in Table 8.4.

Table 8.4 Number of vendors capable of overhauling the airframe of Boeing 747s

Continent	Number of vendors
North America	22
South America	4
Asia	20
Africa	3
Europe	14
Australia	3

8.6 CASE STUDY: RAF MAINTENANCE CONCEPTS AND DEPOT REPAIR

Aircraft maintenance levels in military organizations differ from those of civilian operators. The UK Royal Air Force (RAF) maintenance concept, for an item, is described by a standard terminology defined in Air Publications 100A-01. Location of the work could be as follows:

- *First line*: The maintenance organization immediately responsible for the maintenance and preparation for use of the complete systems or equipment. It normally undertakes tasks like functional testing, replenishment, role changing, minor modification, fault diagnosis and minor repair. In the RAF a flying squadron is typical of a first line organization.
- *Second line*: The maintenance organization responsible for providing maintenance support to specified first line organizations. This includes scheduled maintenance of aircraft and bay maintenance of assemblies. This function is provided in the RAF by the engineering and supply wing of a flying station at which a number of flying squadrons may be based.
- *Third line*: The maintenance organization within the services but

excluding the organizations within the first and second lines. Typical tasks performed at this level are the repairs, partial reconditioning and modifications requiring special skills and equipment, and relatively infrequently used capability which it is not economic to provide generally. It stops short of complete strip, reconditioning and reassembly.

- *Fourth line*: The industrial maintenance organization providing maintenance support beyond second line to the services under contract.

CHAPTER 9

Maintenance policies

The main objective of this chapter is to analyse existing maintenance policies, and their advantages and disadvantages.

With respect to the relationship of the instant of occurrence of failure and the instant of performing the maintenance task, the following maintenance policies exist.

(a) *Failure-based maintenance policy*, FB, where corrective maintenance tasks are initiated by the occurrence of failure, i.e. loss of function or performance.
(b) *Life-based maintenance policy*, LB, where preventive maintenance tasks are performed at predetermined times during operation, i.e. at fixed lengths of operational life.
(c) *Inspection-based maintenance policy*, IB, where conditional maintenance tasks in the form of inspections are performed at fixed intervals of operation, until the performance of a preventive maintenance task is required.
(d) *Examination-based maintenance policy*, EB, where conditional maintenance tasks in the form of examinations are performed in accordance with the monitored condition of the item/system, until the execution of a preventive maintenance task is needed.
(e) *Opportunity-based maintenance policy*, OB, where corrective maintenance is performed on the failed item and preventive maintenance tasks are performed to the remaining items from the designated group of items.

The exact timing of conditional and preventive maintenance tasks is determined by the specific strategy, mainly determined by the operator of the item/system. Thus, these maintenance policies from the maintenance planning point of view are known as the scheduled maintenance, whereas FB and OB maintenance are referred to as unscheduled maintenance.

It is the task of the maintainability analysis to determine what maintenance is required to retain the functionability of the system during its operational life.

9.1 FAILURE-BASED MAINTENANCE POLICY

Failure-based maintenance policy, FB, represents approaches where corrective maintenance tasks are carried out after a failure has occurred, in order to restore the functionability of the item/system considered. Consequently, this approach to maintenance is also known as break-down, post-failure, fire fighting or unscheduled maintenance. A diagrammatical presentation of maintenance procedure for the failure-based maintenance policy is presented in Figure 9.1.

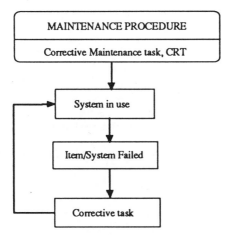

Figure 9.1 Algorithm for failure-based maintenance policy.

Generally speaking, this policy is applicable to items for which the loss of functionability is not significant for either the safety of the user and/or the environment or the economic consequences of failure.

9.1.1 Advantages of failure-based maintenance

The main attraction of this maintenance policy is the full utilization of the operating life of the item under consideration, i.e. the mean duration of utilized life ($MDUL^F$) of the item is identical to the mean duration of functionable life ($MDFL$). Hence, the coefficient of utilization of items considered, denoted as CU^F, will always have a value of 1, thus:

$$CU^F = \frac{MDUL^F}{MDFL} = 1 \qquad (9.1)$$

where:

$$MDUL = MDFL^F = \int_0^\infty D(l)\mathrm{d}l \qquad (9.2)$$

as illustrated in Figure 9.2.

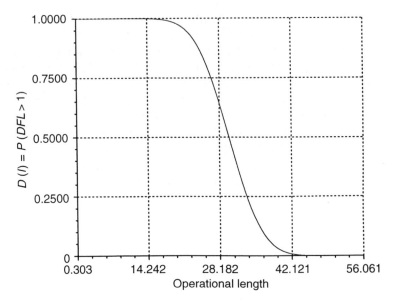

Figure 9.2 Relationship between $MDUL^F$ and $MDFL$ for failure-based maintenance policy.

In practice, this means that the user recovers full monetary investments
in the item/system when the FB maintenance policy is applied.

9.1.2 Disadvantages of failure-based maintenance

Despite the monetary advantage offered by this maintenance policy, it
has some disadvantages, among which the following are the most
important.

- The failure of an item can cause consequential damage to other items
  in the system or to the system itself. Analyses of maintenance costs
  have shown that a repair made after failure will normally be three to
  four times higher than if the corresponding preventive maintenance
  task had been performed (Knezevic, 1995).
- As the instance of occurrence of failure is uncertain, the maintenance
  task cannot be planned, hence longer downtimes, due to unavailability
  of resources (spares, personnel, tools and similar), should be
  expected.

Therefore, this policy can be potentially costly, due to the direct costs
of restoring the functionability of the system caused by failure and the
indirect costs incurred as a result of lost production, reputation and in
some cases lives.

9.2 LIFE-BASED MAINTENANCE POLICY

With a life-based (LB) maintenance policy, preventive maintenance tasks are performed at fixed intervals which are a function of the life distribution of the items considered. As the main aim is to prevent failure and its consequences, this approach to maintenance is very often called preventive maintenance policy. Another name for this policy found in the literature is planned maintenance. The reason for this is the fact that maintenance tasks are performed at a predetermined frequency, which means that it is possible to plan all tasks and fully support them.

A diagrammatical presentation of LB maintenance procedure is presented in Figure 9.3. The frequency of maintenance tasks, FMT^L, is determined even before the item has started functioning. Thus, at the predetermined length of operational life a specified preventive maintenance task takes place. If the item fails between preventive tasks a corrective maintenance task has to be performed, as shown in Figure 9.3.

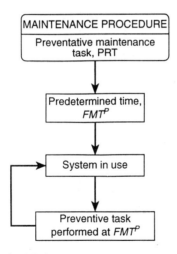

Figure 9.3 Algorithm for life-based maintenance policy.

The LB maintenance policy could be effectively applied to items/systems that meet some of the following requirements:

(a) the probability of occurrence of failure in future is reduced by performing the task;
(b) the total costs of applying this policy are substantially lower than that of the FB maintenance policy;
(c) monitoring of the condition of the item is not technically feasible or it is economically unacceptable.

9.2.1 Advantages of life-based maintenance

One of the main advantages of this maintenance policy is the fact that preventive maintenance tasks are performed at a predetermined instant of time enabling all maintenance support resources to be provided in advance, and potential costly outages avoided.

Another advantage of LB policy is avoidance of occurrence of failure which in some cases could have catastrophic consequences to the user/ operator and environment (Chernobyl, Bhopal, Piper Alpha and similar).

9.2.2 Disadvantages of life-based maintenance

Despite the advantages given above, the LB maintenance policy has several disadvantages that must be recognized and minimized. For example, it could be uneconomical because the majority of items are prematurely replaced, irrespective of their condition. Hence, the coefficient of utilization of the item/system considered, CU^l, has a value less than one, and it is defined as:

$$CU^l = \frac{MDUL^L}{MDFL} << 1$$

where $MDUL^L$ is defined as:

$$MDUL^L = \int_0^{FMT^L} D(l)dl << MDFL$$

as illustrated by Figure 9.4.

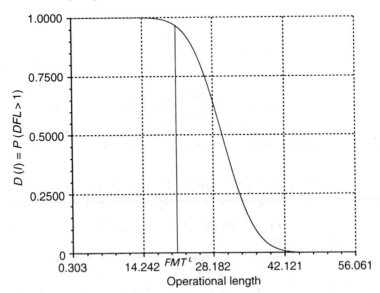

Figure 9.4   Relationship between $MDUL^L$ and $MDFL$ for life-based maintenance policy.

9.3 INSPECTION-BASED MAINTENANCE POLICY

The advantage of this procedure is a provision of better utilization of the item considered than in the case of applying preventive maintenance, with provision of the required level of safety of utility flow.

Inspection is a conditional maintenance task, the result of which is a statement about the condition of the item, i.e. the condition is satisfactory or it is unsatisfactory, which is determined according to the RCI. The common feature of all of these tasks is that the results obtained do not have any effect on the scheduling of next checking. Before the item/system is introduced into service the most suitable frequency of the inspections, FMT^I, has to be determined. Thus, during the operation of the item/system inspections are performed at specified fixed intervals until the critical level is reached, $RCI(FMT^I) > RCI_{cr}$, when prescribed preventive maintenance tasks take place. If the item fails between inspections, corrective maintenance takes place. The algorithm presented in Figure 9.5 shows maintenance procedure in the case where inspection is used as the condition monitoring task.

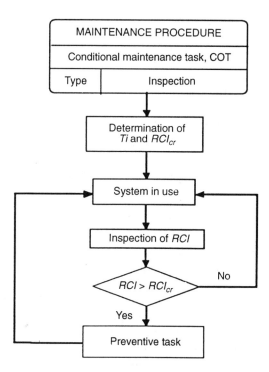

Figure 9.5 Maintenance procedure for inspection-based maintenance policy.

9.3.1   Advantages of inspection-based maintenance

Operating systems with a condition monitoring technique will provide information about the condition of the constituent items. Therefore, maintenance engineers are starting to realize the value of this information. The benefits of condition monitoring can be summarized as follows.

(a) Detection, at the earliest time possible, of deterioration in condition and/or performance of an item/system.
(b) Reduction of system downtime, since maintenance engineers can determine the optimal maintenance interval through the condition of constituent items in the system. This allows better maintenance planning and more efficient use of resources.
(c) Improved safety, since monitoring techniques enable the user to stop the system before a failure occurs.
(d) Increased availability, by being able to keep systems running longer.

9.3.2   Coefficient of utilization

The coefficient of life utilization (CU^l) of the item whose replacement is based on its condition could be determined according to the following expression:

$$CU^l = \frac{MDUL^l}{MDFL} = \frac{\int_0^\infty D_{RCI_{cr}}(l)\,dl}{MDFL} \tag{9.3}$$

9.4   EXAMINATION-BASED MAINTENANCE POLICY

In order to increase the level of utilization of items preventively replaced, and still maintain a low probability of failure during operation, it is necessary to obtain more information about the items' behaviour during the operation process. As the durability function represents a main source of information in the above expression, the only way ahead was to create a new approach to durability which is able to provide a fuller picture of the process of change in condition of the items considered. Fully aware of this, Knezevic (1993) developed a methodology for the determination of durability based on the relevant condition predictor, RCP. This new approach provided additional information about the change in condition of the items considered during their operational life. Subsequently, a new method for the control of maintenance procedures was developed (Knezevic, 1995). Thus, provision of fuller information on the process of change in condition resulted in a higher level of utilization of the items maintaining a low probability of failure during the operation.

It is a dynamic process because the time of the next examination is fully determined by the real condition of the system at the time of examination. Dynamic control of maintenance tasks according to the new model allows each individual system to perform the requested function with the required probability of failure, as in the case of life-based preventive maintenance but with fuller utilization of operating life, hence with a reduction of total cost of operation and production.

As the required level of reliability can only be maintained by applying a preventive maintenance policy, the *RCP* approach to maintenance introduced the critical level of relevant condition predictor RCP_{cr}, above which appropriate maintenance tasks should be performed. The interval between limit and critical values is called the safety interval, and depends on the ability of the operator to measure the condition of the item through *RCP*.

Depending on the numerical value of *RCP*, at any operational length the item under consideration, from the point of view of maintenance could be in one of the following three states:

(a) $RCP_{initial} < RCP(l) < RCP_{cr}$: continue with examinations.
(b) $RCP_{cr} < RCP(l) < RCP_{lim}$: preventive maintenance task required.
(c) $RCP_{lim} < RCP(l)$: corrective maintenance task required, because the failure has already occurred.

In order to minimize interruptions of operation and thereby to increase the availability of the system, the new method does not consider any stoppages until the time to the first examination of the condition of the item, FMT_1^E, expressed in some operating units (kilometres, hours, years, etc.). It is the instant of operational length up to which the required probability of failure-free operation is maintained. The result of the examination is given as a numerical value of the relevant condition predictor, $MRCP(FMT_1^E)$, and it presents the real condition of the item at this instant of time. Regarding the value recorded, the following two conditions are possible:

(a) $MRCP(FMT_1^E) > RCP_{cr}$, which means that a prescribed maintenance task should take place.
(b) $MRCP(FMT_1^E) < RCP_{cr}$, which means that the item can continue to be used.

In (b), the question which immediately arises is: when will the next examination have to be done, in order to preserve the required reliability level? The time to the next examination depends on the difference between the RCP_{cr} and $MRCP(FMT_1^E)$. The greater the difference, the longer is the operational length to the next examination, FMT_2^E.

At the predetermined time of the next examination, FMT_n^E, either condition is possible, and the same procedure should be followed, as shown in Figure 9.6.

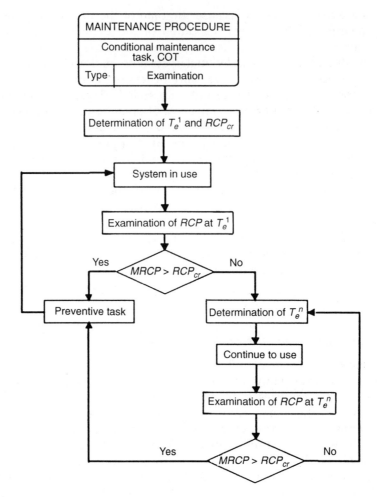

Figure 9.6 Algorithm for the control of the maintenance proces.

9.4.1 Advantages of examination-based maintenance

The following points represent major advantages of the *RCP*-based approach to maintenance.

(a) Provision of the required reliability level of each individual item.
(b) Reduction of maintenance cost as a result of:
(i) extended operating life of each individual item compared with life-based maintenance;
(ii) increased availability of the item by a reduction of the number of inspections in comparison with inspection-based maintenance.

(c) Provision of a plan for maintenance tasks from the point of view of logistic support.

The advantages of the examination-based maintenance policy are:

(a) Fuller utilization of the functionable life of each individual system than in the case of life-based maintenance.
(b) Provision of the required reliability level of each individual system as in the case of life-based maintenance.
(c) Reduction of the total maintenance cost as a result of extending the realizable operating life of the system and provision of a plan for maintenance tasks from the point of view of labour, equipment and spare parts.
(d) Applicability to all engineering systems. The main difficulties are the selection of a relevant condition predictor and the determination of the mathematical description of this $RCP(l)$.

9.5   CASE STUDY: RAF MAINTENANCE ANALYSIS – FIXED LIFE OR ON CONDITION

The support authority is responsible for determining the overall aircraft, weapon or item maintenance policy. Most failures of aircraft items occur in a random manner which is not closely related to the time in service of the items considered. Such failures can only be dealt with by replacing the items as and when they lose functionability. For some items, the probability of failure increases significantly with age or usage. Within aircraft systems, items are classified as either functionally significant or not. If 'the consequences of its failure would affect either the safety of the aircraft, its operational capability or the economics of its operation' the item is considered as functionally significant. The consequence is that such items may have to be replaced or removed for maintenance on the predetermined basis. This results in the RAF adopting a fixed life policy for:

(a) items for which safety considerations require that in-use failures should be eliminated or at least minimized; and
(b) items for which timely replacement or removal for maintenance would significantly increase system availability and/or decrease overall maintenance cost.

A fixed life policy is a costly undertaking so support authorities select only those items which have a failure mode which satisfy all of the following criteria:

(a) It is known or firmly expected that there will be a marked increase in the probability of failure after the item has been in service for some time.

(b) Incipient failures cannot be detected by condition monitoring.
(c) The item is functionally significant in that either the failure mode has safety or operational implications or a fixed life policy is justified because replacement or removal at predetermined intervals would offer a significant improvement in system availability or decrease the overall maintenance costs.

Because it involves maintenance tasks on items which may be fully serviceable and because of all the stringent measures required to monitor and control the items concerned, a fixed life policy is a very expensive option. Consequently the support authority reviews, at the appropriate intervals, the maintenance policy for its range of items. To provide full information of an item's status the support authority may call upon the relevant RAF third line facility or the design authority/ manufacturer to carry out a sampling programme to determine such parameters as the item's mean duration of functionable life, the modes of failure, any pertinent failure trends and the relevant maintenance history of the item. The taileron actuator is an example of an item whose maintenance policy was changed as a result of such a review. It was originally subject to a 900-hour life to overhaul and a finite life of 4000 hours. Following a sampling programme the support authority decided on a change of the maintenance policy to one of 'on condition' but still retained the finite life of 4000 hours. Up to now this item has been operated under this policy for 8 years without experiencing catastrophic failures of an actuator.

Cost of maintenance policy

The main objective of this chapter is to analyse the costs of different maintenance policies, namely:

- Failure-based (FB)
- Life-based (LB)
- Inspection-based (IB)
- Examination-based (EB).

The direct maintenance cost for each of the above-listed maintenance policies will be addressed in this chapter.

10.1   COST OF FAILURE-BASED MAINTENANCE POLICY

In order to assess the direct cost involved with application of FB maintenance policy of an item/system, it is necessary to analyse all cost categories involved. Thus, during the stated operational length, L_{st}, of, say, one year or 3000 hours, the total direct maintenance cost, $CMP^F(L_{st})$, is a function of:

(a) the direct cost of each corrective maintenance task performed, CMT^c;
(b) the number of corrective maintenance tasks, NMT^c, performed during the stated operational length;
(c) the operational length itself, L_{st};
(d) the opportunity cost, or the cost of potential lost revenue, CLR^c.

The generic expression for total direct cost for FB maintenance policy could therefore be expressed as:

$$CMP^F(L_{st}) = f(CMT^c, NMT^c, L_{st}, CLR^c) \tag{10.1}$$

As the analysis of the maintenance cost performed so far clearly demonstrates that the first two elements of the above expression are random variables, it follows that the total maintenance cost for a stated operational length is also a random variable. Thus, the most frequent form in which it is used in maintenance analysis of real-life problems is as the expected value. Consequently, the mean total maintenance cost for a stated operational length, $MCMP^F(L_{st})$, is a function of the expected values of the consisting elements of equation 10.1 which could be rewritten as follows:

$$MCMP^F(L_{st}) = f(MCMT^c, MNMT^c, L_{st}, MCLR^c) \qquad (10.2)$$

Before the final expression for $MCMP^F(L_{st})$ is derived, it is necessary to determine the expressions for each of the consisting elements.

The mean cost of the corrective maintenance task is defined by equation 5.5.

The expected number of maintenance tasks performed during the stated time of operation, in cases of FB maintenance policy, is defined by Table 12.1.

The cost of lost revenue, CLR^c, is a function of the following variables:

$$CLR^c = f(DMT^c, DST^c, HR) \qquad (10.3)$$

where DMT^c represents a random variable known as the duration of maintenance task, which is related to the elapsed active maintenance time. DST^c represents a random variable known as the duration of support task, and it is related to the elapsed support time, which represents a time during which the required maintenance task cannot be successfully completed due to the lack of resources (spares, material, trained personnel, tools, needed specialist equipment, facilities and similar) (Knezevic, 1995). It is necessary to stress that different repair policies for failed items have a significant impact on the DST^c thus:

(a) For items which are discarded after their failures, the random variable DST^d is a function of the logistics delay time, LDT, and the corresponding administrative delay time, ADT^d, thus:

$$DST^d = f(LDT, ADT^d) \qquad (10.4)$$

(b) For items which are repairable after their failures, the random variable DST^r is a function of the turn around time, TAT, and the corresponding administrative delay time, ADT^r, thus:

$$DST^r = f(TAT, ADT^r) \qquad (10.5)$$

HR represents the monetary value of the losses to the operator/user incurred each hour during which the item/system is not available for the revenue-earning operation.

In order to develop the expression for the determination of the cost of lost revenue for FB maintenance the mean values for the consisting elements in equation 10.3 will be used, thus:

$$MCLR^c = (MDMT^c + MDST^c) \times HR \qquad (10.6)$$

Therefore, the mean cost of FB maintenance policy for a stated operational length is equal to the product of the direct maintenance cost for each maintenance task initiated by the failure, increased for the cost of lost revenue, defined by equation 5.5, and the mean number of corrective maintenance tasks performed during the stated operational life, defined by Table 12.1. Now, it is possible to put forward the final expression for the determination of the expected total direct cost for the FB maintenance policy, thus:

$$MCMP^F(L_{st}) = (MCMT^c + MCLR^c) \times MNMT^c(L_{st}) \qquad (10.7)$$

In order to illustrate the model presented for the evaluation of the policy suitability of FB maintenance policy an illustrative example will be used.

Example 10.1

Consider a model of a compressor which consists of three maintenance-significant items. The problem is to determine the optimal maintenance policy which will provide the minimum total cost of maintenance for 240 000 km. The appropriate data are presented in Table 10.1. The estimated hourly rate of revenue is £10.

Table 10.1 Input data for Example 10.1

Characteristics and units	Items		
	1	*2*	*3*
Maintenance task	Replacement	Cleaning	Replacement
Failure distribution	Normal	Normal	Normal
A_f (km)	60 000	80 000	100 000
B_f (km)	18 000	20 000	30 000
C^c_{MSS} (£)	250	30	1 200
C^c_{MTE} (£)	15	3	150
HR^c (£)	500	300	1 000
$MDMT$ (hr)	10	5	300
$MDST$ (hr)	90	45	0

Table 10.2 Obtained values for FB maintenance policy for Example 10.1

Characteristics and units	Items		
	1	*2*	*3*
Maintenance task	Replacement	Cleaning	Replacement
CMT^c (£)	765	333	2 350
CLR^c (£)	1 000	500	3 000
MNF (240 000)[a]	4.293	3.034	2.338
$MCMP^F$ (240 000) (£)	75 710.1	2 527.3	12 508.3

[a]The numerical value for MNF (240 000) is determined according to equation 19.15.

Solution

In the interests of script economy, only the salient details of this case are given in Table 10.2.

10.2   COST OF LIFE-BASED MAINTENANCE POLICY

In order to assess the direct cost involved with application of LB maintenance policy to an item/system, it is necessary to analyse all cost categories involved. Thus, during the stated operational length, L_{st}, of, say, one year or 3000 hours, the total direct maintenance cost, $CMP^L(L_{st})$, is a function of:

(a) the direct cost of each preventive maintenance task performed, CMT^p, at specified frequency, FMT^p;
(b) the direct cost of each corrective maintenance task performed, CMT^c, in the cases where the failure takes place between two preventive tasks;
(c) the frequency of executions of preventive maintenance tasks, FMT^p;
(d) the stated operational length itself, L_{st};
(e) the opportunity cost, or the cost of potential lost revenue, CLR^p.

The generic expression for total direct cost for FB maintenance policy could therefore be expressed as:

$$CMP^L(L_{st}) = f(CMT^p, CMT^c, FMT^p, L_{st}, CLR^p) \qquad (10.8)$$

As the analysis of the maintenance cost performed so far clearly demonstrates that the first two elements of the above expression are random variables, it follows that the total maintenance cost for a stated operational length is also a random variable. Thus, the most frequent form in which it is used in maintenance analysis of real-life problems is as an expected value. Consequently, the mean total maintenance cost for a set operational length, $MCMP^L(L_{st})$, is a function of the expected values of the consisting elements of equation 10.8. Thus, it could be rewritten as follows:

$$MCMP^F(L_{st}) = f(MCMT^p, MCMT^c, FMT^p, L_{st}, MCLR^p) \qquad (10.9)$$

Before the final expression for the $MCMP^F(L_{st})$ is derived, it is necessary to determine the expressions for each of the consisting elements.

The mean costs of preventive and corrective maintenance tasks are defined by equations 5.6 and 5.5.

The frequency of the execution of the preventive maintenance tasks have to be determined according to specific criteria. Models for the determination of the optimal frequency of the execution of preventive maintenance tasks are analysed in Chapter 6.

The cost of lost revenue, CLR^P, is a function of the following variables:

$$CLR^P = f(DMT^P, DST^P, HR^P) \qquad (10.10)$$

where DMT^P represents a random variable known as the duration of maintenance task, which is related to the elapsed active maintenance time. DST^P represents a random variable known as the duration of support task, and it is related to the elapsed support time, which represents a time during which the required maintenance task cannot be successfully completed due to lack of resources (spares, material, trained personnel, tools, needed specialist equipment, facilities and similar) (Knezevic, 1995). It is necessary to stress that in the case of preventive maintenance tasks this cost element could be equal to zero. HR^P represents the monetary value of the losses to the operator/user incurred each hour during which the item/system is not available for the revenue-earning operation due to the execution of the preventive maintenance task. This cost element could be equal to zero, due to the fact that performance of the maintenance task could be planned in advance so that it could be completed during the times when the item/system is not required.

In order to develop the expression for the determination of the cost of lost revenue for LB maintenance, the mean values for the consisting elements in equation 10.10 will be used, thus:

$$MCLR^P = (MDMT^P + MDST^P) \times HR^P \qquad (10.11)$$

Therefore, according to the models presented in Chapter 6, the mean cost of LB maintenance policy for a stated operational length is equal to the product of the unit maintenance cost determined according to the optimization criterion selected, and the stated operational length, thus:

$$MCMP^L(L_{st}) = L_{st} \times UMC^L \qquad (10.12)$$

Example 10.2

Using the data from Example 10.1, determine the cost of LB maintenance policy for the three items considered, during 240 000 km stated operational length.

Solution
In the interests of script economy, only the salient details of this case are provided. Results from the calculation of the frequency of preventive maintenance task, according to equation 6.8, are presented in Table 10.3.

Cost of maintenance policy

Table 10.3   Cost of LB maintenance policy

Item	Maintenance interval, FMT^P (1000 km)				
	20	40	60	80	100
1	5 525	3 909*	5 310	6 418	6 470
2	1 966	1 092*	1 185	1 753	2 158
3	36 896	19 038	14 190	13 258*	13 983

*Indicate optimal values.

Thus the optimal frequencies of LB maintenance are 40 000 km for items 1 and 2 and 80 000 km for item 3.

10.3   COST OF INSPECTION-BASED MAINTENANCE POLICY

In order to assess the direct cost involved with application of IB maintenance policy to an item/system, it is necessary to analyse all cost categories involved. Thus, during the stated operational length L_{st}, of, say, one year or 3000 hours, the total direct maintenance cost, $CMP^I(L_{st})$, is a function of:

(a) the direct cost of each conditional maintenance task performed, CMT^I, at specified frequency, FMT^I;
(b) the direct cost of each corrective maintenance task performed, CMT^c, in cases where the failure takes place between two conditional tasks;
(c) the frequency of executions of inspection as a conditional mainten-ance task, FMT^I;
(d) the stated operational length itself, L_{st};
(e) the opportunity cost, or the cost of potential lost revenue, CLR^I.

The generic expression for total direct cost for FB maintenance policy could therefore be expressed as:

$$CMP^I(L_{st}) = f(CMT^I, CMT^c, FMT^I, L_{st}, CLR^I) \qquad (10.13)$$

As the analysis of the maintenance cost performed so far clearly demonstrates that the first two elements of the above expression are random variables, it follows that the total maintenance cost for a stated operational length is also a random variable. Thus, the most frequent form in which it is used in maintenance analysis of real-life problems is as an expected value. Consequently, the mean total maintenance cost for a set operational length, $MCIM(L_{st})$, is a function of the expected values of the consisting elements of equation 10.13. Thus, it could be rewritten as follows:

$$MCMP^I(L_{st}) = f(MCMT^I, MCMT^c, FMT^I, L_{st}, MCLR^I) \qquad (10.14)$$

Before the final expression for $MCMP^I(L_{st})$ is derived, it is necessary to determine the expressions for each of the consisting elements.

The mean costs of conditional and corrective maintenance tasks are defined by equations 5.9 and 5.5.

The frequency of the execution of the conditional maintenance tasks have to be determined according to specific criteria. Models for the determination of the optimal frequency of the execution of the conditional inspections are analysed in Chapter 6.

The cost of lost revenue, CLR^I, is the function of the following variables:

$$CLR^I = f(DMT^I, DST^I, HR^I) \qquad (10.15)$$

where DMT^I represents a random variable known as the duration of inspection of a maintenance task, which is related to the elapsed active maintenance time. DST^I represents a random variable known as the duration of support task, and it is related to the elapsed support time, which represents a time during which the required maintenance task cannot be successfully completed due to the lack of resources (spares, material, trained personnel, tools, needed specialist equipment, facilities and similar) (Knezevic, 1995). It is necessary to stress that in case of conditional maintenance tasks this cost element could be equal to zero. HR^I represents the monetary value of the losses to the operator/user incurred each hour during which the item/system is not available for the revenue-earning operation due to the execution of the preventive maintenance task. This cost element could be equal to zero, due to the fact that performance of the maintenance task could be planned in advance so that it could be completed during the times when the item/system is not required.

In order to develop the expression for the determination of the cost of lost revenue for IB maintenance, the mean values for the consisting elements in equation 10.15 will be used, thus:

$$MCLR^I = (MDMT^I + MDST^I) \times HR^I \qquad (10.16)$$

Therefore, according to the models presented in Chapter 6, the mean cost of IB maintenance policy for a stated operational length is equal to the product of the unit maintenance cost determined according to the optimization criterion selected, and the stated operational length, thus:

$$MCMP^I(L_{st}) = L_{st} \times UMC^I \qquad (10.17)$$

10.4 COST OF EXAMINATION-BASED MAINTENANCE POLICY

In order to assess the direct cost involved with application of EB maintenance policy to an item/system, it is necessary to analyse all cost categories involved. Thus, during the stated operational length L_{st}, of, say, one year or 3000 hours, the total direct maintenance cost, $CMP^E(L_{st})$, is a function of:

(a) the direct cost of each examination, as a conditional maintenance task performed, CMT^E, at specified frequency, FMT^E;
(b) the direct cost of each corrective maintenance task performed, CMT^c, in cases where a failure takes place between two conditional tasks;
(c) the frequency of executions of examination as a conditional maintenance task, FMT^E;
(d) the stated operational length itself, L_{st};
(e) the opportunity cost, or the cost of potential lost revenue, CLR^E.

The generic expression for total direct cost for FB maintenance policy could therefore be expressed as:

$$CMP^E(L_{st}) = f(CMT^E, CMT^c, FMT^E, L_{st}, CLR^E) \qquad (10.18)$$

As the analysis of the maintenance cost performed so far clearly demonstrates that the first two elements of the above expression are random variables, it follows that the total maintenance cost for a stated operational length is also a random variable. Thus, the most frequent form in which it is used in maintenance analysis of real-life problems is as the expected value. Consequently, the mean total maintenance cost for a set operational length, $MCMP^E(L_{st})$, is a function of the expected values of the consisting elements of equation 10.18. Thus, it could be rewritten as follows:

$$MCMP^E(L_{st}) = f(MCMT^m, MCMT^c, FMT^m, L_{st}, MCLR^E) \qquad (10.19)$$

Before the final expression for the $MCMP^E(L_{st})$ is derived, it is necessary to determine the expressions for each of the consisting elements.

The mean costs of conditional and corrective maintenance tasks are defined by equations 5.9 and 5.5.

The frequency of the execution of the conditional maintenance tasks have to be determined according to specific criteria. Models for the determination of the optimal frequency of the execution of the conditional inspections are analysed in Chapter 6.

The cost of lost revenue, CLR^E, is the function of the following variables:

$$CLR^E = f(DMT^E, DST^E, HR^E) \qquad (10.20)$$

where DMT^E represents a random variable known as the duration of inspection as a maintenance task, which is related to the elapsed active maintenance time. DST^E represents a random variable known as the duration of support task, and it is related to the elapsed support time, which represents a time during which the required maintenance task cannot be successfully completed due to the lack of resources (spares, material, trained personnel, tools, needed specialist equipment, facilities and similar) (Knezevic, 1995). It is necessary to stress that in the case of conditional maintenance tasks this cost element could be equal to zero. HR^E represents the monetary value of the losses to the operator/user incurred each hour during which the item/system is not available for the revenue-earning operation due to the execution of the preventive maintenance task. This cost element could be equal to zero, due to the fact that performance of the maintenance task could be planned in advance so that it could be completed during the times when the item/system is not required.

In order to develop the expression for the determination of the cost of lost revenue for EB maintenance the mean values for the consisting elements in equation 10.20 will be used, thus:

$$MCLR^E = (MDMT^E + MDST^E) \times HR^E \qquad (10.21)$$

Therefore, according to the models presented in Chapter 6, the mean cost of EB maintenance policy for a stated operational length is equal to the product of the unit maintenance cost determined according to the optimization criterion selected, and the stated operational length, thus:

$$MCMP^E(L_{st}) = L_{st} \times UMC^E \qquad (10.22)$$

PART THREE

Maintainability Analysis

CHAPTER 11

Maintainability measures

The probability theory is axiomatic. Fully defined probability problems have unique and precise solutions. Problems are wholly abstract, although they are often used for the probabilistic modelling of the behaviour of the real world.

(Knezevic, 1993)

The objective of this chapter is to define the measures through which maintainability can be quantitatively described, just as functionability is numerically expressed through performance parameters. Earlier in the book it was demonstrated that the duration of maintenance task can only be described in probabilistic terms. Thus, maintainability is fully defined by the random variable DMT and its probability distribution. The most frequently used characteristics of maintainability are:

(a) Probability of task completion.
(b) Mean duration of maintenance task.
(c) Percentual duration of maintenance task.
(d) Variability of duration of maintenance task.
(e) Success of task completion.

A brief definition and description of these characteristics follows.

11.1 PROBABILITY OF TASK COMPLETION

This maintainability measure represents the probability that the maintenance task considered will be successfully completed by a stated time, T_{st}. It is denoted as PTC_{DMT}, and is fully defined by the following expression:

$$PTC_{DMT} = P(DMT \leq T_{st}) = \int_0^{T_{st}} m(t)dt \qquad (11.1)$$

11.2 MEAN DURATION OF MAINTENANCE TASK

This maintainability measure, denoted as $E(DMT)$, represents the expectation of the random variable DMT which can be used for calculation of the characteristic of a maintenance task, thus:

$$MDMT = E(DMT) = \int_0^{\infty} t \times m(t)dt \qquad (11.2)$$

The above characteristic could also be expressed in the following way:

$$MDMT = E(DMT) = \int_0^\infty [1 - M(t)]dt \qquad (11.3)$$

which represents the area below the function which is complementary to the maintainability function.

11.3 DMT_p DURATION OF MAINTENANCE TASK

This maintainability measure, denoted as DMT_p, represents the duration of maintenance task by which a given percentage of maintenance tasks considered will be successfully completed. It is the abscissa of the point whose ordinate presents a given percentage of restoration. Mathematically, DMT_p time can be represented as:

$$DMT_p = t \to \text{ for which } M(t) = P(DMT \leqslant t) = \int_0^t m(t)dt = p$$
$$(11.4)$$

The most frequently used measure of DMT_{90} is time, which presents the restoration time by which 90% of maintenance trials will be completed, as shown in Figure 11.1.

$$DMT_{90} = t \to \text{ for which } M(t) = P(DMT \leqslant t) = \int_0^t m(t)dt = 0.9$$

It is worth noticing that in military-oriented literature (Blanchard, 1976, 1991; Blanchard *et al.*, 1995; Patton, 1983, 1988) and defence

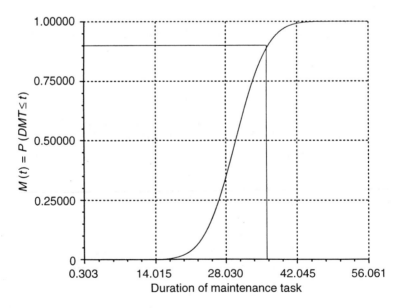

Figure 11.1 DMT_p time.

contracts, the numerical value of DMT_{95} is referred to as maximum repair time and it is denoted as M_{max}, thus $M_{max} = DMT_{95}$.

11.4 VARIABILITY OF DURATION OF MAINTENANCE TASK

In certain cases it is difficult when using only the knowledge of a standard deviation to decide whether the dispersion is particularly large or small, because this will depend on the mean value. In these situations, the coefficient of variation, $CV(DMT)$, defined as:

$$CV(DMT) = \frac{SD(MDMT)}{MDMT} \tag{11.5}$$

is very useful because it gives better information regarding the dispersion. This characteristic of the probability distribution is also known as the variability of the random variable.

11.5 SUCCESS OF TASK COMPLETION

This measure, denoted as $STC(t_1, t_2)$, represents the probability that the trial which has not been completed at time t_1 will be finished by the time t_2. From the point of view of probability this problem represents an example of conditional probability because the task could be completed by t_2 given that it was not at t_1. This measure is fully defined by the following expression:

$$STC(t_1, t_2) = P(DMT \leqslant t_2 | DMT > t_1)$$

Making use of equation 4.2 which defines the maintainability function, $M(t)$, and applying the principles of conditional probability, the above expression for the restoration success could be rewritten as:

$$STC(t_1, t_2) = P(DMT \leqslant t_2 | DMT > t_1) = \frac{M(t_2) - M(t_1)}{1 - M(t_1)} \tag{11.6}$$

Thus, the restoration success is the conditional probability which is fully defined by the above ratio.

In the case when the beginning of the interval coincides with the beginning of the restoration process, $t_1 = 0$, the restoration success is equal to the maintainability function at time t_2, thus:

$$STC(t_1, t_2) = \frac{M(t_2) - M(t_1)}{1 - M(t_1)} = M(t_2)$$

because $M(0) = 0$.

This measure of maintenance provides very useful information for the maintenance planners and managers.

Example 11.1

For the maintenance task, whose duration time could be modelled by the Weibull distribution with parameters $A_m = 29$ minutes, $B_m = 2.9$ and $C_m = 0$, determine:

(a) the probability that the task analysed will be successfully completed within 20 minutes;
(b) the duration time up to which 20% and 95% of the tasks will be successfully completed;
(c) the mean duration of maintenance task, *MDMT*;
(d) the probability that the maintenance task which has not been completed during the first 29 minutes will be finished within the following 10 minutes.

(a) Making use of Table 4.6, the maintainability function for this particular task is modelled by the following expression:

$$PTC_{20} = M(20) = 1 - \exp - \left[\frac{(20 - 0)}{(29 - 0)}\right]^{2.9} = 0.29$$

thus, there is a probability of 0.29 that the task considered will be finished within the first 20 minutes of maintenance.

(b) The DMT_p time represents the restoration time by which a given percentage of maintenance tasks will be completed. For the Weibull distribution this can be calculated using the following equation:

$$t = A_m[- \ln(1 - M(t))^{1/B_m}]$$

$$DMT_{20} = 29[- \ln(1 - 0.2)]^{1/2.9} = 17.29 \text{ minutes}$$

$$DMT_{95} = 29[- \ln(1 - 0.95)]^{1/2.9} = 42.33 \text{ minutes}$$

(c) For the Weibull probability distribution, the numerical value for $E(DMT) = MDMT$ will be:

$$E(DMT) = MDMT = 29 \times \Gamma(1 + \frac{1}{2.9}) = 29 \times 0.892 = 25.87 \text{ minutes}$$

The numerical value for $\Gamma(1 + 1/2.9) = 0.892$ was obtained from Appendix Table 2.

(d) The probability that a maintenance task will be completed by the time t_2, given that it has not been completed by time t_1, can be determined using equation 11.6.

$$P(DMT \leqslant 39 | DMT > 29) = \left[\frac{M(39) - M(29)}{1 - M(29)}\right]$$

Thus, for Weibull probability distribution:

$$M(29) = 1 - \exp - \left[\frac{(29 - 0)}{(29 - 0)}\right]^{2.9} = 1 - 0.3678 = 0.632$$

$$M(39) = 1 - \exp - \left[\frac{(39 - 0)}{(29 - 0)}\right]^{2.9} = 1 - 0.094 = 0.906$$

Therefore:

$$P(DMT \leq 39 | DMT > 29) = \frac{0.906 - 0.632}{1 - 0.632} = 0.745 = 74.5\%$$

which in practice means that there is a probability of 0.74 that the maintenance task which has not been completed within the first 29 minutes will be successfully completed during the remaining 10 minutes.

11.6 MAINTENANCE PERSONNEL DEMAND PER MAINTENANCE TASK

The maintainability measures covered thus far relate to the duration of maintenance task expressed through the probability distribution of the elapsed maintenance times. Although elapsed times are an extremely important factor in relation to the availability, one must also consider the demand for resources required for the task execution, in particular the number of personnel involved. In some instances elapsed times can be reduced by applying additional personnel in the accomplishment of specific tasks. However, this may turn out to be an expensive trade-off, particularly when high skill levels are required to perform tasks which result in less overall clock time. One of the most frequently used maintainability measures is the mean maintenance personnel demand per maintenance task, denoted as $MMPD^{MT}$. This measure could be quantified according to the following expression:

$$MMPD^{MT} = \frac{\sum_{i=1}^{nma} MPSD_i \times MDMA_i}{MDMT} \tag{11.7}$$

where $MPSD_i$ stands for the maintenance personnel demand for the successful completion of the ith maintenance activity, $MDMA_i$ is the mean duration of the ith maintenance activity, and nma represents the total number of activities which make up the task under consideration.

Example 11.2

The maintenance task under consideration consists of four different maintenance activities, as shown in Table 11.1, together with the maintenance personnel demand for each of them.

Table 11.1 Maintenance personnel demand for the task analysed

Activity	Mean duration (min)	Number of personnel
1	30	1
2	120	3
3	45	1
4	5	2

Determine the maintenance personnel demand for the maintenance task considered.

Solution

The required maintainability measure could be obtained by making use of equation 11.7, thus:

$$MMPD^{MT} = \frac{(1 \times 30) + (3 \times 120) + (1 \times 45) + (2 \times 5)}{30 + 120 + 45 + 5} = \frac{445}{200} = 2.225$$

Hence, the mean demand for the maintenance personnel required for the successful completion of the task analysed is just over 2.

CHAPTER 12

Maintainability statistics

Statistics is concerned with the relation of abstract models to actual physical systems. The methods employed by the statistician are arbitrary ways of being reasonable in the application of probability theory to physical situations. The primary tools of statistics are: probability theory, mathematical sophistication and common sense.

(Knezevic, 1993)

Maintainability measures addressed in Chapter 11 are related to the single maintenance task. However, there are large numbers of items the maintenance of which requires two or more different maintenance tasks to be applied. These tasks could be of a:

(a) corrective nature, in the response to different failure modes in which it could fail;
(b) preventive nature, where the maintenance tasks are performed in order to reduce the probability of occurrence of the failure due to a specific failure mechanism (corrosion, fatigue, wear, thermal deformation and similar);
(c) conditional nature, where the tasks are performed in order to assess the condition of the item in order to determine further course of action.

In order to incorporate all maintenance tasks related to an item and different demands for their execution it is necessary to address them within a stated operational interval. The maintenance process, as indicated in Chapter 3, is a flow of the maintenance tasks performed in accordance with the adopted maintenance policy. The flow of specific maintenance tasks related to a single item during stated operational length, L_{st}, is shown in Figure 12.1.

Since it was clearly demonstrated earlier that the maintenance task considered represents a flow of maintenance activities, completion of which can only be described in probabilistic terms, in the case of the maintenance process it could be said that it represents a flow of maintenance tasks. However, the frequency of demand for maintenance tasks is a function of the durability characteristics of the item considered.

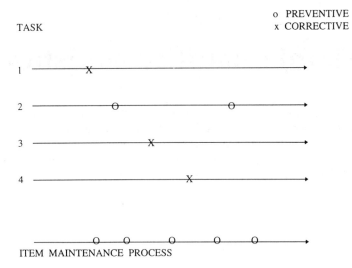

Figure 12.1   Flow of maintenance tasks for a single item.

Consequently, it is obvious that durability of a repairable item and its maintainability characteristics are the main drivers of the maintenance process. The durability factors, expressed through the probability distribution of the duration of the functionable life length, *DFL*, are the basis for determining the frequency of the demand for maintenance. Thus, in this area, durability and maintainability requirements for a given item must be compatible and mutually supportive, in the context of the operational scenario chosen by the operator/user.

12.1   TIME IN MAINTENANCE

Based on the above discussion it is clear that for the item which is exposed to several different maintenance tasks it is necessary to determine the maintainability measures in such a way that all tasks are somehow taken into consideration during the stated operational length. In order to achieve that it is necessary to determine a cumulative time in maintenance during a stated operational length, denoted as $TIM(L_{st})$. Clearly this is a random variable and as such it has to be treated probabilistically. Consequently, the mean time in maintenance during a stated operational length, $MTIM(L_{st})$, could be determined according to the following expression:

$$MTIM(L_{st}) = \sum_{i=1}^{nte} MDMT_i \times MNMT_i(L_{st}) \qquad (12.1)$$

where *nte* represents the number of different maintenance tasks expected to be performed on the item under consideration, which is obtained as the output of the failure mode and effect analysis, and $MNMT(L_{st})$ represents the mean number of tasks expected to be performed during a stated length of operation.

12.1.1 Time in corrective maintenance

Under circumstances where all tasks expected to be performed on an item are of a corrective nature, this maintainability measure is denoted as $MTIM^c(L_{st})$, and it could be calculated according to the following expression:

$$MTIM^c(L_{st}) = \sum_{i=1}^{nct} MDMT_i^c \times MNMT_i^c(L_{st}) \tag{12.2}$$

where *nct* represents the total number of different corrective maintenance tasks expected to be performed on the item under consideration, which is obtained as the output of the failure mode and effect analysis, and $MNMT^c(L_{st})$ represents the mean number of tasks expected to be performed during a stated length of operation, the numerical value of which could be determined according to expressions in Table 12.1 for the corrective maintenance tasks.

Clearly, the above expression is governed by the failure mechanism which drives the demands for a specific maintenance task in one hand, and the operational length considered. Consequently, the mean number of demands for each task could be obtained according to material presented in Chapter 3 where different probability distributions are used to model different failure mechanisms, as shown in Table 12.1.

Table 12.1 Mean number of maintenance tasks for different failure distributions (Knezevic, 1993)

Demand distribution	$MNMT^c(L_{st})$
Exponential	$\dfrac{L_{st}}{A_f}$
Normal	$\sum_{i=1}^{\infty} \Phi^i\left(\dfrac{L_{st} - i \times A_f}{\sqrt{i} \times B_f}\right)$
Lognormal	$\sum_{i=1}^{\infty} C^i(L_{st})$
Weibull	$\sum_{i=1}^{\infty} C^i(L_{st})$

A_f and B_f are scale and shape parameters of probability distributions of a demand random variable, usually it is *DFL*, and $C^i(L_{st})$ is the cumulative distribution function of the cumulative functionable length to the *i*th demand, CFL^i (a full description of which is given in Chapter 19).

Example 12.1

Failure mode and effect analysis has identified three different failure modes of a hypothetical fuel pump. The frequency of occurrence of each mode is modelled by the appropriate probability distribution and its scale and shape parameter, as shown in Table 12.1. Each failure mode could be rectified by performing a specific corrective maintenance task, as shown in Table 12.2. The same table contains the information related to the mean duration of the maintenance tasks required for each failure mode.

Determine the mean time in maintenance of the pump caused by the execution of corrective maintenance tasks during the first 240 000 km of operation.

Table 12.2 Relevant data for item-related maintainability statistics

Failure mode	1	2	3
Probability distribution	Exponential	Exponential	Exponential
Scale parameter (km)	55 950	80 000	102 500
Shape parameter (km)	n/a	n/a	n/a
Maintenance task	Replacement	Cleaning	Overhaul
$MDMT^c$ (hr)	10	5	300

Solution

The expected number of corrective maintenance tasks to be demanded during the stated operational length of 240 000 km could be obtained by making use of Table 12.1, and information provided in Table 12.2. Thus, the calculated values for the $MNMT^c_i(240\ 000)$, for all three maintenance tasks are given in tabular form below.

Maintenance task	Replacement	Cleaning	Overhaul
$MNMT^c(240\ 000)$	4.29	3	2.34

According to equation 12.2, the required maintainability figure of merit could be obtained, thus:

$$MTIM^c(240\ 000) = \sum_{i=1}^{3} MDMT^c_i \times MNMT^c_i(240\ 000)$$

$$= (10 \times 4.29) + (5 \times 3) + (300 \times 2.34)$$
$$= 759.9 \text{ hours}$$

Hence, it is expected that during 240 000 km of operation the pump considered will require around 760 hours of corrective maintenance.

12.1.2 Time in preventive maintenance

In cases where all maintenance tasks related to a single item are of a preventive nature the mean time in maintenance, $MTIM^P(L_{st})$, can be calculated according to the following expression:

$$MTIM^P(L_{st}) = \sum_{i=1}^{npt} MDMT_i^p \times NMT_i^p(L_{st}) \tag{12.3}$$

where npt represents the total number of different preventive maintenance tasks expected to be performed on the item under consideration, and $NMT^P(L_{st})$ represents the number of preventive tasks to be performed during a stated length of operation, the numerical value of which could be determined according to the following expression:

$$NMT^P(L_{st}) = \frac{L_{st}}{FMT^P}$$

Example 12.2

Assume that the pump analysed in Example 12.1 requires two preventive tasks, details of which are given in Table 12.3. Determine the maintainability statistics for the pump considered, with regard to the preventive maintenance tasks only.

Table 12.3 Relevant data for item-related maintainability statistics

Maintenance task	Adjustment	Testing
$MDMT^P$ (hr)	1	2
FMT^P (km)	40 000	80 000

Solution

In the interests of script economy, only the salient details of this case are given below. Thus:

$$MTIM^P(240\ 000) = \sum_{i=1}^{2} MDMT_i^p \times NMT_i^p(240\ 000) = (1 \times 6) + (2 \times 3)$$

$$= 12 \text{ hours}$$

Consequently, during the first 240 000 km the pump considered will require 12 hours of preventive maintenance.

12.1.3   Time in conditional maintenance

In cases where all maintenance tasks related to a single item are of a conditional nature, the mean time in maintenance, $MTIM^m(L_{st})$, can be calculated according to the following expression:

$$MTIM^m(L_{st}) = \sum_{i=1}^{nmt} MDMT_i^m \times NMT_i^m(L_{st}) \qquad (12.4)$$

where *nmt* represents the total number of different conditional maintenance tasks expected to be performed on the item under consideration, as a result of the completion of the failure mode and effect analysis, and $NMT^m(L_{st})$ represents the number of conditional tasks to be performed during a stated length of operation, the numerical value of which could be determined according to the following expression:

$$NMT^m(L_{st}) = \frac{L_{st}}{FMT^m} \, 12$$

Example 12.3

Assume that the pump analysed in Example 12.1 requires two conditional tasks, details of which are given in Table 12.4. Determine the maintainability statistics for the pump considered, regarding conditional maintenance tasks only.

Table 12.4   Obtained values for item-related maintainability statistics

Maintenance task	Inspection	Examination
$MDMT^m$ (hr)	0.5	1
FMT^m (km)	10 000	20 000

Solution
In the interests of script economy, only the salient details of this case are given below. Thus, making use of the information available, the mean time in conditional maintenance could be obtained as follows:

$$MTIM^m(240\ 000) = \sum_{i=1}^{2} MDMT_i^m \times NMT_i^n(240\ 000) = (0.5 \times 24)$$

$$+ (1 \times 12) = 24 \text{ hours}$$

Hence, during the first 240 000 km of operation the pump analysed will require 24 hours of conditional maintenance.

12.1.4 Total time in maintenance

From the point of view of the operator, the mean time in maintenance, $MTIM(L_{st})$, is of prime importance for making business plans and managing their execution. Thus, $MTIM^c(L_{st})$, $MTIM^p(L_{st})$ and $MTIM^m(L_{st})$ are significant factors of maintainability, but all of them are related to specific types of maintenance tasks. Thus, for the items which are exposed to all three types of tasks, i.e. corrective, preventive and conditional, the maintainability measure which takes into consideration the combined impact on the item must be determined. Consequently, the mean time in maintenance of an item which is exposed to a combination of different maintenance tasks, $MTIM_i(L_{st})$, could be determined according to the following expression:

$$MTIM(L_{st}) = MTIM^c(L_{st}) + MTIM^p(L_{st}) + MTIM^m(L_{st}) \quad (12.5)$$

where elements on the right hand side of equation 12.5 are defined by equations 12.2, 12.3 and 12.4 respectively.

It is necessary to stress that the above expression is a function of the cumulative length of stated operational length, L_{st}.

Example 12.4

Making use of Examples 12.1, 12.2 and 12.3, it is clear that three different corrective maintenance tasks, two preventive tasks and two conditional tasks are expected to be demanded and performed during the stated operational interval. Hence, the total time in maintenance of the pump considered during the operational length of 240 000 km could be determined according to equation 12.5, thus:

$$MTIM(240\ 000) = \sum_{i=1}^{3} MDMT_i^c \times MNMT_i^c(240\ 000)$$

$$+ \sum_{i=1}^{2} MDMT_i^p \times NMT_i^p(240\ 000)$$

$$+ \sum_{i=1}^{2} MDMT_i^m \times NMT_i^m(240\ 000)$$

$$= 759.9 + 12 + 30 = 801.9 \text{ hours}$$

Therefore the mean cumulative time in maintenance of the pump analysed during the 240 000 km is just over 800 hours.

In summary it could be said that, for items whose maintenance of functionability by the user/operator requires more than one maintenance task, it is necessary to determine the maintainability statistics in such a way that all tasks related to an item are somehow taken into consideration during the decision-making process at the design stage. There is also the need for realistic quantification of the maintainability necessary for the comparative analysis of two design alternatives or two competing products at the acquisition phase.

The most frequently used maintainability statistics of merit for an item are listed below:

- mean time to restore the functionability of an item;
- maintenance hours per operational unit;
- maintenance personnel demand per maintenance task.

Each of the above figures of merit will now be examined.

12.2 MEAN TIME TO RESTORE AN ITEM

Based on the above discussion it is clear that for the item which is exposed to several different maintenance tasks it is necessary to determine the maintainability statistics in such a way that all tasks are somehow taken into consideration during the stated operational time. The all embracing variable for each item is known as the time to restore, *TTR*. The most frequently used maintainability measure of this type is the mean time to restore for a stated time, denoted as $MTTR(L_{st})$. It represents the mean duration of the maintenance task required to restore the functionability of an item, when two or more different maintenance tasks could be demanded. This measure of maintainability can be calculated according to the following expression:

$$MTTR(L_{st}) = \frac{MTIM(L_{st})}{MNMT(L_{st})} \tag{12.6}$$

where $MTIM(L_{st})$ represents the mean time in maintenance during a stated length of operation, regarding corrective, preventive and conditional maintenance tasks related to the item under consideration, numerical values of which could be obtained according to equations 12.2, 12.3 and 12.4 respectively.

12.2.1 Mean time to corrective restoration

Under circumstances where all tasks expected to be performed on an item are of a corrective nature, the maintainability measure considered is denoted as $MTTR^c(L_{st})$. Making use of equation 12.6, this measure of maintainability can be calculated according to the following expression:

$$MTTR^c(L_{st}) = \frac{MTIM^c(L_{st})}{MNMT^c(L_{st})} \tag{12.7}$$

The full expression of equation 12.7 could be obtained by making use of equation 12.2, thus:

$$MTTR^c(L_{st}) = \frac{\sum_{i=1}^{nct} MDMT_i^c \times MNMT_i^c(L_{st})}{\sum_{i=1}^{nct} MNMT_i^c(L_{st})} \tag{12.8}$$

where *nct* represents the number of different corrective maintenance tasks expected to be performed on the item under consideration.

It is necessary to point out that in cases where all demands for the execution of all corrective maintenance tasks are driven by the exponential probability distribution the above expression becomes independent of length of operation and its final expression has the following form:

$$MTTR^c = \frac{\sum_{i=1}^{nct} \frac{MDMT_i^c}{A_i}}{\sum_{i=1}^{nct} \frac{1}{A_i}} \tag{12.9}$$

In all other cases the numerical value of the mean time to corrective restoration is a function of the length of operation considered, thus $MTTR^c(L_{st})$.

Example 12.5

Failure mode and effect analysis has identified three different failure modes of a hypothetical fuel pump. Each failure mode could be rectified by performing a specific maintenance task, as shown in Table 12.2. The same table contains the information related to the mean duration of the maintenance task required for each failure mode together with the expected number of tasks to be demanded during the stated operational length of 240 000 km.

Determine the mean time to restore the pump by performing corrective maintenance tasks.

Solution

According to equation 12.7, the required maintainability measure could be obtained, thus

$$MTTR^c(240\ 000) = \frac{MTIM^c(240\ 000)}{MNMT^c(240\ 000)} = \frac{(10 \times 4.29) + (5 \times 3) + (300 \times 2.34)}{4.29 + 3 + 2.34}$$

$$= \frac{759.47}{9.63} = 78 \text{ hours}$$

12.2.2 Mean time to preventive restoration

In cases where all maintenance tasks related to a single item are of preventive nature the mean restoration time, $MTTR^P(L_{st})$, can be calculated according to the following expression:

$$MTTR^P(L_{st}) = \frac{MTIM^P(L_{st})}{NMT^P(L_{st})} \tag{12.10}$$

The full expression of equation 12.10 could be obtained by making use of equation 12.3, thus:

$$MTTR^P(L_{st}) = \frac{\sum_{i=1}^{npt} MDMT_i^p \times NMT_i^p(L_{st})}{\sum_{i=1}^{npt} NMT_i^p(L_{st})} \tag{12.11}$$

where *npt* represents the total number of different preventive maintenance tasks expected to be performed on the item under consideration.

Example 12.6

Assume that the pump analysed in Example 12.1 requires two preventive tasks, details of which are given in Table 12.3. Determine the mean time to restore the pump considered, regarding the preventive maintenance tasks only.

Solution
In the interests of script economy, only the salient details of this case are given below. Thus:

$$MTIM^P(240\ 000) = \sum_{i=1}^{2} MDMT_i^p \times NMT_i^p(240\ 000) = (1 \times 6) + (2 \times 3)$$

$$= 12 \text{ hours}$$

$$NMT^P(240\ 000) = \sum_{i=1}^{2} NMT_i^p(240\ 000) = 6 + 3 = 9$$

Finally,

$$MTTR^P(240\ 000) = \frac{12}{9} = 1.33 \text{ hours}$$

Hence, the mean time to preventive restoration of the pump considered during 240 000 km is around 90 minutes.

12.2.3 Mean time to conditional restoration

In cases where all maintenance tasks related to a single item are of conditional nature, the mean restoration time, $MTTR'''(L_{st})$, can be calculated according to the following expression:

$$MTTR'''(L_{st}) = \frac{MTIM'''(L_{st})}{NMT'''(L_{st})} \qquad (12.12)$$

Making use of equation 12.4, equation 12.12 could be rewritten as follows:

$$MTTR'''(L_{st}) = \frac{\sum_{i=1}^{nmt} MDMT_i^m \times NMT_i^m(L_{st})}{\sum_{i=1}^{nmt} NMT_i^m(L_{st})} \qquad (12.13)$$

where *nmt* represents the number of different conditional maintenance tasks expected to be performed on the item under consideration.

Example 12.7

Assume that the pump analysed in Example 12.1 requires two conditional tasks, details of which are given in Table 12.4. Determine the mean time to conditional restoration for the pump considered, regarding the conditional maintenance tasks only.

Solution
In the interests of script economy, only the salient details of this case are given below. Thus:

$$MTIM'''(240\ 000) = \sum_{i=1}^{2} MDMT_i^m \times NMT_i^m(240\ 000) = (0.5 \times 24)$$
$$+ (1 \times 12) = 24 \text{ hours}$$

$$NMT'''(240\ 000) = \sum_{i=1}^{2} NMT_i^m(240\ 000) = 24 + 12 = 36$$

$$MTTR'''(240\ 000) = \frac{24}{36} = 0.66 \text{ hour}$$

Therefore, it could be concluded that the mean time to conditional restoration of the pump analysed is under one hour.

From the point of view of the operator, the mean time to restore the functionability of an item, $MTTR(L_{st})$, is of prime importance for making business plans and management of their execution. Thus, $MTTR^c(L_{st})$, $MTTR^p(L_{st})$ and $MTTR^m(L_{st})$ are significant factors of maintainability, but all of them are related to specific types of maintenance tasks. Thus, for items which are exposed to all three types of tasks (corrective, preventive and conditional) the maintainability measure which takes into consideration combined impact on the item must be determined. Consequently, the mean time to restore the item which is exposed to a combination of different maintenance tasks, $MTTR(L_{st})$, could be determined according to the following expression:

$$MTTR(L_{st}) = \frac{MTIM^c(L_{st}) + MTIM^p(L_{st}) + MTIM_i^m(L_{st})}{MNMT^c(L_{st}) + MNMT^p(L_{st}) + MNMT^m(L_{st})} \quad (12.14)$$

It is necessary to stress that the above expression is a function of the cumulative length of the stated operational length, L_{st}.

Example 12.8

Making use of Examples 12.1, 12.2 and 12.3, it is clear that three different corrective maintenance tasks, two preventive tasks and two conditional are expected to be demanded and performed during a stated operational interval. Determine the mean time to restore the pump considered, for an estimated length of operation of 240 000 km.

Solution
The required maintainability measure could be determined by making use of equation 12.14, thus:

$$MTTR(240\ 000) = \frac{759.9 + 12 + 30}{9.662 + 9 + 36} = 14.66 \text{ hours}$$

Thus, the mean time to restore the functionability of the pump analysed during the 240 000 km is 14.66 hours.

12.3   MAINTENANCE HOURS PER OPERATIONAL UNIT

The maintainability statistics covered thus far relate to the set of maintenance tasks related to an item, during a stated operational length, and are expressed through variables like $TTR(L_{st})$ and $TIM(L_{st})$. Although these statistics are extremely important for maintenance planning and they have direct impact operational availability, in order to fully assess maintainability of an item, one must also consider the ratio between maintenance hours per operational unit, $MHOU(L_{st})$. In current engineering practice, the expected value of this measure could easily be obtained according to the following expression:

$$MHOU(L_{st}) = \frac{MTIM(L_{st})}{L_{st}} \qquad (12.15)$$

It is necessary to stress that operational unit is used as a generic name for the most relevant unit of the operation. Thus, the operational units could be expressed in hours, kilometres, landings, cycles, weeks or any other unit related to the length of operation.

Example 12.9

Based on the information provided in Example 12.4, determine the maintenance hours per kilometre for the pump considered, during the 240 000 km operational period.

Solution
Making use of equation 12.15, the required value could be obtained as follows:

$$MHOU(240\ 000) = \frac{MTIM(240\ 000)}{240\ 000} = \frac{801.47}{240\ 000} = 0.0033\ (hr/km)$$

12.4 MAINTENANCE PERSONNEL DEMAND PER RESTORATION TASK

In order to fully assess maintainability of an item, one must also consider the maintenance resources induced in the process. This is primarily directed towards the demand on maintenance personnel involved in the accomplishment of specific tasks. In order to assess the demand on this most important maintenance resource, the maintenance personnel hours per restoration task, $MMPD^{RT}(L_{st})$, is most frequently used as a maintainability measure, in current engineering practice. The expected value of this measure could easily be obtained according to the following expression:

$$MMPD^{RT}(L_{st}) = \frac{\sum_{i=1}^{nte} [MMPD_i^{MT} \times MNMT_i(L_{st})]}{MNMT(L_{st})} \qquad (12.16)$$

where *nte* represents the total number of maintenance tasks expected, $MMPD_i^{MT}$ is the mean maintenance personnel demand for the *i*th task and $MNMT(L_{st})$ is the mean number of all maintenance tasks expected to be performed on the item considered during the stated operation length.

Example 12.10

Based on the data provided in Table 12.5, determine the maintenance personnel demand per restoration task of the item analysed.

Table 12.5 Maintenance data related to the item analysed

Task	MMPD$_i^{MT}$	MNMT$_i(L_{st})$
1	2.1	4
2	1	2.7
3	3.1	0.78
4	1	4.1

Solution

The required maintainability statistics could be determined by direct application of equation 12.16, thus:

$$MMPD^{RT}(L_{st}) = \frac{(2.1 \times 4) + (1 \times 2.7) + (3.1 \times 0.78) + (1 \times 4.1)}{4 + 2.7 + 0.78 + 4.1} = \frac{17.6}{11.58}$$

$$= 1.5$$

Hence, the item considered requires, on average, 1.5 maintenance personnel for each restoration task performed.

12.5 SYSTEM-BASED MAINTAINABILITY STATISTICS

One of the common perceptions is that maintainability is simply the ability to reach an item to restore it. However, that is only a small aspect. Maintainability is actually just one dimension of system design and a system's maintenance management policy. For example, it could be required from the designer that only three screws are acceptable on a certain partition panel in order to provide a quick and easy access to the item located inside. However, this request has to be placed into global context and it becomes a trade-off. If the item behind that panel needs to be checked once in every 5–6 years, or say 50 000 miles (80 000 km), it does not make much sense to concentrate much intellectual effort and spend project money on quick access. A lot of fasteners and connectors could be tolerated and the item may not be quickly accessible, but all of that has to be traded off against the cost and operational effectiveness of the system. Thus, the length of time which an item spends in a state of functioning (SoFu) provides vital information for all decisions regarding the timing and content of maintenance tasks as well as for all decisions relating to planning and provisioning of resources needed in the most effective way.

Based on the discussion so far, it is obvious that durability of repairable items and their maintainability characteristics are very closely related. The durability factors, expressed through the probability distribution of the duration of the functionable life length, *DFL*, are the basis for determining the frequency of the demand for maintenance. Maintainability deals with the characteristics in system design pertaining to minimizing the resource and time requirements for maintenance of the system when it becomes operational. Thus, in this area, durability and maintainability requirements for a given system must be compatible and mutually supportive, in the context of the operational scenario chosen by the operator/user.

In order to incorporate all maintenance tasks related to a system and different demands for their execution, it is necessary to address them within a stated operational interval. Maintenance process, as indicated in Chapter 2, is a flow of the maintenance tasks performed in accordance to adopted maintenance policy. The flow of specific maintenance tasks related to a system during a stated operational length, L_{st}, is shown in Figure 12.2.

Since it was clearly demonstrated earlier that the maintenance task considered represents a flow of maintenance activities, completion of which can only be described in probabilistic terms, in the case of the maintenance process of a system it could be said that it represents a flow of maintenance tasks. However, the frequency of demand for maintenance tasks is a function of the durability characteristics of the items considered. Consequently, it is obvious that durability of repairable items and its maintainability characteristics are the main drivers of the maintenance process. The durability factors, expressed through the probability distribution of the duration of the functionable life length, *DFL*, of each consisting item are the basis for determining the frequency of the demand for maintenance. Thus, in this area, durability and maintainability requirements for a given item must be compatible and mutually supportive, in the context of the operational scenario chosen by the operator/user.

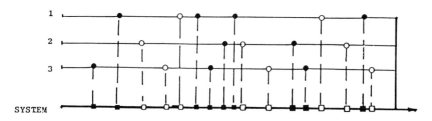

Figure 12.2 Flow of maintenance tasks for a system.

Consequently, the main objective of this chapter is to address the maintainability statistics related to an engineering system, based on the maintainability statistics of its consisting items. This approach to the analysis of a system is known as the 'bottom-up' approach.

The most frequently used maintainability statistics related to the system are:

- mean time to restore functionability of a system;
- maintenance hours per operational hour;
- maintenance personnel demand per restoration task.

Each of the above-listed statistics will be examined in this chapter, and illustrated through several numerical examples.

12.6   MEAN TIME TO RESTORE A SYSTEM

The maintainability statistics of a system considered have to be determined according to the number of consisting items, as well as the number of maintenance tasks associated with each of them. The methodology for the determination of maintainability statistics for one item has been shown above. Applying the same type of analysis it is possible to derive an expression for the determination of the mean duration of a maintenance task for a system according to the maintainability statistics of its consisting items.

Hence, the mean time to restore the system, during a stated operation length, $MTTR_s(L_{st})$, represents the mean or average time required for the successful completion of a maintenance task, among all anticipated maintenance tasks (corrective, preventive and conditional), for the system as a whole, and it can be calculated as:

$$MTTR_s(L_{st}) = \frac{\sum_{i=1}^{nmi} MTIM_i(L_{st})}{\sum_{i=1}^{nmi} MNMT_i(L_{st})} \qquad (12.17)$$

where nmi represents a number of maintenance-significant items within the system, and $MTIM_i(L_{st})$ is the mean time in maintenance of the ith item.

It is necessary to stress that the above expression is valid under the assumption that all items are connected in series from the reliability point of view.

Example 12.11

The system under consideration consists of three maintenance-significant items, A1, A2 and A3. The relevant durability and maintainability characteristics for each item, for 10 hours of operation, are given in Table 12.6.

Table 12.6   Input data for Example 12.11

Item	A1	A2	A3
Failure distribution	Exponential	Exponential	Exponential
Scale parameter (hr)	4050	536	9050
Maintenance task	Corrective	Corrective	Corrective
MDMT (hr)	0.9	0.4	1.0

Determine the mean time to restore the system during the 10 hours of operation.

Solution
In order to determine the required maintainability statistics for the system it is necessary to calculate the mean time in maintenance for each of the consisting items and the corresponding mean numbers of maintenance tasks expected to be demanded during the stated length of operation. In this particular case the required statistics for items are obtained by making use of equation 12.2 and Table 12.1 and the values obtained are shown in Table 12.7.

Table 12.7   Output data for Example 12.11

Item	A1	A2	A3
$MNMT_i^c(10)$	2.46	18.66	1.1
$MTIM_i(10)$	2.214	7.464	1.1

Making use of equation 12.17, the required measure of the system could be obtained, thus:

$$MTTR_s(10) = \frac{2.214 + 7.464 + 1.1}{2.46 + 18.66 + 1.1} = \frac{10.778}{22.22} = 0.485 \text{ hours}$$

Thus, the mean time to restore the system is just under 0.5 hour.

12.7   MAINTENANCE PERSONNEL DEMAND PER MAINTENANCE TASK

The maintainability statistic, covered thus far, relates to the time to restore a system during a stated operational time, expressed through the

$MTTR_s(L_{st})$. Although this statistic is extremely important for the maintenance planning and directly impacts operational availability, one must also consider the maintenance resources induced in the process. This is primarily directed towards the demand on maintenance personnel involved in the accomplishment of specific tasks. In order to assess the demand on this most important maintenance resource, the maintenance personnel hours per restoration task of the system, $MMPD_s^{RT}(L_{st})$, is most frequently used as a maintainability measure, in current engineering practice. The expected value of this statistic could be obtained according to the following expression:

$$MMPD_s^{RT}(L_{st}) = \frac{\sum_{i=1}^{nmi} [MMPD_i^{MT} \times MNMT_i(L_{st})]}{\sum_{i=1}^{nmi} MNMT_i(L_{st})} \qquad (12.18)$$

where nmi represents the total number of maintenance-significant items and $MMPD_i^{MT}$ is the mean maintenance personnel demand for the ith item, and $MNMT_i(L_{st})$ is the mean number of all maintenance tasks expected to be performed on the ith item considered during the stated operation length.

Example 12.12

Based on the data provided in Table 12.8 determine the maintenance personnel demand per restoration task of the system analysed.

Table 12.8 Maintenance data related to the system analysed

Item	$MMPD_i^{MT}$	$MNMT_i(L_{st})$
1	2.1	4.0
2	1.0	2.7
3	1.0	4.1

Solution
The required maintainability statistics could be determined by direct application of equation 12.18, thus:

$$MMPD_s^{RT}(L_{st}) = \frac{(2.1 \times 4) + (1 \times 2.7) + (1 \times 4.1)}{4 + 2.7 + 4.1} = \frac{15.2}{10.8} = 1.4$$

Hence, the system considered requires, on average, 1.4 personnel for each restoration task performed.

12.8   MAINTENANCE HOURS PER SYSTEM OPERATIONAL HOUR

This is another frequently used maintainability statistic at the system level. It is directly related to the durability and maintainability characteristics of the consisting items, as well as to the selected maintenance policy. In Example 1.1 it was stated Pan Am's first B747 has been exposed to 806 000 maintenance hours and, in total, it has flown 80 000 hours. Thus, the ratio between maintenance hours and operational is 10:1.

Hence, the mean maintenance hours to keep a system functionable, during a stated period of time, $MHOH_s(L_{st})$ required for the successful completion of all anticipated maintenance tasks (corrective and preventive) for the system, per operational hour, could be calculated as:

$$MHOH_s(L_{st}) = \frac{MTIM_s(L_{st})}{L_{st}} \tag{12.19}$$

It is necessary to stress that the above expression is valid under the assumption that all items are connected in series from the reliability point of view.

PART FOUR

Maintainability Engineering

CHAPTER 13

Maintainability engineering

There has always been a Chief Pilot on every Boeing model, but 777 is the first Boeing model with a Chief Mechanic. This certainly illustrates the recognition of the importance of the maintenance process to successful airline operation.

(*Airliner*, January–March, 1995)

With the increase of the rate and pace of technological advances, more and more individual companies and whole industries have been and are being forced to abandon complacency and to look into the future. Action, in such cases, usually starts with increased emphasis on research, development and design.

The importance of the maintainability engineering function, MEF, in the design process in a company is directly proportional to the importance of the design function. A strong, competent design function is essential in areas of rapid technological advances, such as aerospace and weapon systems. The design function is also of great importance to producers of consumer goods (such as motor vehicles, office equipment and household appliances), to machine-tool producers, and in many other similar areas. Design organizations are usually strong staff organizations in companies producing these systems. Although of lesser importance in companies producing simple systems or systems of stable, proven design, the design function is an important one in all production industries.

The design function within a company has certain responsibilities to the organization management. Working in the assigned system areas, the design function must create designs which are functional, reliable, maintainable, producible, timely and competitive.

Whenever possible, the designers are expected to use proven design techniques. When design objectives cannot be met by using the proven and familiar design methods, the designers are expected to adapt their methods, borrow design techniques from other industries, or use some of the new state-of-the-art materials and processes which are available. Since designers are usually, by disposition, creative it is often difficult for them to resist trying something new, even though a proven technique is available. Designers are well known for their receptiveness to the efforts of parts and package suppliers' sales engineers, who are selling the outstanding merits of their new system. A major responsibility of the design management is the establishment of a system which makes it appreciably easier for designers to use proven designs than to use unproven designs.

Since it is often impossible to meet all design objectives to the maximum desired extent, designers are frequently required to trade off between them. By requiring unusually tight tolerances or by specifying an exotic material, they may improve reliability at the expense of maintainability. By not fully testing the ability of the design to function under the worst combinations of operational environment, the designers may take a chance on worsening maintainability so that the design disclosure can be released on schedule. Some of these compromises and trade-offs are unavoidable, and it is the design function which has both the information and responsibility to make the required decisions in these cases. Thus, the fact that trade-offs have to take place makes the interaction between all participants within the design function essential, which in practice means that final decisions should not be made without full support obtained by maintainability engineers, as a part of an integrated team.

13.1   ROLE OF THE MAINTAINABILITY ENGINEERING FUNCTION

In order for a design to provide an acceptable inherent maintainability, provisions must be made within the design concept and must continue throughout development to its completion.

However, while performing the maintainability analysis of the design proposed, maintainability personnel may discover and require correction of design features, errors, or omissions which affect feasibility. However, this is not a primary purpose of maintainability analysis. Maintainability is concerned with all design issues which are able to assure that functionability of a feasible design could be easily, safely and economically maintained during operation under specified environments and other operating conditions.

The maintainability engineering function works with the design function in several ways to achieve its objective. Maintainability acts at various times as a helper, and as a conscience of a design team, but very rarely should it act as an inspector (although it is still practice in many organizations).

As a part of an integrated design team, maintainability engineers perform certain analytical and statistical analyses. These include collection, analysis, and feedback of data on development hardware/software. Generally speaking, MEF assists the design team in predicting and measuring inherent maintainability during the various design stages.

Maintainability serves as a conscience to the design function by closely participating in the design progress, in a concurrent manner, towards the specified maintainability goals. In addition, all trade-offs affecting maintainability are very closely examined.

Inevitably, MEF assesses the design output in order to verify the proposed solution before the design effort can progress. Some of the checkpoints for such enforcement of maintainability requirements are the maintainability approval of solutions proposed by the designers, maintainability approval of actual usage of the approved maintenance resources, maintainability approval of design reviews and, finally, maintainability signature approval of the design-disclosure documents (drawings, specifications and procedures).

The maintainability engineering function may also include coordination of the design test programmes, direction of independent maintainability test programmes and actual conduct of all testing, identification and establishment of control systems for portions of the design which have special limitations, preparation of maintainability specifications applicable to suppliers, and the imposition of maintainability requirements on suppliers through review and approval of procurement documents.

13.2   MAINTAINABILITY ENGINEERING FUNCTION OPPORTUNITIES

Generally speaking, the more difficult the design assignment, the larger the maintainability effort that is required, but the larger the impact on design.

The design problems encountered in an aircraft, a major weapon system, or a complex worldwide communications network require substantial design and maintainability efforts. At the same time, if the design is well within the existing technology, is simple, and has ample space, weight and design-time allowances, then a relatively small maintainability effort may be adequate.

Thus, a major maintainability effort (major in the sense of being a substantial fraction of the design effort) is required under the following circumstances:

(a) In the design of any very complex system.
(b) In the design of systems with very high maintainability requirements, particularly when the designers are working within severe space and weight limitations (submarines, spacecraft, racing cars and so forth).

The general rule could be made that the more constraints on the design team and the tighter the constraints are, the greater the required maintainability programme is. One of the major reasons for this relationship is that, under the pressure of these constraints, the design team may intentionally or unintentionally neglect the maintainability requirements. A strong maintainability effort is needed both to assist and to check on the design team in matters affecting maintainability requirements and measures.

During the design development the design team must work out compromises between various requirements. The penalties for design teams not fully meeting required performance, schedule, cost, producibility and other goals are much more immediate and certain than are the penalties for not fully meeting maintainability goals. Consequently, a MEF must be a part of a team in order to call immediate and forceful attention to any deficiencies in the design provisions for maintainability. The presence of a MEF should assure that full consideration is given to maintainability requirements. Clearly, in some cases it may still be necessary to trade off a maintainability requirement for performance or for any other design characteristic, but such a trade-off must be made with full knowledge of all possible consequences.

13.3 MAINTAINABILITY ENGINEERING FUNCTION OBSTACLES

Very few integrated design teams deliberately skimp on provisions for attaining the full maintainability requirements in their creations. However, some of the following dangers are always present:

- oversight
- lack of specific knowledge
- rationalization

and each of these is briefly addressed below.

13.3.1 Oversight

This type of obstacle occurs in cases when the design team fails to take care of one of those innumerable details which make up the completed design.

For example, a design team is fully aware that a special fastener is required for a specific item but fails to indicate it on the drawing. Thus, if this oversight is not caught, a substantial delay in completion of a corresponding maintenance task might occur.

Further, it is very easy to fail to notice that the execution of maintenance tasks will, occasionally, be required to be performed under non-typical conditions, such as low/high temperature, chemical contamination and similar, when the completion of the tasks is significantly more difficult and often impossible.

13.3.2 Lack of specific knowledge

It is fair to say that members of design teams cannot know all there is to be known about everything connected with every design, nor do they have the time to verify every detail. Consequently, all design teams do what they can to check where they believe it necessary, and call in the experts in certain highly specialized areas such as use of specific maintenance resources, ergonomics, safety issues and similar.

For instance, the design teams may specify the use of specific test equipment or a tool which was the best available for the purpose the last time when there was a need for it. However, a new technology that functions better and that performs well in excess of the design requirements may have become available.

13.3.3 Rationalization

Design teams are usually pressed for time, and hence on many occasions there is an honest belief that the design proposed will meet all of the design requirements including the maintainability requirements, but an additional series of tests should be run to be absolutely certain. Waiting for the test results will put the design behind schedule. It is easy for the design team to work themselves into a frame of mind that the tests are not really necessary. This same practice of rationalization includes explaining away test failures with such comments as 'It was a testing error' or 'It was an early design' or 'The real environments will never be nearly that rigorous anyway'.

When the consequences of not achieving the maintainability require-ments are extremely dangerous, expensive in revenue terms, or jeopardize safety, reputation or national security, then maintainability considerations become extremely important. When maintainability characteristics are of high priority, it cannot be left to good intentions or to chance. Thus, there must be an independent check and balance of every maintenance task (including those operations which the maintainability-quality function itself may perform, such as the writing of test procedures) and continuous attention paid to details. Hence, no organization and no person can be considered so good or so omnipotent that its or their output need not receive a searching independent analysis.

13.4 DESIGN METHODS FOR ATTAINING MAINTAINABILITY

Those specific practices design teams may follow to achieve maintain-ability, which would not necessarily be followed if the main concern is in only getting the design to function, are discussed here. They are:

- accessibility
- modularity
- simplicity
- standardization
- foolproofing
- inspectability.

Each of the above-listed practices is briefly addressed below.

13.4.1 Accessibility

All equipment and subassemblies which require regular inspection should be located in such a way that they can be accessed conveniently and are fitted with parts that can be connected rapidly for all mechanical, air, electric and electronic connections.

For example, in the case of the TGV train (see Example 13.1), the roof panels can be rapidly dismounted, and lateral access panels and numerous inspection points allow for all types of progressive inspections and components in a short space of time. The auxiliary equipment in power cars and passenger cars is located so that work positions for maintenance staff are ergonomic and especially in such a way that several specialists can carry out maintenance work simultaneously without disturbing other maintenance staff at work.

Generally speaking, whenever possible, it should not be necessary to remove other items to gain access to items requiring maintenance, especially frequently replaced items. Also, it should be possible to replace or top-up items like lubricates without disassembly.

13.4.2 Modularity

A modular approach is a fundamental guarantee of ease of replacement. This can only be achieved if interface equipment is standard. In such an approach the range of physical orders of magnitude at input and output of each module ensure that no readjustments are required when they are incorporated into a unit of equipment.

For example, SAAB Gripen's RM12 engine is modular in design which makes it easy and quick to inspect and replace only the necessary module.

13.4.3 Simplicity

Generally speaking, the simpler the design, the easier is maintainability. For example, a reduction in the quantity of parts or in the number of different parts used is a standard approach in trying to improve the maintainability.

For instance, no tools at all are required to open and close the service panels on the SAAB Gripen aircraft. All control lights and switches needed during the turnaround time are positioned in the same area, together with connections for communication with the pilot and refuelling. On the same aircraft, a simple, portable mini-hoist is used for loading the external stores as well as for engine replacement.

13.4.4 Standardization

Standard fasteners, connectors, tools and test equipment have usually been thoroughly tested and are less likely to cause problems. Consequently, designers should use standard parts as far as possible, e.g. seals, nuts and bolts, especially high-replacement rate items.

Since typical design teams are by nature creative people, it requires a great deal of restraint to stay with simple designs and continuous use of standard parts, tools and facilities.

13.4.5 Foolproofing

Items which appear to be similar but are not usable in more than one application should be designed to prevent fitting to the wrong assembly. Incorrect assembly should be obvious immediately, not at a later stage such as the fitting of cover plates or during testing. Some of the following considerations should be inbuilt during design:

- If an item is secured with three or more fasteners, stagger their spacing.
- Ensure that shafts which are not symmetrical about all axes cannot be fitted wrongly, either end to end or rotationally.
- Where shafts of similar lengths are used, ensure that they cannot be interchanged, e.g. vary their diameters.
- Avoid using two or more pipe fittings close together with the same end diameters and fittings.
- Where pipes are in close proximity to one another ensure that the run of each pipe relative to the others is easily discernible.
- Flat plates should have their top and bottom faces marked if they need to be installed with a particular face upwards.
- Springs of different rates or lengths within one unit should also have different diameters.

Thus, design teams should make it as difficult as possible to assemble or use their design incorrectly. When possible, cable lengths should be such that only the correct cable can reach the black-box connector. When the design is such that more than one cable can reach a black-box connector, the cable connectors should be of different sizes so that only the connector of the correct cable will fit. When a functional package is to be a space, full maintainability consideration must be given to the problems of removal and replacement by ordinary people under field conditions. If the design is such that it is extremely difficult to remove and replace a line/shop replaceable unit (LRU/SRU), the probability that the maintenance task will not be completed successfully becomes substantial. If the design is such that it is possible to drop an attaching bolt, nut or screw into a vital or inaccessible area, the probability that this will happen, given a number of opportunities, becomes high.

13.4.6 Inspectability

Whenever it is possible, designers should create a design which can be subjected to full non-destructive, functional checkout. For example, a circuit breaker can be functionally checked, whereas such a check on a fuse is destructive. Here the testing advantage must be weighed against the maintainability of the fuse. The ability to inspect important dimensions, joints, seals, surface finishes and other non-functional attributes up to and past the assembly point where they are likely to be degraded is also a very important characteristic of a maintainable design.

Example 13.1

Integration between maintainability engineers and design teams existed from the start of the TGV project (French designed and built high-speed train). A multidisciplinary design team was formed where maintainability engineers played an important and officially acknowledged role. They worked directly with rolling stock design engineers and provided them with the benefit of their experience, thereby avoiding conflicts and delays. As a result, provisions and specifications for maintainability were built into technical documents defining rolling stock. They were based on systematic analyses of past experience and records of all technical provisions hindering maintenance procedures, and tried and proven solutions which should be incorporated in new rolling stock.

However, the SNCF Rolling Stock Department undertook to integrate maintainability to an even greater degree. From the very start of research and for several years, maintenance specialists were deliberately assigned to work with rolling stock research, development and design departments. They ensured that at the blueprint or CAD stage, solutions familiar to maintenance engineers were incorporated in practice in design specifications. At the same time they received and passed on to maintenance departments the outcome of initial research work and diagrams which would be useful for developing maintenance procedures and staff training programmes as well as for defining and setting up installations and equipment required for maintenance.

In this example some of the achievements made by the integrated design team are addressed below:

- Wheels are not subjected to a large degree of wear because wheel materials, geometry, lubrication and the limited forces to which they are subjected, especially with modern brake gear, are such that reprofiling has been pushed back to beyond 450 000 km (roughly 280 000 miles). Even then, wheels are reprofiled simply and at low cost with numerically controlled, pit-mounted wheel lathes.
- Electric commutator motors which required monitoring, maintenance and replacement when they reached their overhaul period have now been replaced by more powerful self-commutated synchronous drive motors for the Atlantic TGV and asynchronous motors for all auxiliaries.
- Visual and instrument monitoring of commutators, brush replacement machining, replacement of worn or damaged commutators and especially rewinding of sections have all been eliminated. Monitoring of the mechanical parts for the new motors is expected to be very simple and reliability of the associated static convertors fully under control.
- Destination indicator panels on the Southeast TGV train are manually controlled mechanical systems: these have been replaced on the Atlantic TGV by microprocessor-based, static liquid crystal displays, which are remote-controlled by radio.
- For the TGV, mechanized cleaning of the external bodyshell might have been problematic because of the contours of trainset ends. Maintainability was achieved by special kinematics built into an automatic train washing machine and by an automatic mechanism controlling the rate of advance of trainsets.
- The innovation of retention toilets has not been a hindrance for operations nor for maintenance: underground automatic evacuation installations have been trouble-free.
- In order to protect components from dirt accumulation, electronic equipment and power circuits on the Atlantic TGV are cooled in sealed units filled with a liquid refrigerant and are thereby protected from direct contact with ventilation air to eliminate pollution and the difficult task of cleaning components.

Failure protection measures built into the design include:

- *Built-in redundancy*: Much of the equipment on Paris–Southeast trainsets features built-in redundancy, e.g. in the power system in command and control circuitry and in technical and passenger comfort auxiliary equipment. Back-up components take over automatically if a failure occurs without causing any disruptions to train operation. This aspect of maintainability is enhanced on the Atlantic

TGV by a facility which stores records of switchovers to back-up components and, for some functions, data is transmitted to the maintenance centre.

- *Automatic monitoring*: On all TGV trains, the main safety functions (fire detection, mechanical damage, instability) are monitored automatically by the train-borne system to safeguard against exceptional catalytic failures and prevent frequent costly checks that are unlikely to yield defects. This equipment, for which fortunately there is only a very minute probability that it might be needed, has its own automatic built-in test facility. To monitor the temperature of roller-bearing axle boxes on the high-speed line, the solution of automatic monitoring at 40 km (25 mile) intervals along the line has proved to be highly appropriate and will be used again on the Atlantic TGV line. Although the term 'hot box detector' has been used, in fact the system is an automatic infrared thermometric network for which data is computerized and processed centrally. In addition to its role as an emergency hot box detector, the system can supply highly useful real-time preventive data on any abnormal changes in axle box temperatures to the maintenance department. To limit the impact of a pantograph failure on the catenary system, an automatic monitoring facility which is currently being tested and will be developed shortly will detect any unusual localized wear on the contact strip.

Some of the measures taken by the design team in order to increase maintainability of the TGV train are given below.

- *Maintainability of articulated trainsets*: The articulated fixed formation structure of TGV trains is particularly well suited to very high speeds, but originally in the design stages there were some who feared that they would not be flexible enough in operation if vehicles had to be withdrawn for repairs. Five years of intensive operation and low-cost maintenance have clarified this debate. There have been very few occasions when a passenger vehicle had to be withdrawn and these have been detected prior to departure for a train service.

    The situation would be no different with a non-articulated train; moreover, in this instance, the many connections between cars and the complexity of inter-car gangways designed for high speeds would make it impossible to withdraw a car easily and prepare another, without creating major delays in the train service.

    By contrast, maintainability of French TGV rolling stock and well-adapted terminal installations make it possible to place the full trainset back in service within a fairly short space of time by replacing a failed component instead of an entire car, even if it is an axle or a truck that has failed. Generally this principle is applied even to the end car which nevertheless is a separate vehicle in the articulated trainset.

    At the Paris–Southeast workshops it takes about one hour to change an axle and an idle truck in a trainset can be changed in no more than 1 hour 30 minutes including all associated operations.

- *Fast and accurate trouble-shooting*: The first generation of TGV trains was already equipped with testers together with diagrams of circuit logic which provided a good standard of trouble-shooting. On Atlantic TGV trainsets, memory functions assigned to the various command and control circuit microprocessors store the values of operating parameters when a failure occurs. This facility provides a graph of failures, guarantees correct fault diagnosis and guards against recurrence of intermittent failures. The train-built computer and various test equipment form a comprehensive computer-aided fault diagnosis system, although it must be said that the number of failures should be minimal.

Preliminary organization measures used include:

- *Documentation*: It would be wrong to overlook an important aspect of maintainability, documentation, which is an essential prerequisite for organization of maintenance. If supporting documents, drawings, technical descriptions, diagrams and functional manuals were not circulated prior to delivery of rolling stock, the equipment would not be maintainable and could not be placed in service without creating a risk. At the SNCF in general and especially for TGV trains, a large proportion of this information was made available well before lead times.
- *Maintenance regulations*: These documents are used to make a preliminary examination of initial maintenance duties and of expected maintenance intervals; maintenance specialists also use them to draw up their own maintenance manuals. Innovation rolling stock like the TGV is designed using components that have been tried and tested extensively for durability and the types of failures which are likely to occur are known. Hence, it is indeed relevant for teams of experienced engineers and technicians to foresee where failures might occur and draw up initial maintenance regulations. Of course, in the early period following commissioning, these regulations are more severe voluntarily, but they are very quickly adapted to the situation in practice and become cost effective within a short space of time.
- *Instruction and training*: This dual-stage preparation of technical documentation specifying maintenance practices is used by future management teams already appointed to take charge of organizing supplementary training for staff selected for the home depot of trainsets. This is followed by a period in which maintenance duties are simulated and then by systematic training to perform these duties on sub-assemblies and on the first few trainsets delivered.

Although it is quite certain that system reliability and component durability play a key role in the quality of service and the economics of the TGV network, it is important to point out that all of these technical and organizational measures for trainset maintainability have been instrumental in ensuring simple and low-cost maintenance and for the reasonable level of unavailability.

Training in the new types of technology used was organized in advance and maintenance staff adapted smoothly to the problems encountered and to changes in qualifications.

Maintainability of maintenance tools has contributed to ease of maintenance work, an element of comfort which is vital to the care and professionalism required for this work.

Maintainability and all of the measures connected with it form a new approach which calls for open and active relations among the partners involved and represents a challenge in terms of qualifications, organization and timescales which the French railroad industry and the SNCF Rolling Stock Department have met successfully for TGV rolling stock.

Regarding the preventive maintenance tasks the following considerations were made:

- Automatic monitoring equipment is designed to meet the need to examine and inspect rolling stock regularly. Testing and fault detection equipment is designed to meet the need to re-establish redundancy promptly when temporary failures occur.
- Items are conveniently located for ease of access.
- Items used for a particular technology are grouped together in functional units corresponding to the same technical speciality.
- Wear has been reduced by lubrication of moving mechanical items (gearing, roller bearings), and in some cases by replacing moving items with a solid-state technology. For example electro-mechanical switching and contact functions have been replaced advantageously by wear-free and maintenance-free static convertor power electronics.
- Ease of cleaning and possibilities for mechanized cleaning are also taken into account in the design of passenger stock for reasons of hygiene, comfort and aesthetics.

Regarding the corrective maintenance tasks, some of the considerations made at the design stage are addressed below:

- Provision was made for testability, which in practice means that the possibility of measuring the orders of magnitude of the physical parameters which are essential for fault detection was allowed for, although it was not functionally necessary. Hence, many of the complex functions incorporated in the TGV include integrated test facilities or a remote fault detection system; these systems may function as fault analysis systems and include a facility to transmit data to repair centres.
- For maintenance tasks involving replacement of failed items, every

provision is made for ensuring safety and swift replacement (snap-on mountings, polarized slots, lifting and handling gears and similar).

- The repair and renewal capacity of structures has been considered, i.e. weldability, dismountability of items and parts vulnerable to impact, wear and ageing.
- Materials and housings were selected with the objective to eliminate problems such as combustion, oxidation and ageing, which for decades represented the major part of repair and renewal work for railroad equipment.

It should be noted that the majority of systems failures have not been caused by the malfunction of some exotic device, the design of which pressed the state of the art. Rather, parts were not made correctly (bogus parts), and in other cases human failures such as failure to torque and secure a fastener properly or failure to install an explosive device properly were the cause. No detail is too minor to cause a problem. High inherent and achieved maintainability are, to a considerable degree, dependent on painstaking attention to detail.

In this chapter the range of design responsibilities has been addressed, with a particular emphasis on maintainability requirements. The reasons why inherent maintainability must be high and why maintainability is needed as an independent check-and-balance function on design to assure that maintainability requirements and considerations get their prompt and proper share of design attention, have been presented. Further, the methods for designing a maintainable system have been explored and the methods, procedures and practices used in achieving and assuring maintainability in design have been reviewed.

In summary, inherent maintainability is the primary responsibility of the design organization with maintainability service as an independent check and balance on the design function, principally to make sure that the design function has given its maintainability responsibility the detailed attention which is necessary.

13.5 MAINTAINABILITY LESSONS LEARNED

As a result of extended research through maintainability literature, design guidelines and personal experience, the following selection of 'lessons learned' has been created as a reminder and guide for maintainability engineers of future systems.

- Use of standard parts should be encouraged as far as possible (especially seals, nuts and bolts, and all other high-replacement-rate items).

- Gaining access to items requiring maintenance should not require removal of other items.
- Lubrication should be possible without disassembly.
- It should be possible to relieve force in powerful springs before they are removed.
- Pipes fitted to items should be in one axis so that the item can be removed in one direction.
- Items which come into contact with tools should not be painted.
- Adequate wall thickness should be provided if a hole in a body is used for lockwire, in order to prevent breakout after repeated use.
- Labels and decals should be hardwearing, not fading, positioned to avoid damage, difficult to peel off, easy to renew, and should follow contours of item without lifting.
- Items not visible after assembly should be restrained to prevent their dislodging during the assembly of other items.
- Positive indication of locking together of items should be provided.
- Location of a bolt/fastener should be shown by an indicator arrow, if its location is not easily visible.
- Pointed bolts should be used when alignment may be difficult.
- Small thread sizes should be avoided as they are prone to damage.
- Good access should be provided to any item which requires torque loading.
- Use of special tools should be avoided.
- Provision of visible indication of correct installation of critical items should be provided.
- Cables needing inspection and repair should be easily accessible.
- Cables should be secured with reusable clips.
- Cables should be kept to a minimum practical length in order to minimize the risk of damage due to excess slack.
- Plugs and sockets should be identified by shape, colour coding, or similar means.
- Possible galvanic reaction between dissimilar metals should be considered (stainless steel and aluminium mating should be avoided).
- Attachment of test equipment should be provided for in-place testing.
- Conformal coating of the printed circuit board (PCB) should be considered, particularly in cases when it is not environmentally sealed.
- Environmental and electromagnetic compatibility seals between mating metal surfaces should be considered, especially for safety critical items.
- Use of existing test equipment should be encouraged where practical.
- Items should be tested in their completed form with no need for subsequent further assembly.

- Adjustments on the item on the test rig should be possible, obviating the need for removal, estimated adjustments and subsequent retests.
- Sufficient clearance to remove and refit a seal without causing damage to it or the item to which it is fitted should be provided.
- Automatic renewal of old seals should be ensured if new seals are needed with a new part.

CHAPTER 14

Maintainability allocation

One of the first tasks of the design process is to translate overall functional requirements for a new system into its physical requirements in relation to performance, power consumption volume, cost, reliability, maintainability, supportability, producibility, weight, development time and similar. Subsequently, those system characteristics have to be decomposed into corresponding requirements related to the constituent modules and items which are located on the levels below, applying a top-down approach.

The process of assigning requirements to individual items to attain the specified system requirements is called the allocation process. The main advantages of an allocation process are listed below. Thus, it:

- Permits system-level requirements to be disseminated to a lower indenture level.
- Enables designers and subcontractors to work to relevant and appropriate targets for their subsystems.
- Reduces wasteful and inappropriate design effort.
- Permits the initial evaluation of the feasibility of achieving system requirements using a particular design approach.

The allocation process is very difficult because it is performed at a very early stage of design when the characteristics of the modules and items located at the lower levels have not been identified. At the same time the allocated values should act as guidelines for their identification and as their specifications.

This allocation problem is complex for several reasons, among which are: the role an item plays for the functionability of the system, the method of accomplishing this function, the complexity of the item, the durability of the item and its physical location within the system. The problem is further complicated by the lack of detailed information on many of these factors early in the system design phase.

Thus, the main objective of this chapter is to review existing methods for maintainability allocation and standards which are used for the allocation of maintainability requirements in day to day engineering practice.

14.1 CONCEPT OF MAINTAINABILITY ALLOCATION

Maintainability measures, like many others, are specified at the product/ system definition phase at the top level. For example, the requirements might be that the $MTTR_s$ should be less than or equal to, say, 45 minutes. Consequently, when the first ideas about a new product/system emerge it is necessary to decompose the system requirements into individual requirements for consisting subsystems, modules, units and items. Once allocated the maintainability measures become the target for the designers and maintainability engineers which should be achieved by their design solutions. Clearly, the maintainability allocation task can only be performed in conjunction with allocation tasks regarding other system requirements, like reliability, durability, stated operational time, availability and similar. A graphical representation of an allocation problem is shown in Figure 14.1.

Over the years, several methods have been developed for maintainability allocation, each attempting to address the goals outlined above. However, it is necessary to stress that as most of the basic durability/ reliability allocation models are based on the assumption that item failures are independent, the failure of any item results in product failure (i.e. the product is composed of items in series), and that the failure rates of the items are constant, the corresponding existing maintainability models are limited by these facts.

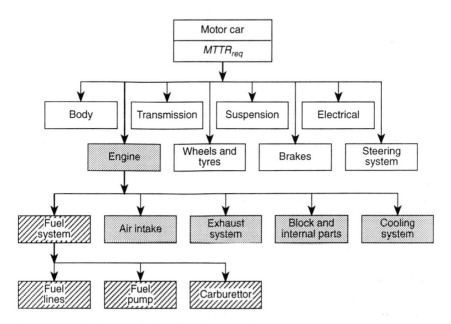

Figure 14.1 Maintainability allocation problem.

Analysis and categorization of existing maintainability allocation methods have been undertaken by Hunt (1992). According to this study, the most frequently used maintainability methods are:

- Defence Standard 00–41
- Military Standard 470B
- Published literature
- British Standard 6548
- Linear programming (LP) method.

Each of the above-listed methods will be examined in this chapter.

14.2 DEFENCE STANDARD 00–41

Defence Standard 00–41 is a guidance document which sets out the 'MoD Practices and Procedures for Reliability and Maintainability' for use in the design, development and production of equipment for United Kingdom Ministry of Defence usage. Issue 2 of this Defence Standard comprised six parts. The practices and procedures for the furtherance of maintainability were set out in part 6.

The general approach to maintainability allocation called for the annotation of maintainability activities and was derived from maintenance flow diagrams with estimates of times to complete those tasks. No specific method of estimation of duration of maintenance tasks was given other than statements like 'the maintainability allocation process shall recognize that items which are least reliable should be most maintainable' and 'that there may be insufficient design data at this [early] stage to reach accurate estimates [of times to perform activities] and in such cases values will have to be assigned on the basis of previous experience and/or engineering judgement'. These estimates were then used in conjunction with allocated reliability figures to deduce the consisting item mean active corrective maintenance time, $MDMT^c$, and hence the system conformance to the maintainability goals.

In June 1993 Issue 3 of this Defence Standard was released. However, it adopts the same approach to the allocation of maintainability requirements as described for the previous issue.

14.3 MILITARY STANDARD 470B

Military Standard 470B (1983) is a guidance document which sets out the Department of Defense requirements for activities to be addressed in a maintainability programme for systems and equipment supplied under United States of America Department of Defense contracts.

Task 202 deals with maintainability allocation and the only guidance given regarding methods to be used for estimating the values of the allocated maintainability parameters is that these estimates may be derived from predictions, data on similar items, experience with similar items or engineering estimates based on personal experience and judgement, and a statement that all allocated maintainability values should be consistent with the mathematical maintainability model developed under task 201 of Military Standard 470B.

Task 201 calls for this maintainability model to consider design characteristics which impact maintainability (e.g. frequency of failure, maintenance time, fault detection probability, etc.), the level of maintenance addressed and the relationship between the specified maintainability parameter and the operational system parameters (e.g. logistics supportability, maintenance manpower, the maintenance plan, etc.). The complexity of the model will vary according to the complexity of the equipment under consideration.

14.4 PUBLISHED LITERATURE

Blanchard (1976, 1991) proposes a method of maintainability allocation incorporating the contribution of consisting item failure rates to the overall system failure rate and a subjective estimate of the item-related maintainability characteristic of the consisting items under consideration, according to the following expression:

$$MTTR_s^c = \frac{\sum_1^k n_i \lambda_i \times MTTR_i^c}{\sum_1^k n_i \lambda_i} \tag{14.1}$$

where $MTTR_s^c$ is the mean active corrective maintenance time (or mean time to repair) for the system, $MTTR_i^c$ is the corresponding measure for the ith consisting item, n_i is the quantity of ith consisting items in the system, λ_i is the constant failure rate for the consisting item (applicable to cases where demand for the maintenance task is exponentially distributed), and k is the number of different types of consisting items in the system.

At this stage in the system life cycle, the inherent characteristics of the system design which contribute to the maintainability characteristics are not known and so the general principle that the items that have a higher contribution to the system failure rate have a lower MTTR target allocation is applied. This general approach may have to be modified to cater for allocated consisting item maintainability targets which are unrealistic due, for example, to cost or technology constraints, and an iterative process of re-allocation and evaluation is employed to arrive at an acceptable solution.

Example 14.1

The required mean time to restore the system under consideration is 0.5 hours. The first indications about the new system show that it consists of three maintenance-significant items, A1, A2 and A3. The allocated durability/reliability characteristics are given in Table 14.1, together with experience-based estimated maintainability characteristics for each item.

Table 14.1 Input data for Example 14.1

	A1	A2	A3
$MDMT_i^c$ (hr)	0.9	0.4	1.0
n_i	1	1	1
λ_i	0.246	1.866	0.11

Based on the data available, determine the maintainability targets for items A1, A2 and A3. Also determine whether the system requirements will be met, $MTTR_s \leqslant 0.5$ hour?

Solution
Making use of equation 14.1, the estimated value for the $MTTR_s$ could be obtained, thus,

$$MTTR_s = \frac{\sum_{i=1}^{nsi} \lambda_i \times MTTR_i^c}{\sum_{i=1}^{nsi} \lambda_i}$$

$$= \frac{(0.246 \times 0.9) + (1.866 \times 0.4) + (0.11 \times 1.0)}{0.246 + 1.866 + 0.11}$$

$$= \frac{0.2214 + 0.7464 + 0.11}{2.222} = \frac{1.0778}{2.222} = 0.485 \text{ hour}$$

As the calculated value for the $MTTR_s$ is less than required, it seems that the overall system target will be met if the predicted values for $MTTR_i$ are achieved. Clearly, in this example, it is assumed that the durability/reliability targets are achievable.

One of the main drawbacks of this approach is its reliance on subjective judgement. The other general approach is to estimate the value using previous or personal knowledge as a guide (i.e. take an educated guess), factor this estimate by a complexity factor (failure rate) and assess the result to see the system impact; if it meets the system requirements with reasonable figures then the estimate is accepted, otherwise it is reiterated until an acceptable solution is obtained. This approach depends largely on the subjective judgement of the individual allocation of the targets (i.e. is not reproducible) or depends on the availability of either previous knowledge/data or details of design characteristics. Although this approach does have its applications, undue reliance on subjective judgements is undesirable, and in addition the information necessary to make these judgements may not be available when the maintainability allocation needs to be performed.

14.5   BRITISH STANDARD 6548

British Standard 6548 part 2 (1992) is a 'Guide to maintainability studies during the design phase' and incorporates IEC 706–2 part 2 section 5 as published by the International Electrotechnical Commission (IEC). BS 6548 part 2 Annex A suggests three methods of maintainability allocation to address the following circumstances:

- new design, no previous knowledge available;
- partially new design, previous knowledge available;
- design with previous knowledge available.

All of the above-mentioned design categories are examined below.

14.5.1   New design

This is used in cases where the new design is analysed and no previous knowledge is available regarding the maintainability characteristics of the various consisting items. Thus, for k consisting items:

$$MTTR_i^c = \frac{MTTR_s^c \times \sum_1^k n_i\lambda_i}{kn_i\lambda_i} \qquad (14.2)$$

where $MTTR_s^c$ is the target mean active corrective maintenance time (or mean time to repair) for the system.

Example 14.2

This example uses a system comprising eight consisting items, A to H, whose proportional failure rate contributions are known, with a requirement for a system $MTTR_s^c$ of 30 minutes and preferably with no maintenance task taking, on average, longer than 120 minutes. In addition there is a minimum feasible maintenance time of 5 minutes due to the need to run a system test prior to any further maintenance task.

Allocated maintainability values to each item in order to achieve a system requirement of 30 minutes by applying BS 6548 Method 1, are shown in Table 14.2.

Table 14.2 Obtained maintainability measures for Example 14.2

Item	No. of items per system, n	Failure rate, λ (10^{-3})	Total failure rate, $n \times \lambda$	BS 6548 Method 1 $MTTR_i^c$
Unit A	1	0.3430	0.3430	10.93
Unit B	1	0.2032	0.2032	18.45
Unit C	1	0.1112	0.1112	33.72
Unit D	1	0.2956	0.2956	12.69
Unit E	1	0.0439	0.0439	85.42
Unit F	1	0.0014	0.0014	2678.57
Unit G	1	0.0001	0.0001	37 500.00
Unit H	1	0.0016	0.0016	2343.75

The use of this allocation method produced results which were rather unrealistic.

14.5.2 Partially new design

In situations where a partially new design is considered, previous knowledge of maintainability characteristics is available for some of the various consisting items. Thus, if the previous knowledge is available for $\ell < k$ consisting items, then *MMTR* for consisting item *j* could be determined according to the following expression:

$$MTTR_j^c = \frac{(MTTR_s^c \times \sum_1^k n_i\lambda_i - \sum_1^\ell n_i\lambda_i \times MTTR_i^c)}{(k - \ell)n_j\lambda_j} \qquad (14.3)$$

where $j = \ell + 1, \ldots, k$. *j* is the number of different types of consisting item in the system for which previous knowledge is not available.

14.5.3 Previous knowledge available

In this case it is expected that knowledge is available for all of the *k* consisting items, i.e. consisting item *i* is expected to show an *MTTR* of M_i. Then:

$$MTTR_s^c = \frac{\sum_1^k MTTR_i^c \times n_i\lambda_i}{\sum_1^k n_i\lambda_i} \qquad (14.4)$$

When $MTTR_i^c$ is less than or equal to the system target value $MTTR_s^c$, the maintainability requirement is met. If the system target value is not met, then the allocation must be revised or other corrective action must be taken.

It should be noted that this method is identical to that proposed by Blanchard in the previous section, except that Blanchard (1976, 1991) calls for the value of $MTTR_i^c$ to be estimated subjectively and iteratively refined.

14.6   LINEAR-PROGRAMMING-BASED ALLOCATION METHOD

An attempt was made by Hunt (1993) to interpret a maintainability allocation problem as a linear programming problem, denoted as LP. The objective function is the mathematical representation of the system maintainability model and the constraints include the system $MTTR_s^c$ requirement, the maximum permissible maintenance time, the minimum feasible maintenance time and the proportional contribution of the consisting items to the overall system maintenance time. Clearly, the allocation of time as the variable concerned (measured in hours or minutes) cannot be negative, therefore the non-negativity constraints are satisfied.

Once the initial LP model has been formulated, based on the algorithm proposed by Hunt (1992, 1993), the constraints can be varied, the model is rerun and the impact of these changes on the other system elements and the achievement of the system requirements is evaluated.

The main advantages of a linear-programming-based approach to maintainability allocation are as follows. It:

(a) Permits durability and maintainability trade-offs to be evaluated, i.e. aids assessment of the impact of trade-offs on the system and subsystem durability and maintainability requirements.
(b) Permits independent variation of constituent item parameters and the assessment of the effect of those variations.
(c) Is flexible and can be adapted for differing scenarios by amending the LP model.
(d) Permits the setting of maximum and minimum values.
(e) Permits modelling of the system such that the system size has no detrimental effect on calculation.
(f) Permits the insertion of known data or previous knowledge about a consisting item.
(g) Minimizes undue initial reliance on subjective estimates while still being flexible enough to accommodate data from these sources if required.
(h) Produces reproducible results.

(i) Is usable with minimal information at an early stage of design and can generate information for the initial setting of targets, the evaluation of support options, etc. This information can then be progressively refined as required.

(j) Permits the consideration of complexity factors (e.g. λ).

This technique and algorithm have been applied to a number of theoretical and practical problems; some examples are contained in Hunt (1992, 1993) and have produced satisfactory results. In addition, this method and algorithm have produced favourable results when compared to current methods, as shown in the example below.

Example 14.3

Using the initial data from Example 14.2 perform maintainability allocation function respecting a requirement for a system $MTTR_s^c$ of 30 minutes with desirably no maintenance task taking, on average, longer than 120 minutes. In addition there is a minimum feasible maintenance time of 5 minutes due to the need to run a system test prior to any further maintenance action.

Table 14.3   Results from linear programming model

Item	No. of items per system, n_i	Failure rate, $\lambda_i(10^{-3})$	Total failure rate, $n_i \times \lambda_i$	Linear programming $MTTR_i^c$
Unit A	1	0.3430	0.3430	17.63
Unit B	1	0.2032	0.2032	29.76
Unit C	1	0.1112	0.1112	54.38
Unit D	1	0.2956	0.2956	20.46
Unit E	1	0.0439	0.0439	123.96
Unit F	1	0.0014	0.0014	120.00
Unit G	1	0.0001	0.0001	120.00
Unit H	1	0.0016	0.0016	120.00

The results obtained from the BS 6548, method 1 calculations, shown in Table 14.2, were lower than the corresponding results from the linear programming model results shown in Table 14.3 for items A, B, C, D and E, while the method 1 results for items F, G and H were vastly in excess of the LP model results. These excessive values were a result of the lack of a means for setting maximum value limits using the current method 1 approach and also contributed to the poor results for consisting items A, B, C, D and E. In addition the current method 1 approach does not permit the setting of minimum value limits which, although it did not have an effect in this example, could, depending on the equipment and the value of the minimum limit, have a significant impact on calculated results.

In summary, use of BS 6548, method 1, as shown in Table 14.2, produced results which were non-optimal and unrealistic, while use of the linear programming model produced results which maximized the amount of time that could be allocated to each consisting item while meeting the overall system requirement.

14.7 CLOSING REMARKS

The process of assigning system maintainability requirements, set up by the customer or market research team, to individual items is known as the maintainability allocation process. The main advantages of an allocation process are listed below:

- Dissemination of system-level maintainability requirements to a lower indenture level.
- Provision of reference points for evaluating the level of consisting item maintainability achievement at specific programme milestones.
- Reduction of wasteful design efforts.
- Provision of the initial evaluation of the feasibility of achieving system maintainability requirements using a particular design approach.
- Establishment of an initial baseline for the maintainability aspects of the proposed design.

Although the allocation methods analysed are potentially beneficial to the maintainability analysis they have drawbacks, some of which are listed below:

(a) The BS 6548 method 1 approach, which is based solely on the number of unique consisting items in a system does not work for large values of n and k. It can be shown that as the system becomes larger so the allocated maintenance time becomes smaller and the targets less realistic and achievable.

(b) Current methods do not permit the setting of minimum value limits: minimum values could be set and input as part of a BS 6548, method 2 calculation, but this then limits the calculation and produces results based upon the assumption that the minimum values set will be achieved and not exceeded. This does not permit any adjustment of this time other than by increasing or decreasing the minimum value(s), performing the calculation again and comparing the results. This process is extremely laborious, especially where the minimum values may be different for each consisting item and achievement of the best solution is unlikely. In addition this process does not lend itself to application for complex time models.

(c) Current methods do not satisfactorily allow the setting of maximum value limits: maximum values could be set and input as part of a BS 6548 method 2 calculation, but again this then limits the calculation and produces results based upon the assumption that the maximum values set will be achieved and not reduced or exceeded. This does not permit any adjustment of this time other than by decreasing or increasing the maximum value(s), performing the calculation again and comparing the results. This process is extremely laborious and the achievement of the best solution is unlikely. In addition this process does not lend itself to application for complex time models especially where the maximum values may be different for each time element.

(d) Despite the advantages of the LP method, there are some disadvantages regarding the use of linear programming. These are primarily concerned with the user interface and can be summarized as follows:

(i) linear programming can be complex and for practical problems the use of software packages and computers is essential;

(ii) altering the LP model requires some knowledge of LP, although generally this need only be of a fairly basic level;

(iii) the LP outputs generated require interpretation and some basic knowledge of LP is required.

CHAPTER 15

Maintainability anthropometric analysis

In dealing with the interface of the operators and maintainers with the system, their biological characteristics and limitations must be considered. Consequently, the main objective of this chapter is to address the maintainability anthropometric analysis of design configuration and to consider human performance limitations related to body size and physical strength. System access, location, layout, weight and similar are analysed, including human characteristics as design parameters. The analysis performed sets up the methodology which is able to verify that the proposed design of the new system will permit maintenance personnel to effectively access, remove, align and install equipment within the workspace confines of the system and its operational maintenance environment.

Anthropometric analysis identifies the requirements to arrange or rearrange equipment location and configuration to provide sufficient access and workspace for maintenance personnel. Also it identifies structure and equipment features that impede task performance by inhibiting or prohibiting body movements by the maintainer.

As with other design characteristics, the system design must be well defined through drawings or sketches before detailed anthropometric analysis can be effective. However, a less detailed analysis during concept development can help to assure early application of anthropometric considerations. Results of the analysis lead to improved designs largely in the areas regarding provisions for equipment access, arrangement, assembly, storage and maintenance task procedures. The benefits of the analysis include shorter durations of maintenance activities/tasks, reduction in maintenance costs, improved supportability systems, improved safety, and a reduction in the number of design iterations/ changes.

15.1 GENERAL DESCRIPTION

An anthropometric analysis compares survey data of system users' body proportions to the workspace and equipment configuration provided for maintenance. Biomechanic considerations of maintenance personnel strength and range of movement are usually included in the analysis.

For example, analysis of defence systems are conducted in accordance with the Human Engineering Design Approach Document-Maintainer (DI-H-7057). In this case, a contract may require application of anthropometric standards for the user population as contained in MIL-STD-1472C (1979). Analysis of commercial systems are often guided by the military standards, and draw on non-military specialized anthropometric survey data (MIL-H-46855B, 1981). Analysis of system equipment must begin in the conceptual and demonstration/validation phases of system development in order to avoid possible engineering changes, which are time consuming and costly.

15.2 IDENTIFICATION OF USER POPULATION

The analysis process begins with identifying the user population and its boundaries, or using the user population determined by performing a use study or market research performed by the producer. For example, in the case of defence systems the user might specify age ranges and specialty codes of male and female aircraft mechanics, as well as type of clothing and environmental considerations. In the case of commercial products it is necessary to address geographical locations of potential markets, which should provide more information about physical characteristics of the future users, together with cultural and social background.

15.3 IDENTIFICATION OF USER ANTHROPOMETRIC DATA

Anthropometric data for military users can be found in several military standards, technical reports and government-sponsored computer databases. The data contain detailed information on dimensions of the human body. The data identify body sizes and descriptive statistics applicable to various percentages of the user population. A common design goal is to provide sufficient workspace to accommodate 5th (small, usually female) to 95th (large, usually male) percentile body proportions where the percentile identifies the percentage of the user population that is within the dimensions described. Some anthropometric data related to standing body dimensions are given in Table 15.1, and corresponding data for seated body dimensions are given in Table 15.2.

Further details regarding anthropometric data can be found in the literature.

Table 15.1 Anthropometric data – standing dimensions

Factors	5th percentile		95th percentile	
	Men	*Women*	*Men*	*Women*
Weight (N)	574	464	916	765
Stature (m)	1.64	1.52	1.85	1.72
Eye height (standing) (m)	1.52	1.42	1.73	1.60
Shoulder (acromiale) height (m)	1.34	1.23	1.54	1.43

Table 15.2 Anthropometric data – seated body dimensions in metres

Factors	5th percentile		95th percentile	
	Men	*Women*	*Men*	*Women*
Vertical arm reach, sitting	1.28		1.48	
Sitting height, erect	0.845	0.784	0.969	0.909
Sitting height, relaxed	0.825	0.769	0.948	0.897
Eye height, sitting erect	0.728	0.687	0.846	0.788
Eye height, sitting relaxed	0.708	0.672	0.825	0.776
Mid-shoulder height	0.571	0.537	0.677	0.625
Shoulder–elbow length	0.338	0.302	0.402	0.362
Elbow–fingertip length	0.443	0.389	0.519	0.457
Elbow rest height	0.175	0.187	0.280	0.269
Knee height, sitting	0.497	0.437	0.587	0.516
Popliteal height	0.406	0.380	0.500	0.441
Buttock–knee length	0.549	0.520	0.643	0.619
Buttock–popliteal length	0.458	0.434	0.545	0.526

15.4 ANALYSIS OF ENGINEERING DRAWINGS

Other very important areas of anthropometric analysis are engineering drawings or sketches of maintenance workspaces and workstations. Graphical representations of the maintainer are used to assess the anthropometric suitability of the workspace.

Drawing-board manikins were used extensively to evaluate engineering drawings or sketches of crew and maintainer workspace prior to the advent of computer-aided design (CAD) tools for anthropometrics. Drawing-board manikins and overlays of human dimensions are still useful tools for small design efforts having few drawings or as quick check devices. For example, manikins can be found in MIL-HDBK-759A (1987), representing 5th and 95th percentile male and 5th percentile female USAF pilot body sizes.

15.5 COMPUTER-AIDED DESIGN TOOLS

Computer-Aided Design (CAD) tools have become widely used for anthropometric analysis. Most of the CAD tools address military operator populations such as flying or weapons personnel (CADAM-ADAM, 1985), and provide maintenance personnel anthropometric data and the capability to stimulate maintainer tasks overlaid on CAD engineering designs. Data for male and female maintenance technicians, the encumbrance of clothing, and the tools to be used are a few of the features that are available. Other CAD tools for anthropometric analysis of maintenance workspace can be useful for specialized purposes. For example, CARD (1987) is a computerized database containing data from a variety of anthropometric surveys.

The Computer Aided Three-dimensional Interactive Application, known as CATIA, of Dessault Systems, created originally for the design of the French military fighter Mirage, is used today as the leading computer-aided engineering software, by the world's leading aerospace companies (Boeing, British Aerospace and similar). CATIA software has been used successfully in maintainability anthropometric analysis by Boeing during the development of their latest passengers aircraft B777.

15.6 SYSTEM MOCK-UP ANALYSIS

Full-scale equipment mock-ups can be used to simulate maintenance tasks using people representing key anthropometric body sizes and wearing the appropriate clothing. Moreover, the simulated tasks can be conducted in an environment and using the support equipment that will surround the system in its real-life environment. Workspace inter-ferences can be measured and photographed and task discontinuities can be identified.

15.7 VALIDATION OF ANTHROPOMETRIC ANALYSIS

The validation phase provides the first opportunity for the actual system users to have access to the equipment and to verify that results of the anthropometric analysis performed during the early design phase are correct. Validation of the anthropometric analysis verifies the effective-ness of the engineering design to provide sufficient access and workspace for maintenance personnel.

15.8 PROTOTYPE OPERATIONAL TRIALS

During prototype operational trials anthropometric analyses are con-ducted when actual maintenance tasks are performed. Anthropometric data are collected to verify suitability of the design and identify any problem areas, as representative maintenance personnel are required to perform access, diagnostic, removal, alignment, instalment and similar tasks.

15.9 DEMONSTRATION AND OPERATIONAL TESTING

Demonstration and operational testing of a system provides the first opportunity to conduct an anthropometric analysis using actual operational scenarios. In many situations, this is also the first opportunity to use a sample of end-user maintenance personnel. In cases where the needs to refine the workspace are identified, the results of the operational testing may be used to modify the final production specifications.

15.10 CLOSING REMARKS

Anthropometric analyses are required for the procurement of all military systems that require maintenance or servicing, by the majority of West European and North American ministries of defence. Similar, but less contractually formal analyses of anthropometric characteristics are performed by commercial aerospace, automotive, heavy equipment, railway and ship-building industries.

Anthropometric considerations should be addressed, beginning early in the conceptual phase and continued throughout design and development to the extent allowed by the degree of design maturity in the programme. Engineering drawings must have sufficient detail and maturity before the detailed analysis can be accomplished to validate achievement of the anthropometric requirements. Analysis of critical subsystems must occur early in the conceptual development phase of the programme. For example, defence agencies in NATO countries, involved in human engineering efforts, require some ongoing level of anthropometric analysis throughout procurement prior to manufacturing, to ensure that maintenance personnel anthropometrics and biomechanic characteristics have been considered in the design.

Anthropometric analysis should be performed on all types of equipment and equipment layouts. The analyses are usually performed on the assembled or packaged configuration of the equipment.

Anthropometric analysis for maintenance processes can be extensive when performed for all subsystems within a complex system. In some cases, analysis of only subsystems related to critical maintenance tasks or mission phases may be appropriate for early design efforts. The analyses are conducted down to the level of line replaceable units

(LRUs) and line replaceable modules (LRMs). The analyses usually involve several factors including:

- Anthropometric characteristics of the user population.
- System location, layout and weight.
- Access, removal, installation and alignment task procedures.
- Area, height, congestion and limitations of the workspace.
- Human engineering features facilitating maintenance (i.e. handles, steps, viewing ports, instruction panels, etc.).
- Maintainer reach envelopes and visual angles.
- Tools, support equipment and special devices.
- Hazards, precautions, protective clothing, escape.

An anthropometric analysis begins based on anthropometric descriptions of the user population, functional descriptions of subsystems, engineering design layouts, assembly layouts, and maintainer task analysis. The information is supplemented by:

- Maintenance concepts for each subsystem.
- Maintenance scenarios.
- Logistics support analysis record (LSAR) documentation.
- Support equipment designs.

The output of the anthropometric analysis is an assessment of the anthropometric suitability of the engineering design and is reported to the responsible design groups. For example, in the case of military procurement in the US the analysis also becomes a portion of the Human Engineering Design Approach Document-Maintainer (HEDAD-M). The HEDAD-M is a contract required by military specification (MIL-H-46855B, 1983) and described in a data item description (DI-H-7057). Major findings can be addressed by formal trade studies and programme design changes.

Several cautions should be recognized when performing anthropometric analysis and these include:

(a) Fewer user populations are represented by anthropometric data. Often specialized anthropometric surveys will have to be conducted. It is worth stressing that available defence data for aviators and some other military occupations may not represent other types of users.
(b) The designs of a system evolve and are refined, often changing anthropometric considerations. Thus, the validity of the findings should be ensured by conducting multiple analyses.
(c) Wiring, coolant, air and fuel lines are easy to overlook when examining the layout of items. These items and their fasteners may have serious impact on accessibility and maintainability.
(d) Detailed task analysis data are not always available for early efforts. Overlooking key task procedures can lead to overestimating access-

ibility and underestimating mean durations of maintenance tasks and consequently the values of *MTTR* for items and systems.

Finally, it is expected that anthropometric analyses are performed by human factors engineers having education, training and experience in anthropometry, task analysis, statistics, engineering design processes and computer applications.

CHAPTER 16

Testability analysis

Good troubleshooting is nothing more than good deductive reasoning. At the centre of that reasoning is a careful collection and evaluation of physical evidence. Unfortunately, many aircraft devices use computer chips to provide a function formerly fulfilled by substantial mechanical parts or subsystems. Consequently, troubleshooting, in the traditional sense of searching for physical evidence of failure, is hindered. You can't troubleshoot a computer chip by looking for physical evidence of failure. A broken chip does not look any different than a healthy one. Although it can be argued that broken chips occasionally make smoke, evidence of malfunction is seldom readily apparent. Broken chips do not leak, vibrate, or make noise. Bad software within them does not leave puddles or stains as evidence of its misbehaviour. Ones and zeros falling off the end of a connector pin are difficult to see.

(J. Hessburg, Chief Mechanic New Airplanes, Boeing, in *Airliner*, January–March 1995)

The main objective of this chapter is to address the aspect of maintainability that has received significant attention in recent years by system designers, namely the use of automatic diagnostic systems. These systems include:

- internal or integrated diagnostic systems, referred to as built-in-test (BIT);
- built-in-test-equipment (BITE);
- external diagnostic systems, referred to as automatic test equipment (ATE), test sets, or off-line test equipment.

Each of the above-listed systems will be analysed in this chapter.

16.1 BUILT-IN-TEST (BIT)

According to the brochure for the Microwave Landing System P-SCAN 2000, produced by Siemens Plessey Radar Ltd, it posseses:

A very comprehensive built-in test (BIT) sub-system embedded within the equipment. The BIT sub-system is processor controlled and is fully isolated by hardware interlocks from the safety critical parts of the system. All faults are identified to the BIT subsystem and the LRU reponsible identified. Fault finding can be accomplish locally or remotely following menu driven software prompts. A remote control and status unit (RCSU) and remote status unit (RSU) are available for installation in the technical equipment room and ATC tower as required. A remote maintenance monitoring system (RMMS) is available for connection via public or private switched telephone lines to allow interrogation of the Monitor and to initiate diagnostics.

(Siemens Plessey Electronic Systems Ltd, 1990, UK)

Built-in-test, BIT, generally refers to the fault and error detection functions of a system whose primary purpose is to protect the operator and system from bad or misleading display or commands. This fault-detection function is therefore part of the basic operational function of the system.

As technology advances continue to increase the capability and complexity of modern systems, there is a necessity of relying more and more on the use of automatic diagnostics, i.e. BIT, as a means of attaining the required level of failure detection capability. The need for BIT is driven by operational availability requirements which do not permit the lengthy durations of maintenance activities associated with detecting and isolating failure modes in microcircuit technology equipment. It is also necessary to stress that because BIT operates within the basic system and at the same functioning speed, it therefore provides the capability to detect and isolate failures which conventional test equipment and techniques could not. Also, a well-designed BIT system can substantially reduce the need for highly trained field-level maintenance personnel by permitting less skilled personnel to locate failures and channel suspect hardware to specialized repair facilities which are equipped to diagnose and/or repair defective hardware. However, BIT is not a comprehensive solution to all system maintenance problems but rather a necessary tool required to deal with the complexity of modern electronic systems.

One of the more complex tasks inherent in the acquisition of modern systems is the development of realistic and meaningful operational requirements and their subsequent conversion into understandable and achievable contractual specifications. Before discussing this topic in more detail, typical performance measures or figures of merit which are used to specify BIT performance are addressed.

The most frequently used figures of merit are:

- Percentage detection, BPD_p: the percentage of all faults or failures that the BIT system detects.

- Percentage isolation, BPI_p: the percentage of detected faults or failures that the system will isolate to a specified level of assembly.
- Automatic fault isolation capability, $AFIC$: the product of percentage isolation and percentage detection, thus:

$$AFIC = BPD_p \times BPI_p \qquad (16.1)$$

- Percentage of false alarms, $BPFA_p$: the percentage of BIT-indicated faults where in all cases no failure is found to exist.
- Percentage of false removals, $BPFR_p$: the percentage of units removed as indicated by BIT whose condition was found to be 'satisfactory' at the next higher maintenance level.

It is necessary to point out that definitions for each of the above parameters leave room for a considerable span of interpretations. For example, the percentage detection could refer to failure modes as well as the percentage of all failures that could potentially occur. The detection capability could apply across the failure spectrum (mechanical systems, instrumentation, connections and software) or it could refer to diagnostic capability applicable only to, say, electronic hardware systems.

Early BIT systems were frequently designed to isolate fault, to the module level. This resulted in BIT systems as complex as, and frequently less reliable than, the basic system. The current trend is to isolate to the subsystem or box level, based on BIT's ability to detect abnormal output signal patterns. Intermediate and depot-level maintenance facilities will frequently use BIT or external diagnostic equipment to isolate to the board or piece-part level.

The percentage of false alarms is a difficult parameter to measure accurately because an initial fault detection followed by an analysis indicating that no fault exists can signify several different occurrences, such as:

- The BIT system erroneously detected a fault.
- An intermittent out-of-tolerance condition exists somewhere.
- A failure exists but cannot be readily reproduced in a maintenance environment.

The percentage of false removals can be a more difficult problem to address, because it may be caused by:

- Incorrect BIT logic.
- Wiring or connection problems which manifest themselves as faulty equipment.
- Improper match of tolerances between the BIT and test equipment at the next higher maintenance level.

The resolution of each type of false alarm and false removal requires a substantially different response. From a logistic viewpoint, false alarms often lead to false removals creating unnecessary demands on supply and maintenance systems. Of potentially more concern is the fact that false alarms and removals create a lack of confidence in the BIT system to the point where maintenance or operations personnel may ignore certain fault detection indications. Under these conditions, the BIT system in particular and the maintenance concept in general can neither mature nor provide the support required to meet mission requirements.

The specification of BIT performance must be tailored to the specific system under consideration as well as the available funds and, most importantly, the overall mission requirements. This tailoring activity must include a comprehensive definition of BIT capability based upon the figures of merit presented above.

Example 16.1

This example is based on DoD document 3235.1-H, March 1982.

System XYZ is composed of five line replaceable units (LRUs) with the following BIT and system performance characteristics:

- System: Five (5) LRUs
- *MDFL*: 50 flying hours
- Time period of interest: 5000 flying hours
- Percentage detection: 90%
- Percentage isolation: 90% (to the LRU level)
- False alarm rate: 5% (of all BIT indications)
- *MTTR* (with BIT): 2 hours (includes failures which have been both detected and isolated)
- *MTTR* (non-BIT): 5 hours (includes failures which have not been isolated but may have been detected)

Making use of the data available determine:

(a) What is the expected number of failures during the 5000 flying hours?
(b) What is the expected number of failures to be detected by BIT?
(c) What is the expected number of failures to be isolated to an LRU?
(d) What is the automatic fault isolation capability (AFIC)?
(e) What is the expected number of false alarm indications to occur during the 5000 flight hours?
(f) What is the expected time in corrective maintenance required to repair the 80 detected/isolated failures?
(g) What is the expected time in corrective maintenance required to repair the remaining 20 non-BIT detected/isolated failures?
(h) Assuming that manual or non-BIT maintenance time is required to

resolve the false alarm indications, what total non-BIT corrective maintenance time is required for the 5000 flying hour period?
(i) What is the expected total time in maintenance during the 5000 hours?

Solution

(a) $MNF(5000 \text{ hr}) = \dfrac{5000}{50} = 100 \text{ failures}$

(b) $MNF_{det}(5000) = MNF(5000) \times BPD_p = 100 \times 0.9 = 90$ BIT-detected failures

(c) $MNF_{isol}(5000) = 5$ detected failures $\times$ 90% isolation $\simeq 80$ failures

(d) $AFIC = BPD_p \times BPI_p(LRU \ level) = 0.9 \times 0.9 = 0.81 = 81\%$

(e) Total BIT indications = true failure detections + false alarms, and:

$$x = (BPD_p \times MNF(5000)) + (\text{false alarm rate})(\text{total BIT indications})$$

$$= (0.90)(100) + (0.05)(x) = 94.7$$

Therefore, false alarms = total BIT indications − true indications = 94.74 − 90 = 4.74 ≃ 5

(f) Time = 80 failures $\times$ 2 hours ($MTTR$ with BIT) = 160 hours

(g) Time = 20 failures $\times$ 5 hours ($MTTR$ non-BIT) = 100 hours

(h) Total (non-BIT) time = non-BIT failure repair time + false alarm maintenance time = $(20 + 5) \times 5 = 125$ hours

(i) The mean time in maintenance could be determined in the following way:

$$MTIM(5000) = \text{BIT maintenance} + \text{non-BIT maintenance} =$$
$$160 + 125 = 285 \text{ hours}$$

Thus, even with a relatively high AFIC of 81% the non-BIT-oriented corrective maintenance represents 44% of the total anticipated corrective maintenance hours.

The information presented in this example is simplified. Also it is necessary to stress that in the above example it was assumed that the BIT AFIC was going to be 81%. If, in fact, the AFIC is 81%, then 56% of the maintenance effort will be oriented towards BIT-detected/isolated failures. If the true AFIC is found to be lower, it will be necessary to re-evaluate the overall effectiveness of the entire maintenance and logistics programmes as well as total mission effectiveness.

The development and evaluation of BIT and diagnostics has tradition- ally been an activity that has chronologically followed basic system efforts. The argument usually presented is that 'the basic system has to be **designed** and **evaluated** before we know what the BIT is supposed to test'. This argument has some basis in fact, but there are significant drawbacks associated with lengthy schedule between basic systems and BIT design and testing. For example, design considerations relating to the basic system such as partitioning and subsystem layout determine to a large extent the required BIT design. The BIT design is also driven by the prediction and occurrence of basic system failure modes which BIT is expected to address. Consequently, the two design efforts cannot be conducted in isolation from one another.

From an evaluation viewpoint, conducting the BIT evaluation after the basic system tests are completed may preclude BIT improvement options from being incorporated because the already finalized basic system design would be substantially impacted. Likewise, an inadequate evaluation of BIT which leads to an erroneous confirmation of its capabilities (AFIC) will result in a substantial impact to system operation effectiveness.

The design of BIT is based upon two assumptions regarding the reliability of the basic weapon system: accurate identification of failure modes and correct estimation of the frequency of occurrence of the failure modes. However, if either of the assumptions is proven incorrect by test or operational experience, the resultant BIT performance is likely to be inadequate or at least less effective than anticipated.

Example 16.2

This example is based on the previous one. Assume that an unforeseen failure mode is observed in the third LRU every 250 flying hours, on average. Assess the impact of this development on the non-BIT maintenance.

Solution

$$MNF_{new}(5000) = \frac{5000}{250} = 20$$

Maintenance time associated with these new non-BIT-detected failures could be determined as follows:

$$20 \times 5 \text{ hours/failure} = 100 \text{ hours}$$

Therefore:

Total maintenance hours = 160 + 125 + 50 = 335 hours

and:

Total non-BIT maintenance = 125 + 100 = 225 hours or 67% of total maintenance.

BIT(detected/isolated) maintenance = 160 hours = 42%

This represents 47% of total maintenance.

Consequently, the discovery of one unforeseen, non-BIT-detectable failure has a relatively significant impact on the comparable magnitude of the two maintenance percentages (Table 16.1).

Table 16.1 Percentage of total maintenance hours

	BIT	Non-BIT
Previous estimate (Example 16.1)	56	44
Current estimate (Example 16.2)	47	67

Digital systems have proved to be reliable because of their self-test ability. However, this also causes problems in workshops when reported faults cannot be confirmed on test. Operators have learnt that some faults are transient, and are not repeatable. Possible reasons for this could be:

- differences between predicted and real environment;
- operator's maintenance practices;
- system design problems.

Results of one survey of a Boeing 737-300 FMCS which has covered 914 000 flight hours and recorded 384 diagnostic events are shown in Table 16.2.

Table 16.2 Results of Boeing 737-300 BIT-related survey

Category	Number	%
Failure confirmed by shop test	177	46
Failure confirmed by BIT	123	32
Software design problem	35	9
Insufficient data	26	7
Wrong LRU rejected	23	6
Total	384	100

Normally, manufacturers count failures as those confirmed by shop test. However, this could cause disagreements between operators and suppliers over the unconfirmed removal rate, with the implications that it is due to maintenance errors from the operator.

One of the solutions is to count all internal failures detected by the more accurate BIT system. The data in the table has 177 (46%) removals confirmed by shop test, but a failure confirmed by BIT in 123 more (32%) cases. Thus *MDFL* is only a measure of hard faults and not an indication of the usability of the system. The traditional manufacturer's response might be something like 'half of them are not wrong on test, so they are not our problem', whereas the operator's response to this might have been 'half of them are removed because we cannot find out what is wrong with them on the system'. Other lines of argument from the operator centre on the observation that, although the unit might pass a shop test, its replacement demonstrably cleared an aircraft system fault – with the implication that the shop is not adequate.

Mean time between unscheduled removals (*MTBUR*) is a better measure of total performance. It is affected by maintenance skills, but it also lets out the cases where transient faults are cleared by switching off and on again, to the manufacturer's benefit. *MTBUR* is easily measured from the operator's records, but only the operator may know what it is since suppliers only receive the rejected units.

What is needed is a failure recording system which includes confirmation by the BIT on one hand and recognition of design errors on the other. This would encourage designers to rely on BIT and its integrity. Faults indicated by BIT would have to be corrected; if BIT itself goes wrong, then improve the BIT accuracy.

Digital avionics specialists distinguish the following three types of failure:

(a) 'Hard': those that persist through several attempts to start the system.
(b) 'Soft': those failure symptoms that disappear when power is recycled or when a BIT test is run.
(c) 'Intermittent': those where the fault disappears without any obvious operator action.

Clearly, soft failures are of the greatest concern for the operators of the systems with digital avionics; they may be a combination of hardware faults and software errors. Intermittent faults are more of a nuisance than a danger, but can consume too much maintenance time in attempts to locate them. These intermittent and soft failures in avionics are the source of erratic maintenance practices. A common practice when maintenance staff are unsure of the cause of a fault involving avionics is to work through the system. Typically, the transducer is first to be changed, then the amplifer/processor, then the indicator, and so on until the fault disappears.

16.2 BUILT-IN-TEST-EQUIPMENT (BITE)

Built-In-Test-Equipment, commonly known as BITE, is a common term in the industry, referring to the part of the system which performs the maintenance function. In most digital avionics the equipment part of

BITE includes some hardware and much software. For software purposes this is an important distinction. The maintenance function, or BITE, is classed as non-essential for safety, unlike the fault detection function (BIT), which is an integrated part of a system classed as essential or critical and which must be certified to the same standard as that system.

Typically, the system to be tested is connected to the BITE through an interface unit. This is essentially a routing system so that the stimulus and measurement devices can be connected to the system under test. Stimuli fall basically into two types, namely analog and digital logic signals. Logic signals are applied to the system under test to simulate the condition of various machinery state inputs. Up to 24 bits may be applied for a given test, or any part of the 24 bit word applied for fast test over the sequence. All inputs can be fed with a logic '1' or '0' signal or be left open circuit, depending upon the requirement of the selected test. A range of fixed frequency signals and a variable frequency generator are available as inputs to the system. Progammable d.c. signals as well as defined d.c. voltage levels can be selected and applied. Signals returned from the system under test are measured to check whether they fall within the programmed limits. Both analog and digital logic signals are processed in parallel and the result displayed as a pass or a fail. The type of test is specified, namely, whether an analog voltage, logic level or combination of these are to be designed to the specific level of tolerance. Normally, the operator controls and monitors the test sequence via the control and display panel. Control signals are fed to the central test sequence and back to the display panel. The operator is informed by the display of the result of the test and, if required, the most probable corrective task is indicated.

The BITE system self-test facility may be periodically exercised to ensure that it is functional prior to the testing of a suspect system. The system to be tested is connected to the test system and identified by coding of the connections. Once the system has been connected, 'system test' is selected and the test sequence is started by pressing the select test button. In response, the central control then sets up the first test in the sequence such that appropriate stimulus and measurement devices are connected to the system under test. The measurement is taken and the result is compared to alarm levels and a 'pass' or 'fail' determined and displayed to the operator. The operator selects the tests in sequence until a failure is indicated or the sequence is completed. When a failure occurs, the indicated faulty module is replaced and the test connectors are removed and the system may be re-selected for control. The control panel is mounted in the equipment control console adjacent to the systems requiring test. The system test socket carries connections such that a system may be fully tested. When a connection is made from this socket to any system or the 'self-test' socket, the BITE system will automatically carry out the tests on the selected system or on the test equipment itself depending upon the test mode selected.

The mode switch is a four-position rotary switch which selects the appropriate mode of operation:

(a) Off.
(b) Self-test 1 (lamps and display checks): by depressing the select test button, the displays are automatically checked. The test number starts at '00' and runs up to the full test number of count, and any discrepancy will be indicated to the operator.
(c) Self-test 2 (stimulus and measurement checks): in this a check of the stimulus and measurement devices is carried out. The test sequence is then identical to that for a normal system. The self-test facility is normally concealed under a flag so that operation of the panel is not confused.
(d) Systems test: with the mode switch in this position, and with the system to be tested linked to the system test socket, the system test is carried out. When system test is selected, the 'off-line' lamp is illuminated confirming that the system under test is muted. The first test is selected by pressing the 'select test' button, the result of test (i.e. pass or fail) is displayed and the affected module or minimodule is indicated. Testing is continued by pressing the 'select test' button again and observing the results.

The BITE fault record stored inside an LRU is often the only useful data to assist in failure investigation. It is therefore important that shop staff record and track BITE records of all LRUs that are received. BITE data must be credible if maintenance work is to be based on it.

Avionics suppliers must keep records of equipment repairs and BITE data. Memory for BITE data recording is now at low cost and big enough to record much useful data, such as phase of flight, aircraft and route, location of fault, state of the system at that time and so on. The event recording frequency of BITE systems is programmable by users, a useful feature if learnt.

16.3 THE CONCEPT OF DATABUS

Before the advent of the databus, all connections were point to point. If a particular aircraft signal, for example airspeed, was needed as input to a function, such as navigation, a dedicated pair of wires was needed to carry the signal. However, today as aircraft systems became more extensive, the value of transferring data between systems to achieve some performance optimization became apparent. This in turn led to aircraft wiring becoming increasingly complex due to the many wires and connectors needed, with the attendant failures and maintenance actions. In order to reduce weight, and improve reliability, engineers began to devise means of putting more than one signal on a particular wire.

A simple example of a databus is in multi-channel radio control of model aeroplanes. Here the medium is a radio frequency carrier, and the signals are a string of pulses whose length is proportional to the analog signal being transmitted. A synchronizing pulse initiates the sequence, and a signal is identified by its place in the pulse train. This is called time-sequenced addressing, and leads to the concepts of priority, refresh rate and integrity. In this example, all signals have equal priority, and are refreshed (new data transmitted) at the same rate. Consequently, the system has implicit low integrity since there is no immediate means of knowing, at the transmitter, whether any data has been received at the receiver, and whether that data is correct. These parameters of priority, refresh rate and integrity are used to categorize different classes of databus.

The most frequently used databuses are:

- *ARINC 429*: This was the first databus to be adopted universally in the commercial aircraft market. It has been incorporated into the design of the Boeing 757 and 767, and the Airbus A310. It resulted from studies initiated in the mid-1970s to specify a standard commercial aircraft databus. The advantage of standardization was already proving itself in the industry with, for example, the ARINC 561 definition of the characteristics of an Inertial Navigation System (INS), and the ARINC 571 definition of the data to and from an INS.

- *MIL-STD-1553*: This was developed largely in the same timescale as ARINC 429, and first had widespread adoption in F-16 aircraft. It differs from ARINC 429 in that it is multi-source, multi-sink and bi-directional. This means that units on the bus, referred to as remote terminals, may both transmit and receive data. In order to avoid the potential conflicts that could arise when sources independently wish to put data on the bus, 1553 implements a central bus controller.

- *ARINC 629*: Development of this databus began in 1977 with studies by Boeing of a Digital Autonomous Terminal Access Communication (DATAC) system to supersede ARINC 429. By 1985 DATAC system requirements were designed, and encouraging results were obtained with a prototype implementation on a NASA aeroplane. Since the objectives for the bus were similar in many ways to those of MIL-STD-1553, there are many similarities between the two systems. The main difference between the two is that in ARINC 629 the bus control function is distributed rather than centralized as discussed below.

One of the advantages of a databus is that it can act as a conduit for aggregating maintenance data on a complete aircraft into a central maintenance facility. The maintainer benefits by having a single access port for this data. A consequence of this ability for one system to report on failures of another because they are on the same databus is that the box reporting the failure may well itself be healthy. Maintainers will need to ensure that they do not 'kill the messenger', and condemn as faulty LRUs that are correctly reporting the faults of others. In this instance reporting a fault is evidence of a box's health.

Set against these advantages is the fact that more sophisticated hardware is needed to access a databus in order to extract maintenance information; the days of finding all faults with a 'scope and meter' are history for many modern systems. However, the increasing adoption of databuses should ensure that adequate test equipment at reasonable prices is available to the maintainer.

16.4 CASE STUDY: ADVANCED MAINTENANCE TECHNIQUES IN THE BOEING 777

A new generation of commercial avionics systems was introduced into service on the Boeing 777 airplane (first commercial flight on 7 June 1995, between London Heathrow and Washington Dullas). The Aircraft Information Management System (AIMS) is the first commercial avionics system based on integrated modular avionics (IMA) technology and provides the Boeing 777 with a quantum leap forward in avionics capabilities. AIMS not only provides an unparalleled level of systems integration and functionability, but also offers significant cost of ownership benefits to the airlines, particularly in the area of avionics maintenance. Information in this section is based on Johnson (1993).

While there are many significant maintenance enhancement features provided by AIMS, the focus of this analysis is on AIMS platform BITE and its affect on hardware and software fault isolation. Although there is no question that the transition to digital avionics from analog systems, which began in the late 1970s, has resulted in dramatic improvements in equipment reliability, the ratio of unscheduled removals to confirmed failures ($MTBUR/MTBF$) has not varied significantly. In complex avionics computers it is not unusual to see $MTBUR/MTBF$ ratios in the range of 0.33 to 0.50. Increased complexity and functional capabilities of avionics systems, combined with the corresponding increase in software contained in these systems, have contributed to keeping this ratio relatively unchanged. These same trends have made it increasingly difficult to isolate anomalous software events from intermittent hardware faults and correlate to the resulting flight deck effects (FDE).

The Aircraft Information Management System for the Boeing 777 airplane is developed around the principles of integrated modular avionics (IMA). The concept behind IMA is to provide a system that allows the integration of multiple systems functions into a set of shared common hardware and software resources (e.g. processor, memory, input/output (I/O) and operating system).

The AIMS system for the Boeing 777 is a highly integrated avionics architecture that incorporates the following airplane functions:

- flight management
- displays
- onboard maintenance
- airplane condition monitoring
- communication management
- information management.

The development of AIMS focused on two primary objectives:

(a) enhance functionality and performance;
(b) reduce airline cost of ownership.

Enhanced system functionality with a corresponding decrease in cost per function is a result of the transition from a federated architecture to integrated modular avionics architecture. At the centre of this IMA architecture is the AIMS cabinet. The AIMS cabinet provides a hardware and software platform that allows multiple avionics functions, such as flight management systems (FMSs) or displays, to use shared resources. The AIMS cabinet allows functions of different system certification levels (critical, essential and non-essential) to operate using shared processor, memory, I/O and software resources. Each function is allocated to one or more software partitions for execution in the AIMS cabinet. Undesired interaction between partitions is prevented through rigorous implementation of time and space partitioning.

The entire 777 AIMS system was designed from the onset with maintenance and airline cost of ownership in mind. Examples of maintenance features incorporated into AIMS are:

(a) improved dispatch reliability through fault-tolerant designs and deferred maintenance capabilities;
(b) reduced spares cost through common hardware part number components;
(c) onboard maintenance system (OMS) with airplane-wide flight deck effects correlation of BITE-detected events;
(d) the use of liquid crystal display (LCD) flat panel displays to replace traditional cathode-ray tubes (CRTs), resulting in reduced weight and power with increased reliability;
(e) improved *MTBUR/MTBF* ratio through isolation of hardware and software faults.

The key objective in the development of AIMS platform BITE was to provide a step increase in the *MTBUR/MTBF* ratio. The basic AIMS architecture and BITE software represent a major evolution in the ability to isolate and contain software anomalies and hardware faults and provides unparalleled transparent recovery from most single-events upset events. AIMS lock step processing architecture and high integrity monitoring approach will detect virtually any hardware fault condition. While all of these items will contribute significantly to an improved *MTBUR/MTBF* ratio, the maximum benefit will not be achieved without proper training and coordination with airline maintenance personnel and changes to existing maintenance practices.

The recognition that software anomalies exist in complex avionics systems, combined with an unprecedented ability to isolate software and hardware faults, requires a new thought process on how to deal with these conditions in airline operations. New procedures are required to handle the case where the OMS correlates a flight deck effect to a software fault event. Under these conditions, removing a core processing module (CPM) will result in a no fault found condition since an actual hardware fault does not exist. Airlines and certification authorities will need to agree on practices that allow the maintenance personnel to download the BITE memory for analysis and then run the return to service a test via the maintenance access team. If the test passes, the LRM can be returned to service safely without the need to remove and/or replace the LRM. Not only will this allow the airlines to keep the hardware in service but it will also provide very valuable data in isolating the root cause of any software anomaly that does occur. The only time an LRM should be replaced is when an actual hardware fault is logged in the BITE history and the maintenance procedure requires the LRM to be removed for that fault condition.

Introduction into service of the Boeing 777 and introduction of AIMS represents a significant area of new technology introduction to airline maintenance operations. A certain level of familiarity and confidence needs to be established within maintenance and flight personnel before significant gains are observed in the *MTBUR/MTBF* ratio. However, the AIMS platform BITE architecture and its extensive capabilities will lead to an extremely rapid product maturity cycle. The BITE architecture has already demonstrated tremendous benefits in the development cycle over previous BITE architectures and manufacturers believe that these benefits will very quickly carry over into airline operations.

CHAPTER 17

Condition monitoring techniques

Condition monitoring techniques are devices used to monitor, detect and diagnose the condition of systems monitored. The objective of condition monitoring techniques is therefore to provide information with respect to the actual condition of the system and the change in that condition. It is important to understand the behaviour of failure that an item exhibits so that the most effective monitoring techniques can be chosen. The decision by which the condition monitoring techniques are selected depends greatly on the type of system used and, in the end, is driven by economic and/or safety decisions. Once the decision is made as to which techniques are to be used, it is possible to define the system that will be needed to carry out condition monitoring. With the increased interest in condition monitoring in recent years there are a number of developments regarding the techniques used to monitor the condition of systems. The sensors, instrumentation, recording and analysis devices have been dramatically improved. This development has made it possible to get more reliable information on the condition of the system. Once condition monitoring sensors have been installed and data are being collected, it is necessary to have reliable methods of interpreting the data to detect when faults are occurring. Effective conditional maintenance tasks require a large number of measurements to be taken at intervals that assure recognition of change in condition of an item/system in sufficient time before any corrective action. The volume of data necessary to accurately determine condition of an item/ system required an excessive amount of time to arrive at a form that could be interpreted. Consequently, the demand has been for the development of tools to diagnose the condition of the system and to allow identification of failure prediction. The diagnostics for systems have now developed to a point where the information available is such that the human requires computer assistance to gain the maximum benefit. Artificial intelligence, AI, seems to offer this advance. Artificial intelligence provides many powerful techniques to manipulate large amounts of data. Artificial intelligence techniques, such as expert systems, neural networks, fuzzy logic and knowledge-based systems, have been applied to all engineering fields. In recent years, these techniques have been applied in the discipline of monitoring and diagnosis systems. However, expert systems have received the most attention of all the above artificial intelligence techniques. The idea of using expert systems technology to develop software support tools for system maintenance has been around for over a decade. Expert systems extend the power of the computer beyond the usual mathematical and statistical functions by using dialogue and logic to determine various possible courses of action or outcome. An expert system has several advantages over a human expert, such as:

(a) It can process information much faster and therefore the time for maintenance diagnosis can be reduced.
(b) It can analyse situations objectively and it will not forget any relevant fact, therefore the probability of making a wrong diagnosis could be reduced.
(c) Furthermore, it can detect incipient failures through its on-line monitoring of the condition parameters of the system.

Today, there are a great variety of different monitoring techniques which are well established and widely used to assist and improve conditional maintenance tasks. Generally speaking, condition monitoring techniques can be divided into different categories according to different criteria.

Some of the most frequently used condition monitoring techniques are:

● vibration
● tribology
● performance
● visual
● non-destructive

each of which is briefly analysed in this chapter.

17.1 VIBRATION MONITORING

This type of condition monitoring technique is based on the fact that rotating machines such as pumps, compressors, gearboxes, turbines, etc. produce vibration as machines deteriorate. If any of this system starts to fail, its vibration levels change, and vibration monitoring is all about detecting and analysing these changes. Often only the overall vibration level is being measured and analysed in order to monitor the general condition of the system. Changes in vibration levels can be used as an indicator for impending incipient failure and can sometimes be used in defining the possible cause of the malfunction. Therefore, measuring and analysing the vibration level gives a good indication of the machine's condition and can be used with confidence in a condition-based maintenance programme either as a continuously monitored parameter or in a periodic programme. Vibration can be characterized in terms of three parameters, namely displacement, velocity and acceleration. Based on these parameters, there are basically three vibration-measuring transducers which can be used by maintenance engineers to gather data on rotating machines. The three transducers being considered are: displacement transducer, velocity transducer and accelerometer. Today there are a wide range of instruments available which vary from simple to specialized, and a number of computer-based systems are used for vibration monitoring. Some of these are on-line systems with permanently mounted sensors. Others rely on manual measurements. Regardless of whether they are on-line or off-line, all these systems can analyse measurements, store data, trend and plot results.

Vibration monitoring equipment usually consists of three main items: a means of data acquisition, which is effected by the use of one or more suitable transducers located on the machine to be monitored, some form of signal processing, which is displayed by the vibration signal either as a time series and/or as a frequency spectrum, and a method by which the condition of the machine being monitored can be assessed. Most vibration analyses rely on one or more trending and analysis techniques. These techniques include broadband trending, narrowband trending and signature analysis. Monitoring the condition of rotating machinery based on vibration is becoming increasingly important and is probably one of the best developed techniques within condition-based maintenance.

However, major problems while measuring vibrations are the large amount of information to be processed and the small number of people who are experts at analysing particular vibration data to interpret the machine's condition. These limitations are overcome by using artificial intelligence techniques to automate the interpretation of vibration data. Recently, there has been considerable interest in the application of expert systems to vibration monitoring of systems such as gas turbines, gearboxes and so forth.

17.2 TRIBOLOGY MONITORING

Generally, oil samples are analysed in the laboratory using several different methods. The information from the analysis is useful in determining whether a machine, for example an engine, is experiencing abnormal wear or whether the lubricant is degraded. Tribology monitoring has several techniques which can be used for the execution of conditional maintenance tasks.

17.2.1 Lubricating oil analysis

In lubricating oil analysis, samples of lubricating oil are analysed in order to determine whether they still meet the lubricating requirements. The results of the analysis can be used to determine the lubricant life and therefore when it should be changed or upgraded to meet the specification requirements. Therefore, lubricating oil analysis can be used to schedule oil change intervals based on the actual condition of the oil which means, when the oil condition reaches an unacceptable state, it will be replaced to maintain satisfactory system operation. Results of the analysis may also form the basis of the decision to change the type of the oil for performance improvements. Thus, lubricating oil analysis cannot be used as a tool for determining the operating condition of the system, but it is an important aid to condition-based maintenance.

17.2.2 Wear particle analysis

This is a very important technique used to provide an indication of a change in the condition of the system and also to help to determine the cause of the failure. Wear debris monitoring techniques have been recognized as a highly reliable method for the detection of degraded operating condition for almost all oil-lubricated systems, because the change in the rate of debris collection indicates a change in the condition of the system. The particles contained in lubricating oil carry detailed and important information about the condition of the machine. This information may be deduced from particle shape, size distribution and composition of solids. Various techniques for condition monitoring of oil-lubricated systems are applied to understand the wear processes occurring and to establish a suitable method which could be applicable to detect and diagnose abnormal system condition. Mobley (1990) classifies wear particle analysis into two basic stages: analysis of solids content such as quantity, size and composition of machine lubricant, and analysis of the type of wear.

However, the main limitations in using tribology analysis as a conditional maintenance task are: the equipment cost, acquiring accurate samples and interpretation of data obtained.

17.3 PERFORMANCE MONITORING

Performance monitoring is a method which directly monitors the way that items or systems are performing their intended function. Performance parameters, sometimes referred to as process parameters, include efficiency, temperature, pressure, flow, speed, etc. Data on these parameters are normally collected as part of the operational routine for monitoring system performance. The value of these data can be further exploited to serve as indicators of the system's condition. Therefore, performance monitoring has become an established procedure in many plants. Monitoring overall plant performance can be an effective tool for detecting system faults. Typically, analysis consists simply of trending measurement data over time. A fault condition is recognized when certain limit values are exceeded. It is common in large plants that most systems have installed instrumentation such as thermometers, pressure gauges, etc. which are required to measure the parameters that indicate the actual operating condition of the system. The amount of information that they produce is collected in two ways, by manual or microprocessor-based systems, then the data are analysed in order to prepare outputs such as drawings, listings and in some cases work orders for the personnel who are responsible for the actual maintenance task. Instead of studying the results of performance monitoring systems and decision making, which normally would have to be done by highly skilled personnel, a great deal of time and money can be saved by immediately undertaking the maintenance steps suggested by expert systems. Advantages of such intelligent diagnosis systems are to help in interpreting the vase amount of performance parameters data that conventional condition monitoring systems provide to maintenance personnel. Also, they are appropriate to situations where a large number of interrelated decisions must be taken to make sense of a complex set of data.

17.4   VISUAL INSPECTION

Despite the many sophisticated inspection methods available, an aided visual inspection is still important. An experienced inspector can visually detect many defects such as leaks, loose mountains, surface cracks, etc. A wide range of tools are available to help implement visual inspection, such as mirrors, lenses, telescopes and so forth. Hence, visual inspection is widely used throughout industry as a simple, quick and relatively inexpensive form of monitoring condition, and it often plays an important part in condition-based maintenance.

17.5   NON-DESTRUCTIVE TESTING TECHNIQUES

In this book condition monitoring focuses on the techniques which directly measure the condition of an item or a system. Some of the non-destructive testing techniques can be used to monitor the deterioration of direct condition of the system in service. The range of non-destructive testing techniques available is so wide that only a selected few will be described here.

17.5.1 Magnetic particle inspection

This method is used to detect and locate surface or near-surface discontinuities in ferro-magnetic materials by the generation of a magnetic current within an item. During planned overhaul maintenance, items such as shafts, flywheels, steam turbine blades, etc. can be inspected for cracks. This can be done using magnetic particle inspection. This type of technique is not a quantitative analysis, but a skilled operator may be able to give a reasonable estimate of crack depth.

17.5.2 Eddy current testing

This method is based on the principles of electromagnetism. Whenever a magnetic material is present in an electromagnetic field and some movement is caused between them, an electrical current is induced in the metallic material, and is known as an eddy current. The magnitude and phase of the current induced is determined by the presence of discontinuities in the material such as cracks, bubbles and similar. Eddy current techniques can be used to detect surface and subsurface defects within an item, and can also be used for measuring the thickness of either conducting or non-conducting coating on ferrous or non-ferrous base materials. Such techniques also provide information about structural features such as metallurgical conditions and physical properties.

17.5.3 Acoustic emission

High frequency waves are emitted when strain energy is rapidly released as a consequence of structural changes taking place within a material, such as crack growth and plastic deformation. Material transformations result in the generation of acoustic signals which can be detected and analysed and, therefore, it is possible to obtain information on the location and structural significance of such phenomena. This technique is able to detect the location of the internal material transformation; however, it is unable to identify the magnitude of the crack or any other material transformation, and therefore it needs to be coupled with other techniques, for example, ultrasonic. One of the applications of this technique is inspection of pressure vessels in nuclear plants (International Atomic Energy Association, 1993). Acoustic monitoring is heavily dependent on setting precise baseline conditions. Once the system's acoustic signature is established, results can then be compared to the baseline. Changes from the baselines represent changing system conditions.

17.5.4 Thermography

This method is based on the practice of making pictures from thermal radiation emitted by objects within the infrared range of the spectrum.

Therefore, this technique is based on the principle that some operating systems give off heat. Thermography uses instrumentation designed to measure emissions of infrared energy as a means of determining the operating condition of the system. The amount of radiated heat changes as system operating conditions change. Thermography has been used in condition monitoring in the form of various applications. Perhaps the biggest use of this technique is in detecting faults in electrical circuits, for example in detecting poor electrical connections in transformer and electrical switch gear. It may also be used in detecting the operational life of furnace walls and lining of pressure vessels in petrochemical and steel-making industries.

17.5.5   Radiography

Surface and subsurface discontinuities caused by fatigue, inclusion, stress corrosion, etc. can be photographed using very-short-wavelength electromagnetic radiation, namely X-rays or gamma rays. This technique is an effective way to detect internal cracks, imperfections, non-homogeneities, etc., but determination of the estimated size and position of any discontinuity requires highly specialized personnel. Disadvantages of this technique are: it requires access to two opposite sides of an object, and it tends to be an expensive technique, compared with other non-destructive test methods. This technique has been further developed and methods like neutron radiography, stereo-radiography and micro-radiography have been applied in practice.

17.5.6   Ultrasonic inspection

This is one of the most widely used methods of non-destructive testing. This method is suitable for detection, identification and size assessment for a wide variety of both surface and subsurface defects in metallic material, provided that there is access to one surface. This technique can be used in routine inspection of aircraft, rail vehicles, etc., in searches for incipient fatigue cracks. It can also be used for accurate thickness measurements which can be made using ultrasonic pulse-echo techniques. Therefore, this degree of accuracy permits the monitoring of corrosion by noting small changes in wall thickness. Some ultrasonic equipment, which is made specifically for thickness measurement, provides a direct digital read-out of wall thickness. This type of monitoring technique seems to be the most suitable for monitoring the direct condition parameter. This technique is difficult to use while inspecting parts that are rough, irregular in shape, or not homogeneous.

17.5.7 Liquid penetrans

This particular method is mainly used for detecting discontinuities open to the surface, such as cracks, porosity, disbonds and delaminations.

17.6 CASE STUDY: AIRCRAFT ENGINE CONDITION MONITORING

Until the late 1960s, aircraft engines were usually overhauled after a set period of operating hours agreed with the airworthiness authorities, known as the engine's 'overhaul life'. This operating period between overhauls was increased progressively throughout an engine's service life as more experience was gained and modifications increased the life of its constituent items.

Before a more sophisticated approach was adopted, engine health monitoring consisted mainly of checks on items such as oil filters, oil pressure, operating temperatures and vibration levels, plus examination of any debris which adhered to magnetic chip detectors (MCDs) in the oil system. Records of MCD samples and oil consumption gave advance warning of internal wear leading to potential failures. This type of data was recorded manually and the analysis of debris collected on MCDs required the assistance of a qualified inspector. Such approaches relied, and still do, on subjective judgement of the maintainer in charge (built up over years of experience).

Turbojet and turbofan engines operated by airlines soon achieved much longer overhaul lives than the earlier piston types. It was recognized that regularly overhauling an engine after a set number of running hours was inefficient and costly, and that it was more sensible and economic to overhaul only when the condition of its parts indicated that this was necessary.

The move to on-condition maintenance and repair became standard at the beginning of the 1970s with the introduction of the large turbofans for new wide-body airlines such as the Boeing 747 and Lockheed TriStar. These engines were designed from the start in modular form, enabling sections of an engine to be removed and repaired when necessary, independently of the rest of the power unit. The change reflected the way new engines were being designed and also the rapid progress in monitoring techniques, which now enabled large quantities of data to be collected and recorded electronically during the course of a flight. They incorporated new health monitoring features which permitted the internal condition of key engine parts to be inspected without removing the engine from the airframe and stripping it down, at regular intervals. The most important feature was the extensive use of ports which enabled the condition of engine terminals to be inspected on site using a borescope.

The monitoring of engines in flight has made great progress in the last decade. Early attempts were aborted because of the poor reliability of the sensors and measuring equipment used. There were also problems in analysing the vast amount of data collected. For many years the warning devices in aircraft were less reliable than the equipment they were monitoring. For example, pilots were accustomed to false fire warnings. Today the situation has changed. Although sensors are subjected to vibration and a wide variation in temperature and pressure, modern transducers are much more reliable and accurate than those of the past. Digital systems have led to an increase in the number of parameters being measured, providing the opportunity for condition monitoring in greater depth.

Engine monitoring techniques have enabled airlines to achieve outstanding usage of their equipment, with engines now being run for very lengthy periods before removal (Table 17.1). For example, six of the Rolls-Royce RB211-535E4 engines operated by America West have achieved well over 18 000 hours on the wing without removal, covering more than eight million miles.

Modern fuel systems, with full authority digital engine control (FADEC), make it easier to monitor engines via the digital databus used on the latest airliners, electronic equipment (BITE) and a carefully chosen level of redundancy. The BITE can identify internal electronic failure and obviously false sensor signals (such as a shaft apparently running at 150% of its maximum speed), and can identify problems for maintenance personnel. The redundancy allows the equipment to continue to function to a satisfactory level, even with some failures in the system. The failures can then be fixed at a time convenient to the operator.

In condition monitoring today, data are gathered on a sampling basis when the equipment is in use and recorded for subsequent analysis and

Table 17.1 RB211 stays on a wing (data to end September 1994)

Engine	Operator	Hours
-22B	TWA	20 004
-524B4	Delta	27 523*
-524D4	Qantas	19 038
-524G/H	British Airways	21 500
-535E4	LTU Sud	22 256

*denotes still in service. The 27 523 hours on wing of RB211-524B4 represents the record.

interpretation. Measurements are made only to provide 'snapshots' of performance and engine behaviour at selected stages of a flight, after lift-off, during the climb and cruise. These techniques avoid the costly collection of masses of unnecessary information and thousands of person-hours needed for subsequent analysis.

Sampled data can be recorded in flight on cartridges or floppy disks which are transferred onto ground systems when the aircraft lands. A further enhancement to this system enables information to be transmitted in flight to ground stations and analysed, so that any maintenance action can be taken immediately after the aircraft lands. This is particularly valuable in long-haul operations when an aircraft may be away from base for several days.

Ground-based computer systems can now analyse the data gathered in flight and report 'by exception', signalling only when events are occurring or trends in engine behaviour are moving close to limits, indicating that maintenance action is necessary.

For example, Rolls-Royce's UK engine manufacture has developed a ground-based software system to process data from engine condition monitoring inputs. It provides operators with clear and concise information to help in maintenance decisions.

COMPASS (COndition Monitoring and Performance Analysis Software System) was developed by Rolls-Royce in cooperation with Lufthansa, initially as a system to monitor V2500 engines. The system acts as a 'neutral host', which means it can be used with the analytical routines of other suppliers, to monitor the condition of engines, auxiliary power units, airframes and other systems, when suitable diagnostics are installed.

COMPASS is an interactive system which allows the user to interrogate, investigate and review engine data at any time. It outputs standard trend plots that help interpret engine data. The system is in service in two forms. The version for mainframe computers is used by major operators such as British Airways, Cathay Pacific and Lufthansa, some of which employ it to analyse information transmitted from aircraft in flight. Operators of the V2500 and RB211-524G can install additional engine sensors to provide the necessary data to analyse the performance of individual engine items.

COMPASS PC is for operators with fleets of up to 20 aircraft which use advanced desktop computers for post-flight analysis of a more limited volume of engine data. This version of COMPASS has been widely adopted and is in service with 24 operators, mainly those which have B757 aircraft powered by RB211-535E4 turbofan engines.

As with the mainframe version, COMPASS PC is designed as a 'quiet' system which produces alert messages only when a problem is detected. Data can be fed into the system manually or automatically via an input data file.

Steady advances are being made in monitoring techniques. For example, borescope views of engines are now being transmitted to specialist staff at Rolls-Royce. This enables operators to receive expert advice on the internal condition of their engines without removing them from the aircraft.

The Trent turbofan incorporates more advanced monitoring features than earlier engines. Its FADEC system monitors more functions than the electronic fuel-control system installed on earlier engines of the RB211 family.

The FADEC's features include self-test of its electronic units and maintenance messages transmitted via its electronic control system. Engine performance trends are noted and there is provision for monitoring the individual performance of engine items.

These features are in addition to the normal use of magnetic chip detectors on the Trent's master and individual oil scavenge lines; extensive access for borescope inspection of its rotating components and combustor monitoring of engine vibration levels; on-wing trim balancing; and the standard flight deck engine indications.

Condition monitoring has moved from using maintenance personnel expertise for the analysis and interpretation of data to the position today where computer systems convert data into useful information. A skilled engineer is still required to translate this into recommended maintenance actions.

Future stages will remove the need for a human to define the rules for interpretation and will move to a system which can generate and learn its own rules. The techniques to achieve this are known by the term 'artificial intelligence', or AI.

The aim is to incorporate the procedures and thought processes of human experts into a computer system, freeing humans to concentrate on areas less suitable for automation. Some airlines have been investigating this approach for several years and others are studying the use of programming methods for aircraft and engine condition monitoring and maintenance.

One feature is the use of 'neural net' software which has the ability to be 'trained' to recognize significant changes in a stream of input data by the differences in the way the network responds. It is known as a neural net because it is thought to replicate some aspects of human thought processes.

A neural system would receive corrected and filtered output from the COMPASS system, plus fault messages from the FADEC. After training with large amounts of no-fault data, the neural net will process the input to produce a 'fault' or 'no fault' assessment. In addition, where a 'fault' condition is present it gives an assessment of the probability of particular problems occurring.

This is linked to an 'expert system', which uses the neural net outputs and a set of rules to present the user with a diagnosis. The system interacts with a database to store the latest diagnosis and refer to previous ones if this is necessary. It could also work with a 'hypertest' package to present the user with text and diagrams which complement the diagnosis. These diagrams would be derived from documents, such as interpretation and maintenance guides and fault-isolation guides.

All condition monitoring systems depend on an effective interface with maintenance personnel, as well as their faith in the reliability of the system. In an airline, users requiring output from the system may include flight operations, powerplant engineering, maintenance engineering and those responsible for engine repair and testing.

An automated system must be designed carefully so that its output is suitable for rapid use by personnel who undertake line maintenance – not just by the computer and thermodynamic experts who conceived the system and brought it into use.

Aircraft engine monitoring has made great progress during the last 25 years. Ahead lies greater use of automation to ensure increasingly efficient and reliable operation of engines. Engine monitoring is just one of the techniques which help airline engineering staff to remain invisible to the passengers.

CHAPTER 18

Prediction of maintainability measures

The only way to solve the problem would be to guess the outline, the shape, the quality of answer . . . We have no excuse that there are not enough experiments, it has nothing to do with experiments . . . We should not even have to look at experiments . . . It is like looking in the back of the book for the answer.

(Richard Feynman, in Gleick, 1993, p. 303)

Experience tells us that the biggest opportunities to make an impact on maintainability characteristics are at the design stage. Consequently, the biggest challenge for the maintainability engineers is to quickly and accurately predict the maintainability measures of the future maintenance task at the early stage of design, when changes and modifications are possible at almost no extra cost.

This is a very difficult prediction task due to complex interaction between the sequences of activities within each task and the arrangements for the sharing of maintenance resources. Thus, the main objective of this chapter is to present a methodology for the fast and accurate prediction of maintainability measures, at design stage, for the maintenance tasks of the future systems based on the corresponding measures related to consisting maintenance activities.

The biggest challenge facing maintainability engineers is to predict maintainability measures related to maintenance tasks of:

- future products at the early stage of design;
- the benefits of modifications on existing items/systems.

This chapter responds to this challenge by proposing a new methodology for the fast and accurate prediction of maintainability measures and the identification of resources needed for the successful completion of maintenance tasks considered (Knezevic, 1995). The proposed method is based on the maintainability measures related to the consisting maintenance activities, and the maintenance activities block diagram which is applicable to the maintenance task whose consisting activities are performed: simultaneously, sequentially and combined. The method presented could be successfully used at the very early stage of design when most of the information available is based on the previous experience, as well as at the stage when design is completed and tests are performed in order to generate maintainability data for the adopted configuration of the system.

Example 18.1

In order to illustrate the above statement, the maintenance task analysed in Chapter 3, will be used here. It is related to the replacement of a wheel on a small passenger car, performed by the engineering students at Exeter University, UK (Knezevic, 1995).

Statistical analysis of the data observed shows that the maintenance task considered could be modelled by the Weibull probability distribution, with the scale parameter $A_m = 350$ s, and the shape parameter, $B_m = 3.4$ (see Chapter 22). Consequently, the maintainability function, for this specific task, has the following form:

$$M(t) = P(DMT \leq t) = 1 - \exp[-(\frac{t}{350})^{3.4}] \qquad (18.1)$$

Making use of equations 4.2 and 11.4, the numerical values for other maintainability measures were obtained and they are shown in Table 18.1, where DMT_{10}, DMT_{50} and DMT_{90} are the lengths of downtimes, caused by the replacement of the wheel, during which, 10, 50 and 90% of users will have successfully completed this task.

Table 18.1 Maintainability measures in seconds for the task analysed

MDMT	DMT_{10}	DMT_{50}	DMT_{90}
314.4	180.6	314.2	447.3

It is necessary to stress that the maintainability measures listed are obtained for a maintenance task performed on a system which is already in existence, i.e. according to real-life test. At this stage of the system life cycle it is only possible to quantify the maintainability characteristics, but the opportunities for their considerable improvement are practically reduced to zero.

18.1 MAINTENANCE ACTIVITY BLOCK DIAGRAM

According to the methodology developed in this chapter, each maintenance task will be considered as a set of consisting maintenance activities. In order to analyse the maintainability characteristics of a maintenance task, the concept of a maintenance activity block diagram, MABD, will be introduced. This is a diagrammatical representation of the maintenance task where each of the consisting maintenance activities is represented by a box. The relationship between boxes is determined by the order in which each of them has to be executed. The structure of a block diagram for a particular maintenance task is primarily inherited from design although in some cases it could be altered by adopted maintenance policy. The time needed for the completion of each activity is irrelevant for the size of the box.

Based on the sequence in which maintenance activities are performed, according to Knezevic (1994) all maintenance tasks could be classified and defined as:

(a) *Simultaneous maintenance task*: which represents a set of mutually independent maintenance activities, all of which are performed concurrently.
(b) *Sequential maintenance task*: which represents a set of mutually dependent maintenance activities, all of which are performed in the predetermined order.
(c) *Combined maintenance task*: which represents a set of maintenance activities, some of which are performed in sequence and some simultaneously.

Regardless of the type of the maintenance task the following symbols will be used here in order to derive the expressions for the prediction of maintainability characteristics:

DMA_i, random variable which stands for the duration of maintenance activity, i.
$M_i^A(t)$, cumulative probability of completion of ith maintenance activity.
nca, the number of consisting maintenance activities.

In Chapter 11 several maintainability characteristics were defined, but all of them are related to a single random variable, DMT. The methodology presented provides a facility for the prediction of these characteristics, based on the corresponding characteristics of the consisting maintenance activities. Thus, the time needed for the completion of each maintenance task is defined by the random variable DMT, and the corresponding time related to each maintenance activity is defined by the random variable DMA. Analysing equations 4.1, 4.2,

4.3 and 4.4, it is clear that all of them are fully defined by the maintainability function. Thus, the remaining part of the chapter will concentrate on the methodology for the derivation of the expressions for the maintainability function only.

18.1.1 Simultaneous maintenance tasks

A simultaneous maintenance task represents a set of mutually independent maintenance activities, all of which are performed concurrently.

The above definition fully describes the relationship between component activities and clearly states that all activities are starting at the same instance of time and that they are performed simultaneously but independent of each other. The maintenance task is completed when all component activities have been completed (Figure 18.1).

This type of maintenance is to be found with equipment where several different maintenance activities, from the content point of view, are performed.

In order to illustrate the simultaneous maintenance task a typical pit stop of a Formula 1 racing car will be analysed below. During the course of a race, cars will make one or more stops at their pits. Each stop consists of the preventive replacement of a set of four tyres, cleaning the windscreen, and fuel refilling. As the race is only fought on the track and lost in the pit, all activities related to the pit stop are performed simultaneously. Swiftness of this task can occasionally mean the difference between first and second place. Perhaps the best known example, in recent times, was the case of British Formula 1 racing driver

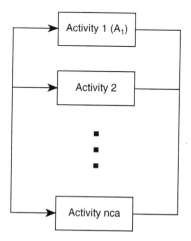

Figure 18.1 MABD for simultaneous maintenance task.

Nigel Mansel losing the wheel of his Renault–Williams car, in the pit-lane at Estoril during the Portuguese Grand Prix in 1991.

18.1.2　Sequential maintenance tasks

A sequential maintenance task represents a set of mutually dependent maintenance activities, all of which are performed in the predetermined order.

The above definition fully describes the relationship between component activities and clearly states that each subsequent activity starts after the completion of the previous one. Thus, none of the subsequent activities can be performed before the completion of the previous one. The maintenance task is completed when the last activity has been completed (Figure 18.2).

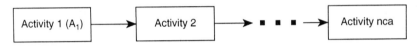

Figure 18.2　MABD for sequential maintenance task.

A typical example of the sequential maintenance task is the replacement of the wheel of the car, where the activities have to be performed in the strictly determined sequence (see example in Chapter 3). Clearly, it is impossible to remove the wheel before the nuts/screws are undone and the wheel lifted off the ground.

18.1.3　Combined maintenance tasks

A combined maintenance task represents a set of maintenance activities, some of which are performed in sequence and some simultaneously.

The above definition fully describes the relationship between component activities and clearly states that activities are performed in combined order (Figure 18.3).

Most maintenance tasks belong to this category, especially today when engineering systems are becoming more complex and consequently their maintenance tasks require higher levels of specialization.

A typical example of the combined maintenance task is the 6000 mile (10 000 km) service to motor vehicles required by their producers. This task consists of activities which are related to the engine, transmission, brakes, body panels, electrical items and so forth. Thus, many of the required activities could be performed simultaneously and the whole

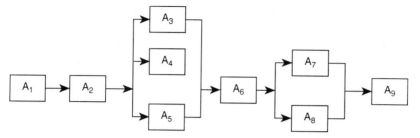

Figure 18.3 MABD for combined maintenance task.

task is finished when all of the consisting activities are successfully completed by the corresponding specialist members of the team.

18.2 MAINTAINABILITY MEASURES FOR SIMULTANEOUS MAINTENANCE TASKS

The definition of a simultaneous task clearly states that all activities are starting at the same instance of time and they are performed simultaneously but independently of each other. The maintenance task is completed when all consisting activities have been completed.

Maintainability measures of the task whose consisting activities are performed simultaneously can be derived from the corresponding measures of consisting activities. Thus, the maintainability function of the maintenance task represents the probability that the task considered will be successfully completed by a certain instance of time, t, as defined by equation 4.2. At the same time, $M(t)$ could be represented as an intersection of events whose probabilities of occurrence are defined by the cumulative probabilities, $M_i^A = P(DMA_i \leq t)$. As the task under consideration will be completed if, and only if, all the comprising activities are completed before or at the stated instant of time t, in the case that random variables DMA_i, where $i = 1$, nca, represent independent events, the maintainability function of the task could also be represented as:

$$
\begin{aligned}
M(t) &= P(DMT \leq t) \\
&= P(DMA_1 \leq t \cap DMA_2 \leq t \cap DMA_3 \leq t \cap \ldots \cap DMA_{nca} \leq t) \\
&= P(DMA_1 \leq t) \times P(DMA_2 \leq t) \times \ldots \times P(DMA_{nca} \leq t) \\
&= \prod_{i=1}^{nca} P(DMA_i \leq t) \\
&= \prod_{i=1}^{nca} M_i^A(t)
\end{aligned}
\tag{18.2}
$$

It is necessary to underline that the above expression could be very complex. The reason for this is the fact that the activity completion functions, $M_i^A(t)$ where $i = 1$, nca, can be defined through any of the theoretical probability distributions (exponential, normal, lognormal, Weibull and similar) and in the majority of cases their product cannot be expressed with any of the well-known distribution functions.

Example 18.2

A maintenance task consists of four activities, A_1, A_2, A_3, A_4. The maintainability functions of all consisting activities are defined by the normal distribution with parameters $A = 40$ minutes and $B = 15$. Determine the DMT_{90} of the maintenance task, in cases where the activities are performed simultaneously.

Solution
For the simultaneous task, the maintainability function can be derived as follows:

$$M(t) = P(DMA_1 \leqslant t) \times P(DMA_2 \leqslant t) \times P(DMA_3 \leqslant t) \times P(DMA_4 \leqslant t)$$

For TTR_{90}, the individual activity probabilities must multiply to give 0.9 (because all four activities share the same maintainability function).

$$0.9 = P(DMA_1 \leqslant t) \times P(DMA_2 \leqslant t) \times P(DMA_3 \leqslant t) \times P(DMA_4 \leqslant t)$$

Each individual probability is thus $\sqrt[4]{0.9} = 0.974$.
In order to find the time (t) by which 97.4% of maintenance tasks are accomplished, the standardized normal variable is used, which is $z = 1.9442$.
Thus the DMT_{90} for the maintenance task is:

$$1.9442 = \frac{t - 40}{15} \rightarrow t = 69.16 \text{ minutes}$$

18.3 MAINTAINABILITY MEASURES FOR SEQUENTIAL MAINTENANCE TASKS

The definition for the sequential task fully describes the relationship between consisting activities and clearly states that each subsequent activity starts after the completion of the previous one. Thus, none of the subsequent activities can be performed before the successful completion of the previous one. The maintenance task is completed when the last activity has been completed.

Maintainability measures of a maintenance task whose consisting activities are performed sequentially, from the point of view of maintainability, can be derived from the maintainability measures of its consisting activities. Thus, the maintainability function of the maintenance task, $M(t)$, whose consisting activities are performed in a predetermined sequence represents the probability that the functionability will be restored by certain instance of time, t, could also be represented as a sum of events of independent random variables DMA_i where $i = 1, nca$. According to the principles of renewal theory (Knezevic, 1995), the maintainability function of the task is equal to the *nca*th convolution of consisting activities, thus:

$$M(t) = P(DMT \leqslant t)$$
$$= P(DMA_1 + DMA_2 + DMA_3 + \ldots + DMA_{nca} \leqslant t) \qquad (18.3)$$
$$= M^{A_{nca}}(t)$$

where $M^{A_i}(t)$ represents the ith convolution of consisting maintenance activities, $M^{A_i}(t)$, which could be determined according to the following expression (Knezevic, 1995):

$$M^{A_i} = \int_0^t M^{A_{i-1}}(t - x)dM^{A_i}(x) \qquad (18.4)$$

where $i = 1, nca$ and $M^{A_i}(t) = M^A{}_1(t)$.

It is necessary to underline that the above expression could be very complex. The reason for this is the fact that the maintainability functions of consisting activities, $A_i(t)$ where $i = 1, nca$, can be defined through any of the theoretical probability distributions and in the majority of cases their convolution cannot be calculated analytically with ease.

Example 18.3

A maintenance task consists of four activities, A_1, A_2, A_3 and A_4. The maintainability functions of all consisting activities are defined by the normal distribution with parameters $A = 40$ minutes and $B = 15$. Determine the DMT_{90} of the maintenance task, in cases where the activities are performed sequentially.

Solution
For each task, the maintainability function will be the same, i.e. $M^{A_1}(t) = M^{A_2}(t) = M^{A_3}(t) = M^{A_4}(t)$ and, for the normal distribution, the maintainability function is derived using the following expression:

$$M(t) = \Phi \left(\frac{t - A}{B} \right)$$

The parameters A and B for the overall task maintainability function are calculated as follows:

$$A = 4 \times 40 = 160$$
$$B = \sqrt{15^2 + 15^2 + 15^2 + 15^2} = 30$$

Thus:

$$M(t) = \Phi \left(\frac{t - 160}{30} \right)$$

and for our example:

$$0.9 = \Phi \left(\frac{t - 160}{30} \right)$$

From the standardized normal distribution tables (see Appendix Table 1):

$$z = 1.2819$$

Thus:

$$1.2819 = \frac{t - 160}{30} \rightarrow t = 198.45 \text{ minutes}$$

18.4 MAINTAINABILITY MEASURES FOR COMBINED MAINTENANCE TASKS

As the definition for a combined maintenance task suggests, it is a combination of maintenance activities, some of which are performed simultaneously and some of them following a predetermined sequence. Thus, the maintainability function for the combined maintenance task depends on the maintenance activities block diagram and the activity completion functions under consideration.

Most maintenance tasks belong to this category, especially today when engineering systems have become more complex and, consequently, their maintenance tasks require higher level of specialization.

18.5 CASE STUDY

In order to demonstrate the practicality of the methodology proposed, an example related to the maintenance task of changing a wheel on a small passenger car will be used (Knezevic, 1995). The list of specified activities which need to be completed, and their sequence, are shown in Table 3.1.

According to the experience gained from the previous model of the car under consideration, and the layout of the present design, the predicted values for the mean time to complete each of 11 consisting activities, $MDMA_i$, could be generated. At this stage of design, the exact type of the probability distribution which could be used to represent each activitiy is not known. Hence, it is not unreasonable to assume that all maintenance activities could be modelled by the normal distribution. Numerical values for the standard deviation, $SDDMA_i$, for each task are reflecting the spread of data among all potential users, their physical and mental differences, as well as the influence of climate, solar radiation, rain, sun and many other factors which might make impact on it. The experience-based values, which take into consideration the variability of the factors which define the environment under which the task is performed, are given in Table 18.2.

The main objective of this exercise is to derive the expressions for the maintainability measures of the task analysed based on predicted activity completion functions of the $M^{A_i}(t)$, where $i = 1, \ldots, 11$.

Table 18.2 Predicted values for consisting activities in seconds

Activity	$MDMA_i$	$SDDMA_i$
1	45	15
2	15	3
3	60	20
4	10	3
5	20	7
6	20	7
7	60	20
8	10	3
9	20	7
10	10	3
11	60	20
Task	330	40

Based on the task description stated above, and the types of maintenance tasks given above, it is not difficult to conclude that the task considered is a sequential maintenance task. Consequently, its maintainability function could be obtained by applying equation 18.3. Thus, in this particular example, it will have the following form:

$$M(t) = M^{A_{11}}(t)$$
$$= \int_0^t \frac{1}{B^{11} \sqrt{2\pi}} \exp \left[-\frac{1}{2} \left(\frac{t - MDMA^{11}}{SDDMA^{11}} \right)^2 \right] dt$$
$$= \int_0^t \frac{1}{42 \sqrt{2\pi}} \exp \left[-\frac{1}{2} \left(\frac{t - 330}{40} \right)^2 \right] dt$$
$$= \Phi \left(\frac{t - 330}{40} \right)$$

where $MDMA^{11} = \sum_{i=1}^{11} MDMA_i$, and $SDDMA^{11} = \sqrt{\sum_{i=1}^{11} SDDMA_i^2}$.

Once the expression for the maintainability function has been obtained, all other maintainability measures could be determined according to equations 4.2 and 4.3.

In order to compare the values obtained from the real-life test and the predicted values obtained by the proposed methodology, the numerical values for the maintainability measures are given in Table 18.3.

The similarity between results shown in Tables 18.1 and 18.3 clearly illustrates the accuracy and the usefulness of the methodology proposed for the prediction of the maintainability measures of related maintenance tasks for the future systems at a very early stage of design. It is necessary to stress that the predicted values are obtained within several seconds of calculation without any additional cost.

The second major advantage of the methodology proposed is the possibility for the quantitative evaluation of the different design options to the maintainability measures.

In the above example it is possible to quantitatively examine the

Table 18.3 Predicted maintainability measures in seconds for the task analysed

MDMT	DMT_{10}	DMT_{50}	DMT_{90}
330	275.5	320	379.5

changes in the maintainability characteristics by making the design changes discussed below under Alternatives 1 to 4.

Alternative 1:

Place a spare wheel in the engine compartment. This will certainly influence activities 1 and 11, because it would not be necessary to remove the contents of the boot and remove the shelf in order to access the spare wheel. Let us assume that the new configuration will reduce $MDMA_1$ to 20 seconds and $MDMA_{11}$ to 15 seconds. The consequences of this design change to $M(t)$ are shown in Table 18.4.

Table 18.4 Predicted values for consisting activities in seconds

Activity	$MDMA_i$	$SDDMA_i$
1	20	7
2	15	5
3	60	20
4	10	3
5	20	7
6	20	7
7	60	20
8	10	3
9	20	7
10	10	3
11	15	5
Task	260	32.5

$$M(t) = \int_0^t \frac{1}{32.5 \sqrt{2\pi}} \exp\left[-\frac{1}{2}\left(\frac{t - 260}{32.5} \right)^2 \right] dt = \Phi\left(\frac{t - 260}{32.5} \right)$$

Alternative 2:

Have a wheel attached to the hub by one central wheel nut. This should decrease the lengths of activities 3, 6, 7 and 9, because there will be only one nut instead of four bolts. Assume that the new configuration will reduce the $MDMA_3$ to 30 seconds, $MDMA_6$ to 10 seconds, $MDMA_7$ to 30 seconds and $MDMA_9$ to 9 seconds. The predicted values of $MDMT$ and $SDTT$ for the new configuration are given in Table 18.5.

$$M(t) = \int_0^t \frac{1}{30.5 \sqrt{2\pi}} \exp\left[-\frac{1}{2}\left(\frac{t - 239}{30.5} \right)^2 \right] dt = \Phi\left(\frac{t - 239}{30.5} \right)$$

Table 18.5 Predicted values for consisting
activities in seconds

Activity	$MDMA_i$	$SDDMA_i$
1	45	15
2	15	5
3	30	10
4	10	3
5	20	7
6	10	3
7	30	10
8	10	3
9	9	3
10	10	3
11	50	20
Task	239	30.5

Alternative 3:

Keep original configuration, but use a hydraulic jack instead of a
mechanical one. This change will affect activities 5 and 8 in the following
way: $MDMA_5 = 10$, $MDMA_8 = 3$. The new values for the mean time
to replace the wheel and the standard deviation are given in Table 18.6.

Table 18.6 Predicted values for consisting
activities in seconds

Activity	$MDMA_i$	$SDDMA_i$
1	45	15
2	15	5
3	60	20
4	10	3
5	10	3
6	20	7
7	60	20
8	3	1
9	20	7
10	10	3
11	50	20
Task	303	39.5

$$M(t) = \int_0^t \frac{1}{39.5 \sqrt{2\pi}} \exp\left[-\frac{1}{2}\left(\frac{t-303}{39.5}\right)^2 \right] dt = \Phi\left(\frac{t-303}{39.5}\right)$$

Alternative 4:

Combine possibilities 1 and 3. The results of these changes are given in Table 8.7.

Table 18.7 Predicted values for consisting activities in seconds

Activity	$MDMA_i$	$SDDMA_i$
1	20	6
2	15	3
3	60	20
4	10	3
5	10	3
6	20	7
7	60	20
8	3	1
9	20	7
10	10	3
11	15	5
Task	243	31.6

$$M(t) = \int_0^t \frac{1}{31.6\sqrt{2\pi}} \exp\left[-\frac{1}{2}\left(\frac{t-243}{31.6}\right)^2 \right] dt = \Phi\left(\frac{t-243}{31.6}\right)$$

Table 18.8 Derived distribution parameters for possibilities examined

Alternative	0	1	2	3	4
MDMT	330	260	239	303	243
SDDMT	40	32.5	30.5	39.5	31.6
DMT_{10}	275.5	215.8	239.4	259.8	200.5
DMT_{50}	330	260	239	303	243
DMT_{90}	379.5	299.2	316.4	361.2	282
DMT_{95}	392.7	312.7	327.2	375.4	292.8

The alternatives are compared in Table 18.8 where 0 stands for the baseline design, and 1, 2, 3 and 4 for the four possible design changes. Maintainability functions for all five configurations are given in Figure 18.4.

Clearly, the proposed methodology enables a design team to quickly quantify the consequences of the design solution chosen as well as the consequences of the possible design changes to the maintainability measures of the maintenance task under consideration, at a very early stage when the changes are implementable at almost no extra cost.

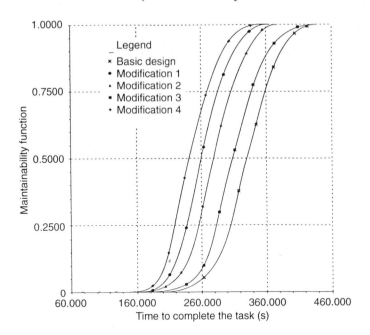

Figure 18.4 Maintainability functions for possibilities examined.

18.6 ANALYSIS OF COMPLEX MAINTENANCE TASKS

Only single maintenance tasks have been addressed so far. However, in many cases it is necessary to perform several independent maintenance tasks at one time. Consequently, a maintenance task which consists of several mutually non-related tasks is called a complex maintenance task.

According to the methodology proposed so far, based on the timing and the sequence in which individual tasks are performed, the following types of complex tasks are possible.

(a) *Simultaneous complex task*: Mutually independent maintenance tasks are performed concurrently. This, in practice, means that each of the consisting tasks are started at the same instance of time and are performed simultaneously but independently of each other. The complex task is completed when all consisting tasks are successfully completed. A typical example of a complex maintenance task is the pit stop of a racing car, where tyre replacement, fuel refilling and windscreen cleaning are performed simultaneously. Clearly this situation requires more maintenance resources and complex logistics, but it minimizes the consequences caused by the downtime. Hence, the replacements are performed simultaneously and the whole task is finished when all of the consisting tasks are successfully completed by the corresponding specialist members of the team, or teams, involved.

(b) *Sequential complex task*: Mutually independent tasks are performed in the predetermined order. This clearly states that the beginning of each subsequent task starts after the successful completion of the previous one. Thus, none of the subsequent tasks can be performed before the completion of the previous ones. The example of the racing car could be used again, but in this case tasks are performed in the following order: tyre replacement followed by refuelling and finished with windscreen cleaning. Clearly, in this situation, the interruption of the operations process would be considerably longer but the demand for maintenance resources would be much less and the logistics simpler.

(c) *Combined complex task*: Mutually independent maintenance tasks are performed in the combined order (sequential and simultaneous). Thus, the relationship between consisting replacement tasks has to be predetermined according to the permissible downtime and availability of the maintenance resources. A typical example of the combined replacement task is the overhaul of an engine or gearbox, where some of the tasks have to be performed in a specific sequence whereas others could be performed simultaneously in order to reduce total time in repair.

Regardless of the type of replacement task, the following symbols will be used here:

DMT_i random variable which stands for the duration of the ith maintenance task.

$M_i(t)$, maintainability function for the ith maintenance task.

The methodology proposed in this part provides a tool for the prediction of the maintainability measures related to all three types of complex maintenance task. Thus, the maintainability measures for each task are fully defined by the probability distribution of the random variable DMT_i, and corresponding measures for the complex task are defined by the probability distribution of the random variable DMT_c, which defines the whole task.

18.6.1 Maintainability function of a simultaneous complex task

The maintainability function for the simultaneous complex task can be derived from the maintainability functions related to the consisting tasks. Thus, the maintainability function related to the simultaneous complex task, $M(t)$, represents the probability that all consisting tasks will be successfully completed by the stated instant of time, t. The maintainability function for the simultaneous complex task, according to the methodology proposed, can be represented as:

$$M(t) = P(DMT_c \leq t)$$
$$= P(DMT_1 \leq t \cap DMT_2 \leq t \cap DMT_3 \leq t \cap \ldots \cap DMT_{nct} \leq t)$$

where *nct* is the number of consisting items. The above expression mathematically describes the complex task whose consisting tasks are performed simultaneously, as an intersection of events at a stated instant of time whose maintainability functions are defined as $M_i(t) = P(DMT_i \leq t)$. It clearly states that the complex task under consideration will be completed if, and only if, all of the consisting tasks have been successfully completed before or at the specified instant of time, *t*.

In case where random variables DMT_i, $i = 1$, *nct* represent independent events, the above expression becomes:

$$M(t) = P(DMT_1 \leq t) \times P(DMT_2 \leq t) \times \ldots \times P(DMT_{nct} \leq t)$$
$$= \prod_{i=1}^{nct} P(DMT_i \leq t)$$

The expression on the right hand side is a maintainability function related to the replacement of the *i*th item. Consequently, the maintainability task which comprises the replacements of any number of consisting items can be calculated according to the following equation:

$$M(t) = \prod_{i=1}^{nct} M_i(t) \tag{18.5}$$

where $M_i(t)$ is the maintainability function of the *i*th task.

It is necessary to underline that the above expression could be very complex. The reason for this is the fact that the maintainability functions of the comprising tasks, $M_i(t)$, where $i = 1$, *nct*, can be defined through any of the well-known theoretical probability distributions, products of which cannot be expressed by any of them. Consequently, their numerical values have to be calculated by numerical methods. This calculation is considerably easier today with extensive use of modern computers.

18.6.2 Maintainability function of a sequential complex task

The maintainability function for the sequential replacement strategy of items from the group, according to the proposed methodology, can be represented as:

$$M(t) = P(DMT_c \leq t)$$
$$= P(DMT_1 + DMT_2 + DMT_3 + \ldots + DMT_{ncr} \leq t)$$

The above expression mathematically describes the sequential replacement of consisting items through a sum of independent random variables, DMT_i where $i = 1$, nct. It clearly states that the task under consideration will be completed only when the successful completion of comprising replacement tasks are performed in a specified order. Making use of the principles of the renewal theory of Cox (1962), the above expression could be rewritten as:

$$M(t) = M^{nct}(t) \qquad (18.6)$$

where $M^{nct}(t)$ represents the nctth convolution of consisting maintainability functions. The convolution of maintainability of function for the ith replacement activity, $M^i(t)$, could be determined according to the following expression of Cox (1962):

$$Mi(t) = \int_0^t M^{i-1}(t - t_i)\mathrm{d}M_i(t_i) \qquad (18.7)$$

where $i = 1$, nct and $M^1(t) = M_1(t)$.

It is necessary to underline that the above expression could be very complex and difficult to use. The reason for this is the fact that the maintainability functions related to the replacement of consisting items can be defined through any of the theoretical probability distributions and in the majority of cases their convolution cannot be expressed by any of the well-known distribution functions.

18.6.3 Maintainability function of a combined complex task

As the definition for a combined complex maintenance task suggests, it is the combination of tasks, some of which are performed simultaneously and some being performed in a predetermined sequence. Thus, the maintainability function for the combined complex task depends on the configuration selected.

18.7 ILLUSTRATIVE EXAMPLE

In order to illustrate the applicability and practicality of the methodology proposed, a hypothetical example will be used. Managing the operation of a racing car is a very exciting business, from a maintenance point of view, because the consequences caused by down times initiated by the failures could be fatal for the final success of the whole process. In this particular case the preventive replacement of a set of four tyres, cleaning the windscreen, and fuel refilling during a pit stop will be analysed. According to data available (Knezevic, 1995) the maintenance tasks analysed and resources needed for their completion are as listed in Table 18.9, where MR30, MR45 and MR2 are sets of resources needed for the completion of the corresponding maintenance tasks.

Making use of equation 18.5, the maintainability function for the simultaneous group replacement strategy has been derived and plotted in Figure 18.5 as a solid line. The corresponding function for the sequential group replacement strategy is shown in the same figure as a broken line.

Calculated numerical values for the mean time to group replacement, $MDMT_c$, and the percentual replacement times, DMT_{c_p}, for both

Table 18.9 Maintainability data for tasks analysed

Maintenance task	Distribution	Parameters	Resources
Tyre replacement	Weibull	$A_f = 45, B_f = 2.7$	MR30
Windscreen cleaning	Normal	$A_f = 8, B_f = 2$	MR45
Fuel refilling	Lognormal	$A_f = 3.67, B_f = 0.19$	MR2

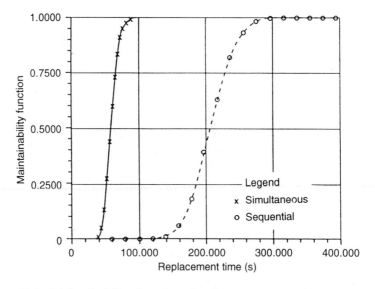

Figure 18.5 Maintainability functions for illustrative example.

Table 18.10 Maintainability data (seconds) for group replacement strategies

Strategy	MDMT	DMT_{10}	DMT_{50}	DMT_{90}
Simultaneous	58.3	45.5	57.8	73.18
Sequential	209.1	167.3	207.5	252.6

replacement strategies are given in Table 18.10, where DMT_{10}, DMT_{50} and DMT_{90} are the lengths of down times, caused by the group replacement, up to which 10, 50 and 90% of maintenance tasks will be successfully completed.

There are several possible combined replacement strategies. For instance, simultaneous replacement of the front tyres, sequentially followed by simultaneous replacement of the rear tyres, sequentially followed by simultaneous fuel refilling and windscreen cleaning, is one of the possibilities. Clearly, the maintainability measures for all combinations possible will be between corresponding values of simultaneous and sequential strategy.

The procedure for the determination of maintainability measures for any configuration of the combined group replacement strategies is identical to the above.

It is necessary to notice that the demand for the maintenance resources varies considerably between different group replacement strategies.

18.8 CONCLUSION

The main objective of this chapter was to demonstrate the methodology for the prediction of the operations/production downtime for the group replacement maintenance policy. The method proposed is based on the maintainability measures related to the individual replacement of consisting items from the group.

The method proposed is applicable to replacement tasks whose consisting items are replaced simultaneously, sequentially and combined. Thus, it is a generic model for the prediction of maintainability measures, which in turn represents vital information needed by maintenance managers for the selection of the most appropriate maintenance policy and strategy for the successful support of the operation/production process.

It is necessary to stress that the method presented could be successfully used at the planning stage of operations/production process when the information available is based on previous experience only.

The numerical examples used clearly illustrate the ability of the proposed methodology to quantify the consequences of the decisions made related to the possible group replacement strategies on the length of the downtime and the magnitude of the revenue lost during these inoperable periods.

CHAPTER 19

Prediction of maintainability statistics

There are a large number of human-made systems whose functionability has to be maintained during the utilization process by the user. The process during which the ability of the system to perform a function is maintained is known as the **maintenance process**, and it is defined (Knezevic, 1995) as:

> a flow of maintenance tasks performed by the user in order to maintain the functionability of the system during its operational life, in accordance to the adopted maintenance policy.

In the analysis of the maintenance process of restorable items, the main interest is determination of the number of maintenance tasks to be performed during a stated operating length, $NMT(L_{st})$. Thus, the main objective of this chapter is to address the methods for the prediction of the number of maintenance tasks to be demanded during the stated length of operation and consequently the frequency of demands for the execution of maintenance tasks.

19.1 CONCEPT OF CUMULATIVE FUNCTIONABLE LENGTH

Renewal theory has been developed into investigation of some general results in the theory of probability connected with sums of independent non-negative random variables (Cox, 1962). These results are extremely useful in modelling maintenance processes (Knezevic, 1995).

Let us suppose that the intervals between successive occurrences of events of a flow are independent, identically distributed random variables, DFL_1, DFL_2, DFL_3, . . . and that their distributions are denoted by $F_i(l)$, thus

$$F_i(l) = P(DFL_i < t) \qquad i = 1, 2, \ldots \qquad (19.1)$$

The instants of occurrences $l_1 = DFL_1$, $l_2 = DFL_1 + DFL_2$, $l_n = DFL_1 + DFL_2 + \ldots + DFL_n$ constitute a random flow, known as a renewal process (Cox, 1962).

Random variables DFL_1, DFL_2, . . ., in certain cases, could be identically distributed with probability density function $f(l)$. This type of renewal process is known as an ordinary renewal process (Cox, 1962).

The maintenance process of a repairable item could be fully defined by the probability distributions of the random variables known as cumulative functionable length to ith failure and denoted as CFL^i. Like any other random variable, CFL^i is fully defined by its probability distribution through some of the following measures:

(a) *Cumulative failure function, $C^i(l)$*: which represents the cumulative probability that the ith transition to a state of failure (or simply failure) will occur before or at the moment of operation, l, thus:

$$C^i(l) = P(\text{cumulative functionable length to } i\text{th failure} \leq t)$$

or

$$P(CFL^i \leq t) = \int_0^l c^i(l)dl \tag{19.2}$$

where $c^i(l)$ is the probability density function of CFL^i.

(b) *Mean cumulative functionable length to ith failure, $MCFL^i$*: which can be determined by using the expectation of the random variable, CFL^i, thus:

$$MCFL^i = E(CFL^i) = \int_0^\infty l c^i(l)dl \tag{19.3}$$

The same result can be obtained from the following expression:

$$MCFL^i = E(CFL^i) = \int_0^\infty 1 - C^i(l)dl \tag{19.4}$$

19.2 PROBABILITY DISTRIBUTION OF CUMULATIVE FUNCTIONABLE LENGTH

Each random variable, CFL^i, could be represented in the following way:

$$CFL^1 = DFL_1$$
$$CFL^2 = CFL^1 + DFL_2$$
$$CFL^3 = CFL^2 + DFL_3$$
$$.$$
$$.$$
$$.$$
$$CFL^i = CFL^{i-1} + DFL_i$$

In some cases there is no difference between these random variables, namely, DFL_i for any $i > 0$. However, on some occasions there are differences between them caused by some of the following factors:

(a) Failures induced by maintenance of the particular or neighbouring items, caused by deployment of untrained personnel, use of inadequate tools and equipment, wrong spare parts or material, lack of necessary facilities and similar.
(b) Failures caused by storage, handling and transportation, and by damp and corrosion.
(c) Use of items which have not been produced by the original manufacturer, i.e. 'bogus' (local production, change of material).

It is necessary to stress that there are cases where the operational length to the subsequent failures are not independent. The most frequent reason for that is a common mode of failure, which has not been rectified by replacing the failed item. For example, the replacement of the fuse in a faulty electrical installation will not rectify the failure in the system, so each subsequent replaced fuse will fail until the prime cause of the failure is removed. Therefore, the operational length between subsequent failures of the fuse is not related to its ability to maintain functionability, but it is related to the operating conditions which are not as specified by the producer.

As the random variables, CFL^1, CFL^2, CFL^3, . . ., CFL^n are independent probability distributions they could be represented by:

(a) Probability density functions, $c^1(l)$, $c^2(l)$, $c^3(l)$, . . ., $c^n(l)$, as shown in Figure 19.2.
(b) Cumulative distribution functions, $C^1(l)$, $C^2(l)$, $C^3(l)$, . . ., $C^n(l)$, as shown in Figure 19.1 for a hypothetical distribution.

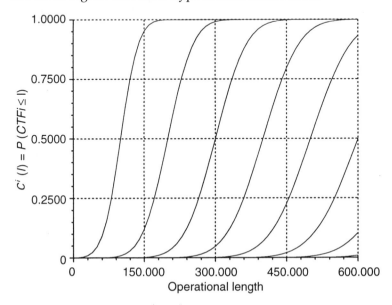

Figure 19.1 Cumulative distribution functions of times to subsequent failures.

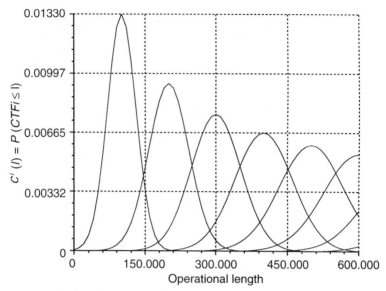

Figure 19.2   Probability density functions of times to subsequent failures.

The cumulative probability function, $C^{i+1}(l)$ of a random variable, CFL^{i+1}, at any instant of operation, represents the probability that the $i + 1$st failure will occur up to cumulative operational length l. In order to derive the expression for the determination of these functions, based on probability distributions of CFL^i, where $i = 1, \ldots, \infty$, it is necessary to analyse the relationship between two subsequent failures. Thus, the probability that the $i + 1$st failure will occur before or at the instant of operational length l, is equal to the occurrence of the complex event, say G, which consists of the following events:

(a) that the ith failure occurred between 0 and, say, $l - x$;
(b) that the subsequent failure, $i + 1$st has taken place in the remaining period x.

Thus, the complex event G could be defined as follows:

$$G = (CFL^i \leqslant l - x \cap DFL_{i+1} \leqslant x)$$

The event G means that the ith failure took place within the interval $[0, l - x]$ and that the $i + 1$st failure has occurred during the remaining length of operation, x. Clearly, the complex event is the intersection of the two sequential, mutually exclusive events. Thus, the probability of occurrence of the event G, $P(G)$, could be determined according to the following expression:

$$P(G) = P(CFL^i \leqslant l - x \cap DFL_{i+1} \leqslant x)$$
$$= P(CFL^i \leqslant l - x) \times P(DFL_{i+1} \leqslant x)$$

As the ith failure could occur at any instant of operating length, the probability of this happening could be numerically determined by the application of the formula for the total probabilities, thus:

$$C^{i+1}(l) = \int_0^l C^i(l - x)\mathrm{d}F_{i+1}(x) \qquad (19.5)$$

It is necessary to point out that $C^0(l) = 1$. Consequently, the cumulative distribution function of the cumulative functionable length l to first failure, $C^1(l)$, is identical to the corresponding function of the duration of the functionable life of the first item $F_1(l)$. Thus, $C^1(l) = F_1(l)$.

Probability density functions of the operating time to the ith failure could be determined in the way shown above, thus:

$$c^{i+1}(l) = \int_0^l c^i(l - x) \times f_{i+1}(x)\mathrm{d}x \qquad (19.6)$$

19.3 WELL-KNOWN DISTRIBUTIONS

In practice considerable difficulties arise when trying to evaluate the numerical values of the function $C^i(l)$, the ith convolution of the function $C^i(l)$, in closed form. The expressions for the prediction of the probability distribution characteristics for the most frequently used well-known theoretical probability distributions will be examined below.

19.3.1 Exponential probability distribution

$$F_1(l) = C^1(l) = 1 - \exp\left[-\frac{l}{A_f}\right]$$

where $A_f = MCFL^1 = MDFL_1$.

$$f_1(l) = c^1(l) = \frac{1}{A_f}\exp\left[-\frac{l}{A_f}\right]$$

The corresponding expression for the random variable CFL^i will be:

$$C^i(l) = 1 - \sum_0^{i-1} \frac{(l/A_f)^i \times \exp\left[-\dfrac{l}{A_f}\right]}{i!}$$

$$c^i(l) = \frac{\dfrac{1}{A_f}\left(\dfrac{l}{A_f}\right)^{i-1} \times \exp\left[-\dfrac{i}{A_f}\right]}{(i-1)!}$$

19.3.2 Normal probability distribution

$$C^i(l) = \Phi \left(\frac{l - A_f^i}{B_f^i} \right) \tag{19.7}$$

The main parameters of the normal distributions are defined as follows:

$$A_f^i = \sum_0^i A_i$$

in the cases where the probability distributions which define times between subsequent failures are defined by the identical expression:

$$A_f^i = i \times A_i$$

The shape parameter, B_f^i, in this case can be calculated according to the following expression:

$$B_f^i = \sqrt{(B^{(i-1)^2} + (B_i)^2}$$

for $i = 1, \ldots, \infty$. If the probability distributions between subsequent failures are identical, the shape parameter, B, could be determined as:

$$B_f^i = \sqrt{i} \times B$$

19.3.3 Lognormal probability distribution

In the cases where the duration of functionable life of each item is modelled by the lognormal distribution, the analytical expressions for the $C^i(l)$ and $c^i(l)$ cannot be derived. Consequently their numerical values could only be determined by applying general expressions.

19.3.4 Weibull probability distribution

For the Weibull probability distributions, the situation is similar to the lognormal distribution, which in practice means that the only way forward is to use a numerical integration method to solve the convolution integrals, which makes calculations considerably more difficult.

Example 19.1

Determine the $C^i(l)$ and $c^i(l)$ for the hypothetical item whose cumulative functionable lengths up to seventh failure are defined by the normal distribution, with parameters as listed in Table 19.1.

Table 19.1 Numerical values for A_f^i and B_f^i

i	1	2	3	4	5	6	7
A_f^i	150	300	450	600	750	900	1050
B_f^i	35.0	49.5	60.6	70.0	78.3	85.7	92.6

Solution

The cumulative distribution function in this case is defined by the following equation:

$$C^i(l) = \Phi \left(\frac{l - A_i}{B_i} \right) \tag{19.8}$$

Table 19.2 lists values of $C^i(l)$, while graphical interpretation of functions $C^i(l)$, $i = 1, \ldots, 7$ are given in Figure 19.3.

Table 19.2

l	$C^1(l)$	$C^2(l)$	$C^3(l)$	$C^4(l)$	$C^5(l)$	$C^6(l)$	$C^7(l)$
0	0.000	0.000	0.000	0.000	0.000	0.000	0.000
100	0.077	0.000	0.000	0.000	0.000	0.000	0.000
200	0.923	0.022	0.000	0.000	0.000	0.000	0.000
300	1.000	0.500	0.007	0.000	0.000	0.000	0.000
400	1.000	0.978	0.205	0.002	0.000	0.000	0.000
500	1.000	1.000	0.795	0.077	0.001	0.000	0.000
600	1.000	1.000	0.993	0.500	0.028	0.000	0.000
700	1.000	1.000	1.000	0.923	0.261	0.010	0.000
800	1.000	1.000	1.000	0.998	0.739	0.122	0.003
900	1.000	1.000	1.000	1.000	0.972	0.500	0.053
1000	1.000	1.000	1.000	1.000	0.999	0.878	0.295

The probability density function of the cumulative time to failure, in this case, is defined as follows:

$$c^i(l) = \frac{\Phi \left(\dfrac{i - A_i}{B_i} \right)}{B_f^i} \tag{19.9}$$

and values for $c^i(l)$ are listed in Table 19.3.

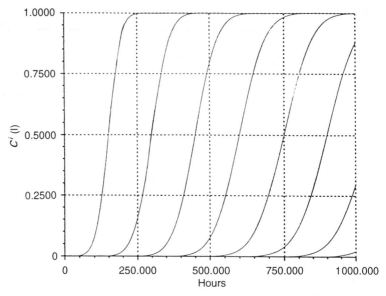

Figure 19.3 Failure functions.

Table 19.3

l	$c^1(l)$	$c^2(l)$	$c^3(l)$	$c^4(l)$	$c^5(l)$	$c^6(l)$	$c^7(l)$
0	0.000	0.000	0.000	0.000	0.000	0.000	0.000
100	0.004	0.000	0.000	0.000	0.000	0.000	0.000
200	0.004	0.001	0.000	0.000	0.000	0.000	0.000
300	0.000	0.008	0.000	0.000	0.000	0.000	0.000
400	0.000	0.001	0.005	0.000	0.000	0.000	0.000
500	0.000	0.000	0.005	0.002	0.000	0.000	0.000
600	0.000	0.000	0.000	0.006	0.001	0.000	0.000
700	0.000	0.000	0.000	0.002	0.004	0.000	0.000
800	0.000	0.000	0.000	0.000	0.004	0.002	0.000
900	0.000	0.000	0.000	0.000	0.001	0.005	0.001
1000	0.000	0.000	0.000	0.000	0.000	0.002	0.004

Making use of equation 19.7, the mean cumulative times to ith failure were calculated. The data obtained are shown in Table 19.4.

Table 19.4 Mean cumulative time to ith failure

i	1	2	3	4	5	6	7
$MCFL^i$	150	300	450	600	750	900	1050

19.4 NUMBER OF CORRECTIVE MAINTENANCE TASKS

In the case of corrective maintenance tasks, the maintenance process could be viewed as the continuous random process which could be defined, at any instant, by the random variable $NMT^c(L_{st})$. This random variable can take non-negative integer values only (the number of maintenance tasks completed during a stated length can be neither a negative nor a real number). Therefore, the number of maintenance tasks which occur up to a stated length should be modelled by a discrete random variable. As any other random variable, the number of maintenance tasks which occur up to a stated length are fully defined by their probability distribution.

In order to define the probability distribution of the random variable $NMT^c(L_{st})$, the following procedure will be used.

The probability that the number of maintenance tasks demanded to be performed up to the stated operational length is greater than i, is equal to the probability that the $i + 1$st failure has occurred within this period of operation, thus:

$$P(NMT^c(L_{st}) > i) = P(CFL^{i+1} < L_{st}) = C^{i+1}(L_{st}) \qquad (19.10)$$

where $C^{i+1}(L_{st})$ is the cumulative distribution function of the cumulative functionable length to the $i + 1$st failure when only the operation process is considered.

The probability that the number of maintenance tasks occurring up to the stated instant of operational length is less than i, is equal to the probability that ith failure has not occurred up to that instant of operation, thus:

$$P(NMT^c(L_{st}) < i) = P(CFL^i > L_{st}) = 1 - C^i(L_{st}) \qquad (19.11)$$

As, at any instant of length, the sum of probabilities of all possible events has to be equal to 1, according to the second axiom of the probability theory, the sum of the following probabilities should obey this law, thus:

$$P(NMT^c(L_{st}) < i) + P(NMT^c(L_{st}) = i) + P(NMT^c(L_{st}) > i) = 1$$

Thus, the expression for the calculation of the probability that the number of maintenance tasks completed during the stated length is equal to i, will have the following form:

$$P(NMT^c(L_{st}) = i) = C^i(L_{st}) - C^{i+1}(L_{st}) \qquad (19.12)$$

It is necessary to say that the probability that the number of maintenance tasks occurred is equal to zero is defined as follows:

$$P[NMT^c(L_{st}) = 0] = C^0(L_{st}) - C^1(L_{st})$$

In order to determine $C^0(L_{st})$, equation 19.11 will be used, thus:

$$P[NMT^c(L_{st}) < 0] = 1 - C^0(L_{st})$$

As it is impossible that $NMT^c(L_{st}) < 0$ it follows that $C^0(L_{st}) = 1$. Consequently,

$$P[NMT^c(L_{st}) = 0] = C^0(L_{st}) - C^1(L_{st}) = 1 - C^1(L_{st}) = 1 - F_1(L_{st}) = D(L_{st})$$

where $D(L_{st})$ is the durability function defined by equation 6.2.

According to equation 19.12 the probabilities of occurrences of subsequent maintenance tasks during the length interval $[0, L_{st}]$ are defined as follows:

$$P(NMT^c(L_{st}) = 1) = C^1(L_{st}) - C^2(L_{st})$$
$$P(NMT^c(L_{st}) = 2) = C^2(L_{st}) - C^3(L_{st})$$
$$.$$
$$.$$
$$.$$
$$P(NMT^c(L_{st}) = i) = C^i(L_{st}) - C^{i+1}(L_{st})$$

Up to now it has been concluded that the number of maintenance tasks completed during a stated length of operation is a non-negative discrete random variable which can take the values between 0 and infinity. Consequently, as any other random variable, it could be represented by a functional method, through the following functions:

- Probability mass function, $h(L_{st}, i)$, defined as:

$$h(L_{st}, i) = P(NMT^c(L_{st}) = i) = C^i(L_{st}) - C^{i+1}(L_{st}) \qquad (19.13)$$

which represents the probability that the number of maintenance tasks completed up to the stated instant of operation is equal to i, where $i = 0, 1, 2, \ldots, \infty$.
- Cumulative distribution function, of the number of maintenance tasks that have occurred, denoted as $H(L_{st}, i)$, which represents the probability that the number of maintenance tasks completed at the stated length will be less than or equal to i:

$$H(L_{st}, i) = P[NMT^c(L_{st}) \le i] = \sum_0^i P[NMT^c(L_{st}) = i] = \sum_0^i h(L_{st}, i)$$

$$(19.14)$$

Graphical representations of the probability functions defined above are given in Figures 19.4 and 19.5 respectively.

The number of maintenance tasks completed during a stated length of operation can also be represented by a parametric method, through the following parameters:

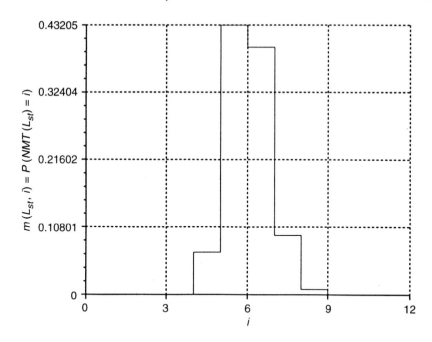

Figure 19.4 The number of maintenance tasks completed at a stated length.

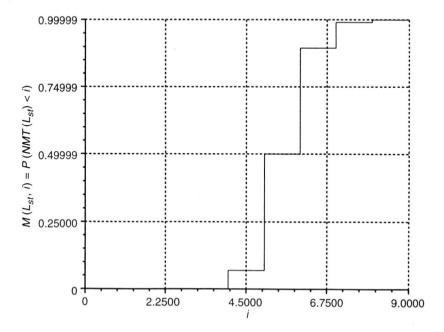

Figure 19.5 The cumulative number of maintenance tasks completed at a stated length.

- The expected value of the random variable $NMT^c(L_{st})$, denoted as $H(L_{st})$, with the mean value, $MNMT^c(L_{st}) = E(NMT^c(L_{st}))$, defined as:

$$MNMT^c(L_{st}) = E(NMT^c(L_{st})) = \sum_{i=1}^{\infty} i \times P[NMT^c(L_{st}) = i]$$

$$= \sum_{i=1}^{\infty} i \times (C^i(L_{st}) - C^{i+1}(L_{st}))$$

$$= 1 \times C^1(L_{st}) - 1 \times C^2(L_{st}) +$$
$$2 \times C^2(L_{st}) - 2 \times C^3(L_{st}) +$$

$$\cdot$$
$$\cdot$$
$$\cdot$$

$$i \times C^i(L_{st}) - i \times C^{i+1}(L_{st})$$

$$= \sum_{i=1}^{\infty} C^i(L_{st}) \tag{19.15}$$

Making use of equation 19.13, expression 19.15 can be rewritten as:

$$H(L_{st}) = MNMT^c(L_{st}) = E(NMT^c(L_{st})) = \sum_{i=1}^{\infty} i \times h(L_{st}, i) \quad (19.16)$$

- The variance of the random variable $NMT^c(L_{st})$, denoted by $VNMT^c(L_{st})$, is defined as:

$$VNMT^c(L_{st}) = \sum_{i=1}^{\infty} (i - MNMT^c(L_{st}))^2 \times P(NMT^c(L_{st}) = i)$$

$$= \sum_{i=1}^{\infty} (i - MNMT^c(L_{st}))^2 \times h(L_{st}, i) \tag{19.17}$$

- The standard deviation of the random variable $NMT^c(L_{st})$, denoted by $SDNMT^c(L_{st})$, is defined as:

$$SDNMT^c(L_{st}) = \sqrt{VNMT^c(L_{st})} \tag{19.18}$$

In practice, considerable difficulties could arise when trying to evaluate the numerical values of the distribution functions in closed form. The expressions for the calculation of the probability distribution characteristics of $NMT^c(L_{st})$ for the most frequently used well-known theoretical probability distributions will be examined below.

19.4.1 Exponentially distributed demand

Making use of the above expressions and general expressions for modelling of maintenance processes, the following equations can be derived for the maintenance process where the demand for the execution of the corrective maintenance tasks is modelled by the exponential probability distribution.

$$h(L_{st}, i) = P(NMT^c(L_{st}) = i) = \frac{(L_{st}/A)^i \times \exp[-L_{st}/A]}{i!} \tag{19.19}$$

$$H(L_{st}, i) = \sum_{0}^{i} h(L_{st}, i) \tag{19.20}$$

$$MNMT^c(L_{st}) = H(L_{st}) = \frac{L_{st}}{A} \tag{19.21}$$

Example 19.2

For the module under consideration, determine the expected number of corrective maintenance tasks to be performed during the period of operation, described by the following operational scenario:

- Number of years: 3
- Number of weeks per year: 30
- Number of days per week: 3
- Number of hours per day: 4

The durability measures of the module considered are modelled by the exponential distribution with scale parameter $A_f = 1000$ hours.

Solution
The stated operational length can be calculated as follows:

$$L_{st} = 3 \times 30 \times 3 \times 4 = 1080 \text{ hours}$$

The expected number of maintenance tasks to be demanded during the stated operational length could be calculated by applying equation 19.21, thus

$$MNMT^c(1080) = H(1000) = \frac{1080}{1000} = 1.08$$

Thus, it is expected to have around one demand for the execution of the corrective maintenance task on the module addressed.

19.4.2 Normally distributed demand

Maintenance process measures for cases where the demand for the execution of the determined corrective maintenance tasks is modelled by the normal probability distribution can be determined according to the following expressions:

$$h(L_{st}, i) = \Phi \left(\frac{L_{st} - A_f^i}{B_f^i} \right) - \Phi \left(\frac{L_{st} - A_f^{i+1}}{B_f^{i+1}} \right) \tag{19.22}$$

$$H(L_{st}, i) = \sum_0^i h(L_{st}, i) \tag{19.23}$$

$$MNMT^c(L_{st}) = H(L_{st}) = \sum_{i=1} \Phi \left(\frac{L_{st} - A_f^i}{B_f^i} \right) \tag{19.24}$$

Example 19.3

For an item whose duration of functionable life is defined by the normal probability distribution with parameters $A_f = 150$ hours and $B_f = 35$, calculate:

(a) the mean number of maintenance tasks during the first 500 and 1000 hours of operation;
(b) the probability mass function of the number of maintenance tasks to be demanded during the first 500 and 1000 hours of operation.

Solution
The cumulative distribution function in this case is defined by equation 9.8.

Probability mass functions for the number of maintenance tasks occurring up to 500 and 1000 hours could be determined by making use of equation 19.22 and data given in Table 19.5. Obtained values are given in Table 19.6.

Table 19.5 Preliminary calculations for Example 19.2

i	1	2	3	4	5	6	7	
A_f^i	150	300	450	600	750	900	1050	
B_f^i	35.00	49.50	60.62	70.00	78.26	85.73	92.60	

L_{st}	$C^1(L_{st})$	$C^2(L_{st})$	$C^3(L_{st})$	$C^4(L_{st})$	$C^5(L_{st})$	$C^6(L_{st})$	$C^7(L_{st})$	$H(L_{st})$
500	1.000	1.000	0.795	0.077	0.001	0.000	0.000	2.873
1000	1.000	1.000	1.000	1.000	0.999	0.878	0.295	6.172

Table 19.6 Probability mass functions for number of failure up to 500 and 1000 hours

i	0	1	2	3	4	5	6	7
$P(NMT^c(500) = i)$	0.0	0.00003	0.2048	0.7186	0.07587	0.0	0.0	0.0
$P(NMT^c(1000) = i)$	0.0	0.0	0.0	0.0007	0.121	0.586	0.273	0.021

19.4.3 Lognormally distributed demand

In cases where the length of operation to the demand for the execution of corrective failure is modelled by the lognormal distribution, the analytical expressions for the probability distribution of $NMT^c(L_{st})$ cannot be derived. Consequently their numerical values can only be determined by applying general expressions. Thus, the expected number of maintenance tasks to be performed during a stated operational length is determined by the following expression:

$$MNMT^c(L_{st}) = H(L_{st}) = \sum_{i=1}^{\infty} C^i(L_{st})$$

19.4.4 Weibull demand

For the Weibull probability distributions, the situation is similar to the lognormal distribution. This in practice means that the only way forward is to use a numerical integration method for the determination of $C^i(L_{st})$ according to equation 19.16 thus:

$$MNMT^c(L_{st}) = H(L_{st}) = \sum_{i=1}^{\infty} C^i(L_{st}) \tag{19.25}$$

PART FIVE

Maintainability Management

CHAPTER 20

Maintainability management

Technical provisions will allow for ease of maintenance only if they are planned from the very outset of rolling stock design stages.
(Paul Monserie, General Engineer Deputy to the SNCF Rolling Stock Director)

In this chapter the primary role of maintainability within the design organization in creating and proving designs with the required inherent maintainability will be considered. Since the obvious is often overlooked and forgotten, some basic points are listed below:

(a) Designers create the design and are responsible for all its character-istics including maintainability.
(b) Each hardware design has an inherent maintainability potential.
(c) The inherent maintainability potential of the design is approached when hardware is produced to the design requirements, but the full inherent maintainability is rarely achieved because the ways and environments under which the system is utilized all tend to result in an achieved or actual maintainability that is less than the inherent maintainability.
(d) Any complex system must start with a design containing very high inherent maintainability characteristics so that there will be a reasonable possibility of placing hardware with the desired or required actual maintainability in the users' hands.
(e) Maintainability personnel must serve as an objective, independent check and balance on the designers, but maintainability personnel must not be permitted to usurp the designer's responsibility for maintainability.

Clearly, the best interests of maintainability during the design process are served by a strong, competent, design function teamed with a strong, competent, maintainability function. It should be stressed that if an organization should be faced with the choice between a strong, competent, design function with a weak maintainability function or a weak design function with a strong, competent, maintainability function, then the better interests of design maintainability would be served by the first condition.

Maintainability design analysis, MDA, is an embracing term which is used to cover many maintainability management functions. Among two of the most important of these functions are maintainability-prediction analysis and maintainability design review.

Maintainability-prediction analysis is a function for assessing the potential maintainability characteristics of a design. It is attempted as soon as the possible design concepts appear. The analysis reports are updated as the design matures. The maintainability-prediction analysis is the major maintainability input to design and to design-review meetings.

A typical initial maintainability-prediction analysis report on a functional system might contain the following sections:

- *Introduction*: which describes the system/item physically and functionally. The use of the system/item is explained, and a picture or sketch is included.
- *Summary of major conclusions and recommendations*: which represents a vital part of the report and should be emphasized by printing it on contrasting paper. It must be remembered that the purpose of the maintainability analysis is to identify design areas needing improvement and to propose those improvements so that corrective action will be taken.
- *Maintainability block diagram*: which shows the location of the maintenance-significant items in the system as well as the major functions of the items themselves.
- *System maintainability estimation*: which leads to a numerical estimate of the maintainability of the system performed by design and maintainability personnel. All assumptions made during the estimation process must be listed. These include required operational scenario, reliability data and durability characteristics, as well as maintenance and inspection policies adopted. The detailed analysis study itself should be an attachment.
- *Items maintainability*: containing sources of failure data of the item, the application of these data to the parts of the system/item and the assumptions used.
- *Failure-mode effect and criticality analysis, FMECA*: where all primary failure modes are identified and described along with their effect and criticality of each on system functionability. Statements should be made on design provisions present to prevent progressive failures, i.e. failures which, in turn, cause other failures.
- *Maintainability analysis*: contains fault-detection and fault-correction information, accessibility of especially limited-life system/items, suggested maintenance requirements, suggested service instructions, and logistic recommendations.

- *Conclusions and recommendations*: which should provide a summary of all recommendations contained in other sections of the analysis report with a reference to the specific section paragraphs where the detailed information is contained. Detailed, specific recommendations for corrective action should also be included.

As the design matures, so will the maintainability in design analysis reports. Later revisions of an analysis report will contain the design-review meeting minutes and information on the resulting actions.

Design analysis is something less than an exact science, but techniques for analysis of maintainability have been quite well worked out. While not an exact figure, the predicted maintainability number resulting from such analysis does provide a rough guide to whether the design is anywhere near the required maintainability.

The maintainability measures/statistics predicted for a particular design as a result of maintainability analysis are of special value in comparing alternative design concepts when the relative inherent maintainability of the designs being compared is the major purpose of the analysis. The maintainability-analysis report often is the only central source of complete, early design description with flow diagrams, schematics, operating scenario, functional descriptions, predicted failure modes, and similar vital information. As such, it serves a valuable auxiliary communication and coordination function. The maintainability-prediction analysis report is, along with the design-disclosure information (drawings, specifications and procedures) itself, a major input to the design review. A third basic input is a set of maintainability design-review checklists completed by the designers.

20.1 MAINTAINABILITY DESIGN REVIEWS

Maintainability design reviews are most successful when conducted within the design organization with the maintainability engineer scheduling and setting up the meetings, taking the initiative, and publishing the minutes. Extensive pre-meeting preparations should be made by using maintainability design-review checklists which are prepared and periodically revised by the maintainability engineer. Whenever possible, design-review checklists should ask questions which require informative answers other than yes, or checklists should ask questions which require informative answers other than yes or no. A minimum of three (conceptual, interim and final) design reviews is required on complex designs. Suppliers who furnish their own design work must also hold their own conceptual, interim and final design reviews. The contractor's design and maintainability personnel should participate in the conceptual and final design reviews.

The process of achieving the required maintainability is easier in some designs than in others. While nearly any design concept can be converted into a maintainable design if enough money, time and effort are expended, the relative ease (comparing two or more design approaches) with which maintainability may be achieved can (and should) be recognized through design reviews. Conceptual design reviews have, of course, a potentially major impact on the design, with successive interim and final reviews having relatively less effect as the design becomes more fixed and less time is available for major changes.

Maintainability design reviews should be combined, wherever possible, with other design reviews such as supportability and reliability to minimize the demand on the designer's time and to resolve conflicting recommendations.

The manager of the design department should normally chair the design-review meeting. As chairperson, the manager makes the final decisions on suggestions and proposed changes. The design-review chairperson is the only person in the meeting who has both the responsibility and authority to make design trade-offs when necessary. The design-review meetings are actually a device for measuring the performance of the designers, and the information gathered provides the chairperson with a better basis of fact for making the necessary trade-offs.

In practice, it is best for the maintainability engineer to give the responsible designers the design-review checklist well in advance of the design-review meeting. The maintainability engineer should go over the checklist with the designers to identify any areas of concern. Ideally, these problem areas should be resolved before the meeting. If the maintainability personnel take exception to the design manager's decision on matters affecting maintainability, the maintainability recourse is through withholding approval or through the management ladder, depending upon the policies of the organization concerned.

The preparation required for a design-review meeting should usually result in most of the deficiencies being cleared up. Tests and analyses that might have been overlooked are made, and the designers will make more use of the various design specialists. In the final analysis, this preparation may, in itself, be the greatest single benefit of design review.

Properly done, the design review can make a major contribution toward 'getting the job done right the first time'. The design review forms a counter force against schedule pressures. Design time requirements are often underestimated. As deadlines approach there is considerable pressure for quick 'intuitive' design solutions without the required searching analysis. A formal design-review programme is a barrier to 'quick and dirty' design solutions, since the designers know they must face their superiors and the maintainability personnel with the correct answers to a searching list of questions. With the design review known to be coming up, it is usually easier for the designers to do the design job thoroughly and correctly the first time.

To summarize, maintainability design analysis is a mathematical, analytical method of estimating or predicting the inherent maintainability of a design. Maintainability analysis should be performed as soon as an interim design and schematics are available. The analysis should be revised every time a significant design change is made.

Maintainability requests for corrective action are made when significant design discrepancies are found. The maintainability analysis report forms a significant input to the design-review meetings, which are a formal, scheduled series of meetings for the thorough, searching review of all aspects of each design with particular emphasis upon the maintainability aspects. Design-review meetings are scheduled and conducted by the design organization with maintainability engineers acting as instigators, recording secretaries, devil's advocates and conscience. When necessary, design trade-offs are made at these meetings, maintainability personnel may refuse to approve or may react through management channels against unacceptable design reviews.

Maintainability analysis and maintainability review may be considered to be the 'inspection and test' of the individual designs for the purpose of measuring and recording conformance to the basic design requirements and conformance to sound design principles and practices. In this sense, maintainability analysis and maintainability review are to the quality of design (inherent maintainability) as inspection and test are to the quality (conformance to design requirements) of the product. Searching and competent maintainability analysis and maintainability review functions are essential in achieving a high inherent maintainability in complex designs. High inherent maintainability is a fundamental requisite for production of hardware which will still demonstrate high maintainability under use conditions.

20.2   IMPACT OF MAINTAINABILITY REQUIREMENTS ON DESIGN

Maintainability requirements, goals and objectives are established by the customer or by competitors' practices. Public organizations (Ministry of Defence, NASA, ESA, CAA, FAA, FIA and similar) are usually quite specific about their maintainability requirements, particularly on weapon systems, space systems, or other complex systems. Large industrial customers such as those in the automotive and airline industries are also very definite in their maintainability requirements upon their suppliers.

However, the maintainability requirements for consumer goods and industrial systems supplied to a number of small users are determined by the producer. Such a determination is based upon both customer reaction and the practices of competitors. It is certainly possible for a company to price iteself out of the market by setting unrealistically high maintainability requirements. However, many more companies/users have found themselves in trouble because they have had a low standard of maintainability requirements, or none at all.

Maintainability requirements, whether imposed or created, are not sacred and should be periodically challenged. This challenging of the need for and validity of any and all design requirements (including maintainability) is the basic element of the value-engineering function. Thus, removal of unnecessary and unrealistic maintainability requirements, during the early stages of a programme, can save substantial time, effort and money as well as increase the amount of resources which can be concentrated on the solution of other design problems.

In the design of a complex system it is necessary to break down the overall maintainability requirement, goal or objective into separate objectives for the elements of the design. This allocation or apportionment is usually made by the design integration function or is made for them by the maintainability function. Generally speaking, design areas of great complexity or in which greatest state-of-the-art advances are required must be given the lowest possible maintainability requirement, while design areas which are straightforward and use thoroughly known and tested design principles will have the most rigorous maintainability requirements.

The allocated maintainability figures should be frequently challenged by comprehensive and responsible design analysis. As the design matures, design management usually makes some adjustments of the maintainability requirements based upon relative progress in the various design areas. Specific maintainability limitations may be imposed upon the design as part of the overall design requirement. In practice, this means that design teams may be prohibited from using certain fasteners, tools, materials, parts, or practices. Similarly, the design teams may also be required by maintainability to follow certain maintainability practices, to use certain tools, equipment and facilities, and to follow certain design practices.

Numerical maintainability requirements, goals, or objectives are a part of the basic design objectives imposed upon the design organization. In addition, maintainability considerations may well result in specific design limitations (dos and don'ts) being imposed upon the design teams. Both these types of maintainability requirement form a part of the design environment within which the design teams must operate.

20.3 MANAGING SYSTEMS MAINTAINABILITY IN FUTURE

Airlines have in the past introduced new aircraft to the fleet with small groups of aircraft engineers. These engineers were based at aircraft manufacturers' main assembly lines to inspect the aircraft during final assembly. These inspectors would arrive just before final assembly began and would begin to inspect selected areas of the aircraft as they made their way along the assembly line. This late arrival to a new aircraft programme meant that all the design work had been completed and the design frozen. In essence, the airline took delivery of an aircraft that the manufacturer believed to be what the customer required. This was not always the case, and a request for a design change, i.e. a MC (master change), at this relatively late stage in production could mean lengthy debate around the following questions.

(a) The time the re-design would take; were there sufficient design staff available?
(b) Could the re-design be incorporated into the production line, without disruption to the aircraft building process?
(c) Could the airline's aircraft be modified before delivery?
(d) Would a retro-fit programme have to be managed for those aircraft that could not be retro-fitted before delivery?

These MCs could be expensive, and the costs incurred would be passed on to the airline requesting the change.

There is also a certain amount of conflict between the aircraft designer and manufacturing processes and the aircraft maintenance engineer. For ease of assembly the manufacturer will install components at convenient times during assembly. This may be convenient to the manufacturer but to the airline engineer this can result in a relatively simple task taking several hours to complete. Hydraulic, pneumatic and electrical systems may need to be broken down for access. Other LRUs (line replaceable units) may also need to be removed to gain the vital access to the LRU that needs to be replaced. Also, once the units have been replaced and the system reconnected there will be hours and hours of function, leak checks and engines running to check the system that has been disturbed.

Example 20.1

This example is based on the paper entitled 'B777 introduction compared to previous aircraft introductions' presented at the one day conference on *The Design and Maintenance of Complex Systems on Modern Aircraft*, in April 1995, organized by The Royal Aeronautical Society.

On previous aircraft introductions the line engineer has had little or no involvement with the design of the aircraft. Normally seeing the aircraft for the first time when it arrives at the airline's engineering base, the airline engineer finds out how maintainable this new aircraft type will be. With the advent of the B777 great emphasis was placed on maintainability, which came from the concept of 'Working Together' and 'Service Readiness Programmes'. As a consequence, the aircraft line engineers became involved from the early days of the design process. With the design now closed, 'Working Together' continues with the airline's involvement in the B777 Cycle Validation Programme.

In 1992 British Airways based a team consisting of one senior manager, two principal airframe engineers, one principal propulsion engineer and one maintainability engineer in Seattle, supported by a London-based project team. This team could draw on representatives from the airline's line engineers to support B777 maintainability reviews. These reviews took place at the prime subcontractors' sites. Some of the real-life experiences of companies involved are cited below.

Garett – Auxiliary Power Unit 331-500: Representatives from British Airways, ANA (Japan), United Airlines (North America) and Boeing attended the APU maintainability reviews. The following is an extract from the August 1992 review:

Design concern regarding the Air Turbine Starter Motor and Air Turbine Starter Control Valve:

(a) When refitting the APU starter motor, it is difficult to align the aligning pin on the gearbox with the hole on the motor flange. An index mark on the flanges (gearbox and motor) would assist alignment.
(b) When spare motors are supplied, a label should be fitted/tied to the shaft directing the maintenance engineer to service the motor with oil before fitting.
(c) On the starter motor and valve, different size clamps are used. All four V-band clamps should be a common type and removable with the same size socket.

The following actions were taken:

(a) Index marks on the motor and adapter flanges are being incorporated to improve alignment.
(b) The vendor has been directed to include a label indicating the motor requires oil servicing before installation.
(c) On production parts, V-band clamps will be removable with the same size socket. The clamps used during demonstration were not production hardware.

Also an aluminium alloy boroscope access plug has been replaced by stainless steel to prevent the plug threads being stripped when being removed, prior to boroscope inspection being carried out.

General Electric – GE90: Six British Airways line engineers were sent to Cincinnati on three occasions for maintainability reviews. Their task was the removal and replacement of all the power plant LRUs, using type designed tooling, to ensure maintainability. Also demonstrated was the fan case removal procedure using specially designed GSE (ground support equipment). Design changes have been incorporated in the engine stand and fan case removal dolly as recommended by the line engineers for ease of operation in the field. Several engine changes have been demonstrated using the GE90 mock-up engine, with the Boeing engine change boot-strap equipment and all the engine change hardware developed for the task.

Boeing Company: The British Airways maintainability engineer, who is permanently based in Seattle, has had a continuing dialogue with the Boeing design team and the B777 Chief Mechanic, Jack Hessburg. The issues cover topics from high intensity radiated field to the more mundane removal of the 13 kg electronic control boxes in the overhead roof space.

A *GE90 engine 3000 cycle validation programme* was carried out at the GE engine test facility at Peebles, Ohio. During this test, at planned engine shut downs all the GE90 LRUs and Boeing Engine Build Unit, EBU, line replaceable units will be removed and replaced in accordance with maintenance manual procedures and the Illustrated Parts Catalogue, IPC. For defect rectification the trouble-shooting charts and Fault Isolation Manual, FIM, will be used, along with the system lock-out procedures in the Mandatory Minimum Equipment List, MMEL. The purpose of using all the manuals and tooling is to verify the correct maintenance practices and procedures.

When the B777 was introduced on its first revenue service in British Airways, the line engineer would already have been involved in maintainability reviews and MMEL steering meetings, and would have reviewed a large proportion of the maintenance manuals. With the cycle validation programme mechanics had gained invaluable hands-on experience and, with the constant 'Working Together' focus on B777 'Service Readiness', British Airways was prepared to introduce the new generation B777 into worldwide airline operation.

20.4 CLOSING REMARKS

It should be noted that the majority of system failures have not been caused by the malfunction of some exotic device whose design pressed the state of the art. Rather, parts were not made correctly (bogus parts), and in other cases there were human failures such as failure to torque and secure a fastener properly or failure to install an explosive device properly. No detail is too minor to cause a problem. High inherent and achieved maintainability are, to a considerable degree, the system of painstaking attention to detail.

In this chapter the range of design responsibilities has been addressed, with a particular emphasis on maintainability requirements. The reasons why inherent maintainability must be high and why maintainability is needed as an independent check-and-balance function on design to assure that maintainability requirements and considerations get their prompt and proper share of design attention, have been presented. Further, the methods for designing a maintainable system have been explored and the methods, procedures and practices used in achieving and assuring maintainability in design have been reviewed.

In summary, inherent maintainability is the primary responsibility of the design organization with maintainability service as an independent check and balance on the design function, principally to make sure that the design function has given its maintainability responsibility the detailed attention which is necessary. In addition, maintainability performs certain functions wherein its work is checked by design for the same reasons.

CHAPTER 21

Systems maintenance model

The primary consideration of all maintenance decisions is neither the failure of a given item nor the frequency of occurrence of the failure, but rather the consequences of that failure upon the system and environment. Analysing all possible consequences of failures of all consisting items within the system, it could be concluded that there are two type of consequences: those affecting safety, and those affecting utility.

Engineering failure analysis provides insight into the type of failures that items/systems are likely to experience. Thus, each item from the list of consisting items within the system should be analysed from the point of view of failure, and, in particular, the consequences of failure must be addressed. The most frequently used engineering tool for performing this task is a failure mode effect and criticality analysis, FMECA. It is a very comprehensive analysis which has a major impact on design in general and reliability and maintainability decisions in particular. Results of this analysis of all consisting items are divided into two groups according to the significance of the consequences of failure. Thus:

(a) *Safety significant items, SSI*: Those items shown by FMECA, to be likely to have hazardous effects and which require special controls in order to achieve an acceptable low probability of individual failure. In the literature, several definitions for safety can be found. According to Niczyporuk (1994) safety is: 'the ability of an object of not causing hazard in relation to life, health and environment'. It is also defined as freedom from unacceptable risk/personal harm, and the risk is the combined effect of the chances of occurrence of some undesirable failure and its consequences in a given system. Therefore, all consisting items of the system whose failure provides a possibility that someone/something could get hurt/damaged or killed/destroyed, as a direct result of it, should be regarded as a safety significant.

(b) *Utility significant items, USI*: Those items which are not safety critical, but where an occurrence of failure is likely to have an effect on the flow of utility, and as such requires a control in order to meet business objectives. Consequences of failure of this type of item could have a significant impact on revenue, cost of maintenance, operational availability, reputation, pride, communal benefit and similar objectives.

A logical diagram for the classification of maintenance-significant items is shown in Figure 21.1. These two categories of items should be separated sufficiently because their influences on maintenance disciplines are quite different.

The occurrence of failures that jeopardize the safety of system/ occupants or environment must be reduced to an acceptable level. Current design practice assures that vital functions are protected by redundancy, fault tolerance, fail tolerance and fail-safe features. This ensures that, if there is failure, a given function will remain available from some other source.

If, however, the loss of a particular function neither endangers the equipment/user nor the environment, then the consequences of failure are evident on utility flow. Thus, the value of maintenance must be measured in utility terms. In some instances, these utility consequences are major, particularly if the failure affects the operational capability of the system. Whenever the system must be removed from service in order to correct the failure, the cost of the failure includes the loss of utility.

The ability to manage utility failure depends largely on the design of the equipment. Today, in the aerospace industry, the predominant strategy is the same as that used to avoid safety-related failures; i.e. inclusion of redundancy, fault tolerance and fail-safe construction

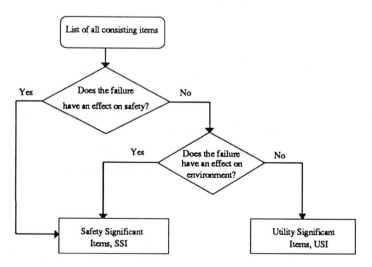

Fig. 21.1 Logical diagram for classification of maintenance-significant items.

beyond that required to certify the design. Clearly, this approach to design has its price; it increases the number of failure possibilities because it adds more items that can fail. It also results in a system that is more complex and integrated, which makes fault isolation more difficult. But the technique also greatly reduces the utility consequences of any single failure; sufficient fault tolerance or redundancy in the design puts initial failures of a system into the utility rather than the safety-related realm.

Utility significant failure should not be a cause for removing the system from operation. Items experiencing utility failure can and should be scheduled into the normal maintenance routine. This is an idea central to justifying excess features in this approach to design. If this is not recognized, then weight, volume, complexity and cost are added without tangible benefit but with obvious penalty.

21.1 SELECTION OF OPTIMAL MAINTENANCE TASKS

Generally speaking, all three types of maintenance tasks (corrective, preventive and conditional) could be applied to every item/system, but only one of them will provide the optimal result regarding the significance of the consequences of failure of the item/system. The most frequently used criteria for assessing the significance of the consequences of occurrence of failure in engineering practice for SSIs are the required reliability level, hazard rate, probability of failure, etc. For USIs the criteria are:

• *Minimum*: maintenance cost, downtime, time to repair.
• *Maximum*: revenue, profit, time between failure, availability.

It is necessary to stress that in some cases the significance of failure could be expressed through two or more of the above-listed criteria.

The selection of the optimal maintenance task, for the maintenance of critical items, is governed by the consequences of failure on the system, and its particular content and scheduling depends on a large number of parameters. In order to analyse the impact of the operational requirements and inherent design characteristics expressed through the reliability, maintainability, availability and similar measures, to the selection of maintenance task, it is necessary to establish the relationship between them. The best way of achieving this is the building of a mathematical model which defines that relationship and provides a basis for all the analysis necessary.

Until recently, and in some cases still, decisions were made by shop-floor engineers based on experience or a subjective feeling. This is understandable to a certain extent since system models changed very rarely and after several years these people had useful and applicable experience. Today, however, when models change quickly and new design solutions are even more complex, it is impossible to rely on experience alone. The given tasks must be solved using established engineering methods. In order to optimize the scope and frequency of maintenance using engineering methods, a suitable objective basis must be created. Without this, decision making is subjective and the optimal solution cannot be found. Subjective decisions, as a rule, lead to cost increases and can also cause undesirable consequences to the system, the user and the environment.

Objective decision making is based on mathematical models which encompass all the factors which have the greatest influence on processes of functioning, change in condition and restoration of the item/system. It is necessary to underline that maintenance models alone do not tell the user what to do, but they do give an indicator for evaluating the degree of risk one takes by making that decision.

An engineering approach to the control of the maintenance process has significant advantages over the traditional 'fire-fighting' way of maintenance management in which all tasks were performed after failure, and also over the 'authorised' way of managing where all maintenance tasks were carried out in accordance with an order based on the personal 'feeling' of someone whose seniority/authority did not permit any sort of analysis of the results obtained.

21.2 MATHEMATICAL MODELS

Models provide answers to questions asked, based on data available, under criteria chosen.

The main objective of every decision is to select the best solution among competing alternatives according to the chosen criterion. This is applicable to decisions related to any activity, task or process. For example, most owners of, say 10-year-old cars with 88 000 miles (140 000 km) on the clock are facing a dilemma, whether to: (a) buy a new car, (b) buy a second-hand car, (c) overhaul the engine in the existing one, (d) hire a car, (e) not have a car at all, and use a taxi when it is needed, or . . .? Cetainly, in real life there are many more alternatives which make a decision even more difficult to make. Thus, the problem is to assess the consequences of each alternative and select that one which seems to be the best according to the information available and selected criteria chosen.

Generally speaking, a better decision requires better information which in return requires a better communication of masses of data and associated rules. In order to achieve this in the area of availability engineering, a deep understanding of several complex disciplines is required. The ideal availability engineer should have a working understanding of every associated engineering discipline, understand the relative importance of special problem areas like physics of failure, workforce, spares, special equipment, facilities, investment, economics, operational scenarios, technical publications, etc., and be able to anticipate the interaction of all these factors so as to find the optimal solution to 'real-life' utilization problems. It should be clear that engineers, in fact human beings, of this description do not exist. Thus, the need exists for some kind of 'magic stick' which will be able to assist the decision maker. As such things do not exist in daily life, mathematical models appear to be able to offer the closest solution.

Models can be used as a tool for communicating masses of data based on complex rules in a structured and disciplined manner. This enables collection and transformation of the input data to be brought into focused attention on various measures of merit known as output. In short, mathematical models, properly executed and applied, can solve the most difficult of the communications problems mentioned above, simplifying the management problem to the point of tractability.

21.3 MIRCE MAINTENANCE MODEL

This model has been developed by Knezevic (1995b) with the objective of selecting the best maintenance policy for maintenance-significant items from the systems point of view. In maintenance literature large numbers of mathematical models can be found, but almost all of them address one item at a time. However, as systems are collections of items, collection of the optimal maintenance policies for consisting items does not always produce the optimal operational effectiveness for the system.

In order to analyse the impact of maintenance on operational effectiveness it is necessary to identify characteristics which best define and quantify this process. As the measures which numerically express these characteristics can take any value from some specified set, it is necessary to treat them as variables.

Variables which define the different characteristics of the item during its operational life, used in the MIRCE maintenance model, are classified in the following categories:

- operational characteristics
- durability characteristics
- reliability characteristics
- maintainability characteristics
- supportability characteristics
- cost characteristics
- safety regulations/standards.

A brief description of each category follows.

21.3.1 Operational characteristics

These characteristics directly depend on the way in which the user of the item plans to operate it. Typical variables include the following.

- Number of years in service, NY.
- Number of operational weeks per year, NWY.
- Number of operational days per week, NDW.
- Number of operational hours per day, NHD.
- Number of operational units, such as landings, miles, cycles, on/off switchings, and similar, per hour, NUH.

All of these variables determine the total length of operational life of the item expressed through the common operational units. Thus, the operational length, L_{st}, for which the maintenance task has to be selected, could be calculated according to the following expression:

$$L_{st} = NY \times NWY \times NDW \times NHD \times NUH \qquad (21.1)$$

Hence, the result obtained from the above expression represents a total number of operational units during the stated operational length, which could be expressed in miles, cycles, years, hours and similar units.

21.3.2 Durability characteristics

The durability characteristics express the ability of the item to maintain functionability when used under the specified condition. The variables which numerically describe this design characteristic are those related to the probability distribution of the random variable known as time to failure. These distributions are fully described by the following parameters.

- Type of distribution of DFL, TD_f.
- The scale parameter, A_f.
- The shape parameter, B_f.

Once the above characteristics are known the durability function, hazard function, $MDFL$, $MdDFL$, DFL_p life and other measures are uniquely defined (Knezevic, 1995b).

21.3.3 Reliability characteristics

These characteristics are related to the system configuration expressed through the reliability block diagram, where all consisting items, based on the impact of their failure to the functionability, are connected in series, parallel, r-out-of-n, and similar categories.

21.3.4 Maintainability characteristics

The maintainability characteristics express the ability of the item to be restored when the maintenance task is performed as specified. The variables which numerically describe this design characteristic are those related to the probability distribution of the random variable known as time to restore. This distribution is fully defined by the following parameters.

- Type of distribution, TD_m.
- The scale of parameter, A_m.
- The shape of parameter, B_m.
- The source parameter, C_m.

Once the above characteristics are known the maintainability function, $M(t)$, $MTTR$, TTR_p and other measures are uniquely defined (Knezevic, 1995b).

It is necessary to remember that time to restore, as a random variable, could have different characteristics for different maintenance policies.

21.3.5 Supportability characteristics

The supportability characteristics express the ability of the item to be supported with required resources for operation and maintenance processes. The variables which numerically describe this characteristic are those related to the probability distribution of the random variable known as time to support. This distribution is fully defined by the following parameters:

- Type of distribution, TD_s.
- The scale parameter, A_s.
- The shape parameter, B_s.

Once the above characteristics are known the supportability function, $MTTS$, $MdTTS$, TTS_P life and other measures are uniquely defined (Knezevic, 1995b).

Obviously, time to support, as a random variable, has different characteristics for different support policies.

21.3.6 Cost characteristics

Brief descriptions of different cost categories are given in Chapter 2. Apart from considering the direct costs (directly related to the performance of each maintenance task) and the indirect costs (related to the administration, customers service, advertising and similar cost centres), in order to select the most suitable maintenance task it is often necessary to determine the cost of lost revenue due to loss of utility flow.

Fixed costs of lost utility flow are related to the costs of salaries of operators/employers, heating, insurance, taxes, facilities, electricity, telephones and similar which are incurred while the item is in a state of failure, SoFa. These costs should not be neglected, because they could be even higher than the other cost categories.

Direct cost of lost utility is usually manifested through the lost revenue and is proportional to the length of the time which the system spends in the SoFa.

21.3.7 Safety regulations/standards

In determining the most beneficial schedule for preventive and conditional maintenance tasks it is necessary to know all safety and environment related requirements, which could be issued by the national, international or profession-specific standards, regulations or recommendations.

21.4 ALGORITHM FOR THE SELECTION OF THE OPTIMAL MAINTENANCE POLICY FOR USIs

The algorithm shown in Figure 21.2 was developed to facilitate the many calculations which are needed before an optimal replacement policy can be formulated for the USIs, based on the MIRCE maintenance system (Knezevic, 1995b). The established algorithm is applicable to all engineering systems.

The combination of the possible maintenance policies and strategies which provides the minimum mean maintenance cost for the stated operational life represents the optimal policy since minimum cost was adopted as the selection criterion.

Example 21.1

Consider a system which consists of three line replaceable units, or LRUs. The problem is to determine the optimal maintenance policy which will provide the minimum total cost of maintenance during the operational period of 240 000 km, for the entire system.

The appropriate durability, maintainability, supportability and cost data are presented in Table 21.1.

Solution

In the interests of script economy, only the final results are provided in Table 21.2, where FBI represents failure-based individual maintenance

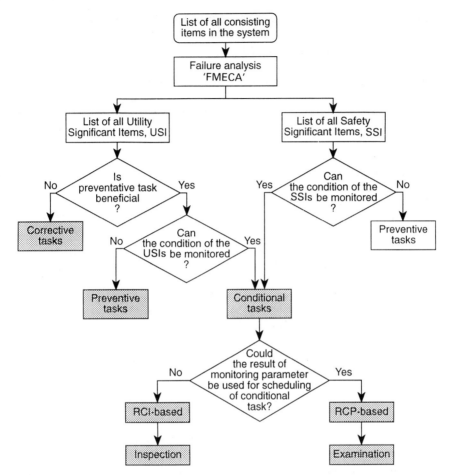

Fig. 21.2 Algorithm for the selection of the optimal part replacement policy for USIs.

Table 21.1 Input parameters for Example 21.1

Items	1	2	3
Maintenance task	Replacement	Cleaning	Replacement
Failure distribution	Normal	Normal	Normal
A_f (km)	60 000	80 000	100 000
B_f (km)	18 000	20 000	30 000
$C^c_{MSS} = C^p_{MSS}$ (£)	265	30	1200
$C^c_{MTE} = C^p_{MTE}$ (£)	15	3	150
HR^c (£)	500	300	1000
HR^p (£)	0	0	0
$MTTR$ (hrs)	10	5	300
$MTTS$ (hrs)	90	45	0

task, FBG represents failure-based group maintenance task, LBI represents life-based individual maintenance task and LBG represents life-based group maintenance task.

Table 21.2 Summary of the possible maintenance policies and strategies for Example 21.1

MS_i	1	2	3	$TCMS^i(T_{st})$
1	FBI	FBI	FBI	22615.4
2	FBG	FBG	FBG	30576.4
3	FBI	FBG	FBG	26700.4
4	FBG	FBI	FBG	29616.7
5	FBG	FBG	FBI	22190.6
6	LBG	LBG	LBI	17660.0
7	LBG	LBI	LBG	20195.0
8	LBI	LBG	LBG	18621.0
9	LBG	LBG	LBG	19885.0
10	LBI	LBI	LBI	18259.0
11	FBI	LBG	LBG	22289.1
12	LBG	FBI	LBG	21627.3
13	LBG	LBG	FBI	16910.3*
14	FBI	FBI	LBI	23365.4
15	FBI	LBI	FBI	21177.4
16	LBI	FBI	FBI	18944.6
17	FBG	FBG	LBI	22940.3
18	FBG	LBI	FBG	28181.5
19	LBI	FBG	FBG	23035.3

*Indicates optimal results.

From the table it can be seen that the optimal maintenance policy and strategy is maintenance strategy 13: Life-based replacement of item 1 and life-based cleaning of item 2 after every 40 000 km and individual failure-based replacement of item 3.

The example given shows that the best choice of the optimal maintenance policy and strategy can reduce the total maintenance cost. The difference between minimum and maximum total average costs of possible replacement strategies, in the above example, is over £13 000 for only one vehicle. For companies with a large number of vehicles, this difference becomes even greater. In this case the proposed methodology provides impressive savings for the same results.

21.5 CASE STUDY: RELIABILITY CENTRED MAINTENANCE

Reliability centred maintenance (RCM) was developed in the aviation industry to determine scheduled maintenance policies for civil aircraft. It has since been adapted for the manufacturing and process industries.

Until the 1960s, scheduled maintenance in the civil airline industry was based on the concept that every item of equipment had a 'right age' at which complete overhaul was necessary to ensure safety and operating reliability. However, the Federal Aviation Agency, frustrated by its inability to control the failure rate of certain types of engine, set up a task force to investigate the capabilities of preventive maintenance which led to extensive research into how equipment fails, in particular age reliability studies, and what conditions must exist for scheduled maintenance to be effective. The research identified six failure patterns which show the conditional probability of failure against operating age for a wide variety of electrical and mechanical items. Thus:

- *Pattern 1*: the well-known bath tub curve.
- *Pattern 2*: demonstrates constant or slowly increasing failure probability with age, ending in a wear-out zone.
- *Pattern 3*: indicates slow increase in probability of failure.
- *Pattern 4*: shows a low failure probability when the item is new then a rapid increase to a constant level.
- *Pattern 5*: exhibits a constant probability of failure at all ages, i.e. a totally random failure pattern.
- *Pattern 6*: starts with burn-in, which drops eventually to a constant or very slowly increasing probability of failure.

The most startling finding from this research, however, was the frequency with which each of these failure patterns occurs in the civil aviation industry. It was shown that 82% of items follow patterns 5 and 6. Thus, these findings contradicted the belief that, as equipment gets older, it is more likely to wear out, the belief which led to the idea that the more often the equipment is overhauled, the better protected it is against failure. The studies showed that this is seldom true, and, moreover, that scheduled overhauls can frequently increase the incidence of failure by introducing burn-in into otherwise stable systems.

The outcome of this research was the development of a method for designing preventive maintenance programmes for aircraft. A rudimentary decision diagram technique was devised first, followed by a handbook on maintenance evaluation, known as MSG-1 (1969). It was used initially to develop the scheduled maintenance programme for the Boeing 747, since judged to have been outstandingly successful.

A later development (MSG-2) was used to develop maintenance programmes for the Lockheed 1011, the Douglas DC-10, the Airbus A300, Concorde, and various military aircraft. The most recent refinement MSG-3 (1992), on which RCM is based, has now become the established process for developing maintenance programmes for aircraft world-wide. The objective throughout has been to assure maximum safety and reliability at the lowest cost. As an indication of the success of

MSG-3, the number of accidents per million take-offs each year has fallen sharply in the civil airline industry.

At the same time, dramatic reductions have been made in the amount of preventive maintenance carried out. Only 66 000 person-hours of structural inspection work are required, for example, on the Boeing 747 for the first 20 000 hours of flying time. Under the traditional maintenance policies 4 million person-hours were required to arrive at the same point for the smaller and less complex Douglas DC-8. By the early 1980s the work being done in the civil airlines was receiving recognition outside the aviation industry.

The implementation of the RCM approach is based on the principle that no preventive maintenance task will be performed unless it can be justified (Knezevic, 1993). The RCM analysis can be divided into two major steps (Anderson and Neri, 1990).

(a) Perform FMECA to identify the SSIs in the system.
(b) Apply the RCM logic analysis process to each safety critical item in order to select the optimum combination of maintenance tasks and suitable intervals for their implementation.

The RCM approach is useful in making decisions about whether a preventive maintenance is needed or not, and whether it will be time-based or condition-based. Therefore, this approach is very valuable in determining the appropriate type of preventive maintenance, but it cannot be used as a tool for deciding optimal interval. Thus, the RCM approach does not contain any basically new method. Rather, it is a more structured way of utilizing the best of several methods and disciplines.

The RCM process, as now developed, possesses three key features:

(a) It recognizes that the inherent reliability of any item is governed by its design and by how it is made, and that no form of maintenance can yield reliability beyond that inherent in the design. An RCM analysis starts by defining the desired performance of each plant in its operating context and ascertains whether the inherent reliability is such that maintenance can deliver that performance. If it cannot, it highlights the problems which are beyond the scope of maintenance and need other actions such as design, modifications, change in operating procedures or raw material changes.
(b) RCM recognizes that the consequences of failure are far more important than their technical characteristics. A structured review of failure consequences focuses attention on the failures which most affect the safety and performance of the plant.
(c) RCM incorporates the latest research on equipment failure patterns into a sophisticated decision algorithm for the selection of preventive maintenance tasks, or the actions which should be taken if no

suitable tasks can be found. The approach recognizes that all forms of maintenance have some value and provides criteria for deciding which is most suitable in any situation.

In daily practice RCM is usually applied by small teams of people normally containing representatives from both maintenance and production, all of whom should have a thorough knowledge of the equipment under consideration. The functions of the equipment together with the associated performance standards are rigorously analysed. All the ways in which the item can fail to meet these standards are then identified and the consequences of each failure are assessed. The final stage, the selection of the most suitable policies, is done with the aid of the decision algorithm.

In many hazardous industries such as nuclear power and petro-chemicals, it is now a requirement for companies to undertake probabilistic risk assessments (PRA), covering all significant hazards on an entire site.

These plants will have many thousands of protective devices designed to prevent major malfunctions – devices such as fire alarm systems, gas detectors, ultimate level switches, pressure relief valves, overload protection, interlocks and stand-by equipment. The majority of these devices do not fail safe, and regular checks are required to confirm that the complete protective system remains operational. As these systems become increasingly sophisticated with multiple protective devices and voting systems, it is very difficult to assess the risk of a multiple failure.

Recent developments in RCM include a method for calculating the level of risk, and for determining how these devices should be maintained. The approach takes into account a number of factors including the consequences of the multiple failure, the mean time between failures of both the protected and protective devices, the number of protective devices and the probability that the maintenance task might cause the failure of the protective device being tested.

The selection of the most appropriate maintenance task is made through the use of the decision algorithm, which takes into account both the technical feasibility of the proposed task and whether it is worth doing. The task selection process always considers the selection of conditional maintenance tasks before any other.

CHAPTER 22

Assessment of maintainability field data

All the well-known distribution functions (exponential, normal, log-normal and Weibull), are derived theoretically and are reproducible only if the experiment is repeated an infinite number of times. On the other hand, looking at examples from everyday engineering it is not difficult to conclude that in the majority of cases the family (and the particular member of the family) of distribution functions which could be used to describe the behaviour of the relevant random variable is unknown. This presents a serious problem for engineers because the empirical data related to the maintainability issues are very limited. This is due to the fact that repetition of experiments is both expensive and time consuming.

Thus, the problem which maintainability engineers face is the selection of the appropriate distribution function to describe the outcome of an experiment using the empirical data obtained from only a few repetitions. Expressed in the language of statistics the task is to make inferences about a population, based on the information contained in a sample.

The main concern of this chapter is therefore the presentation of the methodologies for establishing the relationship between the abstract probability system and probability distributions, created by mathematicians, and the empirical data related to maintainability measures, obtained in day-to-day engineering practice. In other words, we are concerned with the selection of the family of theoretical probability distributions and associated parameters which define the random variable, *DMT*, based on empirical data obtained from the repetition of real-life maintenance tasks trials.

This chapter describes some well-known methods for selection of the most relevant theoretical distribution functions for random variables under consideration. The overall process consists of a few independent but interconnected phases, as shown in Figure 22.1. Each phase will be explained theoretically and illustrated with examples related to the most frequently used theoretical probability distributions, applicable to engineering practice.

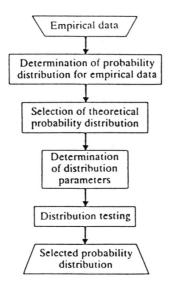

Figure 22.1   Algorithm for the selection of the theoretical distribution function for the empirical data.

22.1   PROBABILITY DISTRIBUTION OF EMPIRICAL DATA

According to the algorithm given in Figure 22.1, the first step is the estimation of the probability distribution of empirical data obtained from the experiment performed. It is necessary to emphasize that the term 'experiment' covers a large number of activities whose aim is to generate and collect raw data (laboratory test, field data, user reports, demonstration test and similar). Regarding the total number of results available there are two distinguishable procedures for the determination of the probability distribution of empirical data:

- ranking method
- relevant frequency method.

The first one is applicable in all cases where the total number of empirical data is less than 50, whereas the second method is recommendable in all other situations. Both methods are described below for a hypothetical random variable, *DMT*.

22.1.1   Number of empirical data less than fifty

According to Bernoulli's theorem (Knezevic, 1993), the probability of occurrence of an event is defined by the ratio between its frequency and the total number of results, n, when this approaches infinity. The problem with the above statement starts in cases where the total number of results is insufficient for the data to be classified into groups. For example, if a specific maintenance task has been repeated four times the result with the lowest value, say t_1, has, from the point of view of maintainability function, the probability of occurrence 0.25. On the other hand, if the same task is repeated eight times, then the result with the lowest value, t_1, represents the cumulative probability of 0.125 (1/8), and the second lowest value represents the cumulative probability of 0.25. It is clear that the difference between the proportion of data represented by the cumulative probabilities in these two cases directly depends on the total number of results. Similarly the cumulative distribution function for the result with the highest numerical value will be 1 (4/4). Obviously, this is not correct because there is always some possibility that a higher value could be obtained if the task considered were repeated more than four times.

Attempts have been made to solve this problem, and the ranking method seems to be the most suitable. The median rank, MR, which represents the median of all possible results which could have occurred in specific rank is the most frequently used ranking method. In some cases the mean rank method can be used as well. Numerical values of median rank for all cases where the total number of results is less than or equal to fifty are presented in Appendix Table 2. According to this table, for $n = 4$ the median rank of the first result is 0.1591 and not 0.25 as might have been expected, the second is 0.38636, the third 0.61364 and the fourth 0.84091.

Apart from using the available tables, the approximate numerical values of the median rank can be determined by using the following expression:

$$MR = \frac{(i - 0.3)}{(n + 0.4)} \qquad (22.1)$$

where i is the order number of each individual result after they have been arranged in ascending order, and n is the total number of results available. The above approximation is accurate to 1% for $n > 5$ and to 0.1% for $n > 50$.

Other than the median rank, the most commonly used are 5 and 95% ranks which correspond to probabilities of 0.05 and 0.95. Appendix Table 2 shows that in the case of four results, 5% rank has a value of 0.01274, which means that a probability of 0.05 exists that the first result will have cumulative probability greater than 0.01279. Thus in only 5% of cases will the first result represent more than 1.27% of the total results. For the same example, rank 0.95 is 0.52713, which means that in

95% of cases, the first results will present less than 52.7% of the total number of results.

As the median rank represents the cumulative proportion of the data corresponding to the rank under consideration, the maintainability function of empirical data, $M'(.)$ can be determined in the same way, thus for the random variable DMT:

$$M'(t_i) = MR(t_i) = \frac{(i - 0.3)}{(n + 0.4)} \tag{22.2}$$

By making use of the above equation the expression for the probability density function of empirical data, $m'(.)$ can be derived as:

$$m'(t_i) = \frac{1}{(n + 0.4) \times (t_{i+1} - t_i)} \tag{22.3}$$

It is necessary to repeat that, in order to apply the above expressions to the empirical data, they must be arranged in ascending order, i.e. $i = 1$ corresponds to the result with the lowest numerical value observed.

In the same way as the probability distribution of the random variable could be represented by the summary measures, the probability distribution of the empirical data could be described by the measures of central tendency and the measures of dispersion. Thus, the mean value of empirical data, $MDMT'$ can be determined by the expression:

$$MDMT' = \sum_{i=1}^{n} \frac{t_i}{n} \tag{22.4}$$

The median, $MdDMT'$ in the case of an odd number of results is equal to the numerical value of the middle result, whereas in the case of even numbers of results, the median is equal to the arithmetical mean of the middle two results.

The standard deviation, $SDDMT'$, can be calculated:

$$SDDMT' = \sqrt{\frac{\sum_{i=1}^{n} (t_i - MDMT')^2}{n - 1}} \tag{22.5}$$

In order to clarify the above statements and make the calculation easier, a form, named M1, was designed. It contains all the steps for the determination of probability characteristics based on the above equations. The form, shown in Figure 22.2, is applicable to all random variables.

Total number of results, $n =$		Order number of results, $i = 1.n$		$B = n + 0.4 =$ .... $+ 0.4 =$ ....
i	x_i	$(x_i - M')^2$	$M'(x_i) = (i - 0.3)/B$	$C_i = x_{i+1} - x_i$	$m'(x_i) = 1/C_i B$

$$\sum_{i+1}^{n}$$

S1	S2

$$M' = S1/n \text{..........} / \text{..........} = \text{..........}$$

$$SD' = \sqrt{S2/(n-1)} = \sqrt{(\text{..........}/(\text{..........}-1)} = \text{..........}$$

Figure 22.2 Form M1.

Example 22.1

The durations of the maintenance task, repeated nine times, are presented below. Calculate and plot the probability characteristics of the empirical data available.

Trial number	1	2	3	4	5	6	7	8	9
Measured value (min.)	84	34	37	99	51	69	26	48	103

Solution

The first step towards the solution is to rearrange the available data into ascending order:

Order number	1	2	3	4	5	6	7	8	9
Trial number	7	2	3	8	5	6	1	4	9
Measured value (min.)	26	34	37	48	51	69	84	99	103

In order to calculate the mean and standard deviation of empirical data equations 22.4 and 22.5 should be used.

Regarding the determination of distribution functions for the empirical data, the following two methods are possible:

(a) Use the data given in Appendix Table 2, in the following way: find the part which is applicable to sample size $n = 9$ and directly read values for $MR(i) = M'(t_i)$ for $i = 1, 9$. For example, $M'(t_1) = 0.06697$ (6.7%). Then the numerical values of $m'(t_i)$ can be calculated by applying equation 22.3.

(b) Use form M1, as shown in Figure 22.3.

Graphical interpretations of the obtained functions are presented in Figure 22.4.

Example 22.1 illustrates a commonly accepted procedure for the calculation of probability distributions for empirical data when the total number of results is less than 50 ($n < 50$). It should be emphasized that extreme care must be taken since a single inaccurate result can distort the picture, by changing the shape of the functions.

22.1.2 Number of empirical data greater than fifty

In contrast to the method above where results had to be ranked, in this case it is possible to use the frequency approach. In order to calculate the probability characteristics of the empirical data the results are classified

Total number of results, n = ...		Order number of results, $i = 1.n$	$B = n + 0.4 = $ .... $+ 0.4 = $ ....

i	x_i	$(x_i - M')^2$	$M'(x_i) = (i - 0.3)/B$	$C_i = x_{i+1} - x_i$	$m'(x_1) = 1/C_i B$
1	26	1240.45	0.074	8	0.013298
2	34	740.90	0.181	3	0.035461
3	37	586.60	0.287	11	0.009671
4	48	179.90	0.394	3	0.035461
5	51	104.40	0.500	18	0.005910
6	69	60.52	0.606	15	0.007092
7	84	518.90	0.713	15	0.007092
8	99	1427.30	0.819	4	0.026596
9	103	1745.50	0.926		
$\sum_{i+1}^{n}$	551	6599.5	$M' = S1/n$ .... 55.1 .... / .... 9 .... = ...	61.22	
	$S1$	$S2$	$SD' = \sqrt{S2/(n-1)} = \sqrt{(....6599.5.....)/(........9........-1)} = ...28.72....$		

Figure 22.3 Form M1 with the data related to Example 22.1.

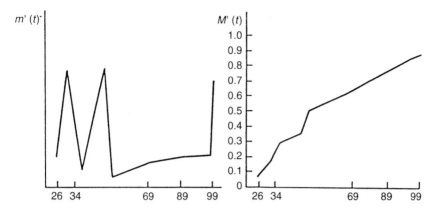

Figure 22.4 $M'(t)$ and $m'(t)$ for empirical data.

into several intervals the number of which, NI, depends on the total number of results, n. The following equation can be used as guidance for determining the suitable number of intervals, thus:

$$NI = 1 + 3.3 \times \log_{10}(n) \qquad (22.6)$$

The length of each interval, LI, is defined by

$$LI = \frac{(t_n - t_1)}{NI} \qquad (22.7)$$

The limits of the intervals can be calculated in the following way:

$$t_{min_i} = t_i + (i - 1) \times LI$$

$$t_{max_i} = t_{min_i} + LI$$

$$t_{m_i} = \frac{t_{max_i} - t_{min_i}}{2}$$

where i is the order number of the interval, $i = 1$, NI; t_{min_i} is the minimum qualifying value for the ith interval; t_{max_i} is the maximum qualifying value for the ith interval and t_{m_i} is the mean value of the ith interval.

Taking into account equation 22.1, the estimated values for the cumulative distribution function of the empirical data considered are defined by the expression:

$$M'(t_{max_i}) = \frac{N(t_{max_i}) - 0.3}{n + 0.4} \qquad (22.8)$$

where $N(t_{max_i})$ is the total number of results whose numerical value is less than or equal to t_{max_i}.

The probability density function of empirical data is defined:

$$m'(t_{m_i}) = \frac{f_i}{n \times LI} \tag{22.9}$$

where f_i is the frequency of the *i*th interval, i.e. it is the number of results whose numerical values are greater than t_{min}, and less than or equal to t_{max_i}.

The mean of empirical data can be obtained as:

$$M' = \frac{\sum\limits_{i=1}^{NI} t_{m_i} \times f_i}{n} \tag{22.10}$$

The standard deviation of available data can be determined:

$$SD' = \sqrt{\frac{\sum\limits_{i=1}^{NI} (t_{m_i} - M')^2 \times f_i}{n}} \tag{22.11}$$

and the same results can be obtained by using the transformed expression:

$$SD' = \sqrt{\sum\limits_{i=1}^{NI} \frac{t_{m_i}^2}{n} \times f_i - M'^2} \tag{22.12}$$

In order to facilitate calculation of these numerical values in everyday practice, the form M2 was designed, which contains all the steps for the determination of the probability distribution for empirical data. This form is illustrated in Figure 22.5.

Example 22.2

Results of 55 observed values of the duration of a specific maintenance task, in minutes, are given below:

3	56	9	24	56	66	67	87	89	99	4
26	76	79	89	45	65	78	88	89	90	92
99	2	3	37	39	77	77	93	21	24	29
32	44	46	5	46	79	99	47	77	79	89
31	78	34	67	86	91	75	33	55	22	44

Calculate the empirical values of the mean and standard deviation of the duration of this maintenance task and plot the corresponding functions: $M'(t)$ and $m'(t)$.

	$x_{min_i} =$ $x_1 + (i+1)$ LI	$x_{max_i} =$ $x_{min_i} + (i+1)$ LI	$x_{m_i} = (x_{min_i} + x_{max_i})/2$	$x^2_{m_i}$	$f_i =$ $N(L_i)$	$x_{m_i} f_i$	$x^2_{m_i} f_i$	$b_i =$ $N(x_{max_i})$	$M'(x_{max_i}) = b_i/n$	$m'(x_{m_i}) = f_i/(n\,LI)$
$\sum_{i+1}^{NI}$						S1	S2			

Total number of results, n = ...

Number of intervals, NI = 1+3.3 log (n) = ...

Length of interval, LI = $(x_n - x_1)/NI$ = ...

$M' = S1/n = ... / .... = ...$

$SD' = \sqrt{S2/n - M'^2} = \sqrt{............ / .... - ....} = ...$

Figure 22.5 Form M2.

	Total number of results, n = ...55...				Number of intervals, $NI = 1+3.3 \log(n) = 1 + 3.3 \log(55) = 6$ ...			Length of interval, $LI = (x_n - x_1)/NI =$ ...16.17...	
i	$x_{min_i} =$ $x_1+(i+1)LI$	$x_{ma_i} =$ $x_{min_i}+(i+1)LI$	$x_{m_i} = (x_{min_i}+$ $x_{max_i})/2$	$x^2_{m_i}$	$f_i =$ $N(L_i)$	$x_{m_i} f_i$	$x^2_{m_i} f_i$	$b_i =$ $N(x_{max_i})$	$M'(x_{max_i}) = b_i/n$	$m'(x_{m_i}) = f_i/(n\,LI)$
1	2.00	18.17	10.08	101.61	6	60.48	609.64	6	0.109090	0.0067460
2	18.17	34.33	26.25	689.06	10	262.50	6890.60	16	0.290909	0.0112440
3	34.33	50.50	42.42	1799.45	8	339.30	14395.60	24	0.436360	0.0089900
4	50.50	66.67	58.58	3431.60	5	292.90	17158.10	29	0.527272	0.0056220
5	66.67	82.83	74.75	5587.50	12	897.00	67050.70	41	0.745454	0.0134930
6	82.83	99.00	90.92	8266.40	14	1272.90	115730.20	55	1.000000	0.015418
$\sum_{i+1}^{NI}$						3125.06	221 834.80			
						$S1$	$S2$			

$$M' = S1/n = \frac{3125.06}{55} = 56.82$$

$$SD' = \sqrt{S2/n - M'^2} = \sqrt{\frac{221\,834.8}{55} - 56.82^2} = 28.37 .$$

Figure 22.6 Form M2 with the data related to Example 22.2.

Solution

The form M2 was used to facilitate the calculation of the probability characteristics of the empirical data available. The results obtained, after they have been classified into six groups, are given in Figure 22.6.

The graphical interpretation of the functions $m(.)$ and $M(.)$ are shown as histograms in Figure 22.7.

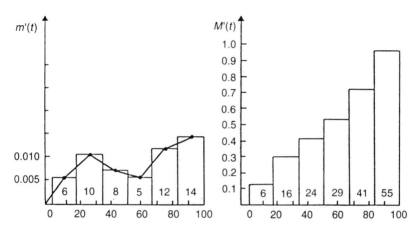

Figure 22.7 *PDF'* and *CDF'* of the empirical data.

22.1.3 Algorithm for calculation

In order to provide a full picture of the method presented for the estimation of probability distribution of available empirical data, the algorithm shown in Figure 22.8 was designed. It shows that the output characteristics of this phase are numerical values of the probability distribution for empirical data, presented in analytical and graphical form. Results obtained using this method must be treated with caution, because they represent only the available empirical data.

22.2 SELECTION OF THEORETICAL DISTRIBUTION

In this phase, selection of the theoretical distribution function which represents the observed empirical data as closely as possible has to be performed. Using the language of statistics this process is known as hypothesis making.

It is very difficult to select one particular family of theoretical probability function with which to associate the probability distribution of empirical data, especially in cases with a small number of results. The main indicator for the selection of the theoretical probability distribution

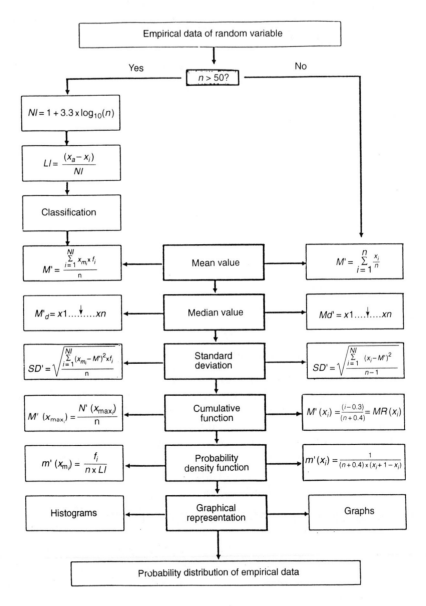

Figure 22.8 Algorithm for calculation of probability distribution of empirical data.

should come from information output from the previous step. That is, one or more of:

- Mean duration of maintenance task, *MDMT'*.
- Median duration of maintenance task, *MdDMT'*.
- Standard deviation of the duration of the maintenance task, *SDDMT'*.
- Probability density function, $m'(t_i)$.
- Maintainability function, $M'(t_i)$.

The following recommendations are provided to assist the decision-making process regarding the selection of the most suitable theoretical distribution for empirical data:

- If the calculated values of the standard deviation and the mean of the duration of the maintenance task considered are relatively close, it is a good indicator that the random variable, *DMT*, obeys exponential distribution.
- As the normal distribution is symmetrical, i.e. mean, median and mode have the same numerical value in the cases where the estimated values of the $M'(.)$ and $Md'(.)$ are reasonably close, the recommended distribution is the normal.
- If the first two conditions are not met, and the plotted probability density function of empirical data is not symmetrical, it is most likely that we are dealing with a lognormal or Weibull distribution.
- If none of the above conditions are satisfied, it is most likely that the observed empirical data does not belong to the same maintenance task, conditions or procedure, i.e. that we are dealing with a mixed distribution. In this case the empirical data should be carefully re-examined and if there is evidence of the mixed distribution, it is necessary to separate mixed results, and then repeat the above procedure for each set of data.

When a hypothesis is made about the family of the theoretical distribution function it is necessary to find the numerical values of the scale parameter *A*, the shape parameter *B* and the source parameter *C*, which together fully define the member of the family. There are two possible ways of determining the parameters of theoretical distribution function for the empirical data: graphical and analytical.

The analytical method requires more calculation time, but gives more accurate results, whereas the graphical method is simpler but leaves room for inaccuracy. In this chapter both methods will be described.

22.3 GRAPHICAL DETERMINATION OF DISTRIBUTION FUNCTION PARAMETERS

The basic tool for the graphical method for determination of the most suitable probability distribution is a special type of graph paper, known

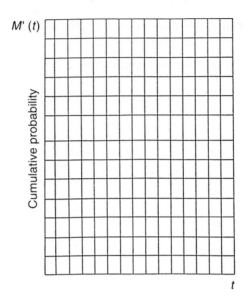

Figure 22.9 Probability paper for a hypothetical probability distribution.

as probability paper. Probability paper for a hypothetical probability distribution is presented in Figure 22.9.

For each theoretical distribution function there is a unique probability paper whose coordinates are designed so that the probability distribution is given as a straight line. Thus, a maintainability function, $M(x)$, defined by parameters A, B and C plotted on the appropriate probability paper, will be a straight line, with positive slope. Using this property, it is possible to plot observed empirical data on the probability paper, as a test of data distribution, and at the same time obtain a graphical assessment of the parameters involved by drawing a straight line to the empirical points.

All types of probability paper are used according to the following procedure:

(a) *Plotting the empirical data*: every point plotted on the probability paper is defined by the pair t_i, $M'(t_i)$, where $i = 1$, n, as shown in Figure 22.10.
(b) *Visual check*: if the points plotted fall in a straight line, it indicates that the empirical data under consideration could be represented with the selected distribution. It should be said that it is unreasonable to expect a perfectly straight line, and that according to some authors the first and last 10% of results can be ignored.
(c) *Fitting the best line*: a straight line should be fitted through the plotted points. The chosen line should be the best representative of all points (see Figure 22.11). This is the weakest part of the graphical

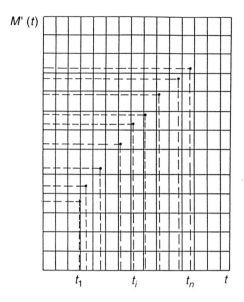

Figure 22.10   Plotted points on probability paper.

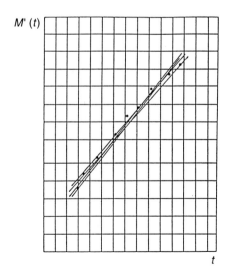

Figure 22.11   Fitting line through the points.

method because the location of the line is subject to the discretion of the person involved.

This method for the selection of a probability function is simple and quick, but carries with it a possibility of error, due to the inaccurate fit of the straight line through the points. In some cases it is possible to fit several best lines as shown in Figure 22.11. In short, subjective decision making leaves room for error.

The procedure for determination of numerical values for the distribution parameters A, B and C are different for each family of distribution. Hence, in the following sections, separate procedures for the exponential, normal, lognormal and Weibull distribution are given.

22.4 ANALYTICAL DETERMINATION OF DISTRIBUTION FUNCTION PARAMETERS

Regardless of the type of probability distribution involved, the procedure is the same for all cases where the analytical method is used. The main idea of this approach is to fit the linear regression line, $y = ax + b$, through empirical data by using the least squares method. This can be done only after the coordinates related to the numerical values of empirical data and their cumulative probabilities, t_i, $M'(t_i)$ where $i = 1, n$ have been linearized. The parameters a and b of the regression line can be determined according to the following expressions:

$$a = \frac{\sum_{i=1}^{n} t_i y_i - \left(\sum_{i=1}^{n} t_i \sum_{i=1}^{n} y_i \right)/n}{\sum_{i=1}^{n} t_i^2 - \left(\sum_{i=1}^{n} t_i \right)^2/n} \tag{22.13}$$

$$b = \sum_{i=1}^{n} y_i/n - a \times \sum_{i=1}^{n} t_i/n \tag{22.14}$$

In order to determine the strength of the relationship between the empirical data and the regression line it is necessary to calculate the coefficient of determination, CD, applying the following expression:

$$CD = \frac{\left(\sum_{i=1}^{n} t_i \times y_i - \left(\sum_{i=1}^{n} t_i \times \sum_{i=1}^{n} y_i \right)/n \right)^2}{\left(\sum_{i=1}^{n} t_i^2 - \left(\sum_{i=1}^{n} t_i \right)^2/n \right)\left(\sum_{i=1}^{n} y_i^2 - \left(\sum_{i=1}^{n} y_i \right)^2/n \right)} \tag{22.15}$$

Given that the value obtained for CD is satisfactory, it is necessary to find the relationship between parameters a and b of the regression line, and parameters A and B or C of the selected probability distribution. As this relationship is specific to each type of theoretical distribution, the appropriate expressions will be given below and illustrated by numerical examples.

In spite of the fact that this is an analytical method it is still possible to plot all the points defined by the empirical data on the coordinate system with equally spaced axes. Thus, there is no need for the use of special probability paper. As the plotting of data does not cause a great deal of trouble and serves as a good tool for checking the distribution of the points, this exercise is recommended.

For situations where there are 50 or less empirical data values, columns 4, 5, and 6 of Appendix Table 2 give the numerical values of the coordinates, t_i, together with values of their sum, $S3$, and the sum of their square values, $S4$. Appendix Table 3 should be used if more than 50 data values are available.

22.5 EXPONENTIAL DISTRIBUTION

22.5.1 Analytical method

In order to linearize coordinates, the following expression should be used:

$$t_{e_i} = \ln(1 - M'(t_i)) \tag{22.16}$$

$$Y_{e_i} = t_i \tag{22.17}$$

Calculated values for t_{e_i} where the total number of empirical data is less than 50 can be found in Appendix Table 2 under the heading $Xe(i)$, otherwise consult Appendix Table 3. The numerical value of the parameter A, the single parameter which fully defines the exponential distribution, can be found according to the following expression:

$$A = |a| + b \tag{22.18}$$

Example 22.3

The results tabulated represent the inspection time, in hours, observed for 14 aeroplanes of a particular make and type.

i	1	2	3	4	5	6	7	8	9	10	11	12	13	14
t_i	102	209	14	57	54	32	67	134	152	27	230	66	61	34

Check the hypothesis of exponential distribution using the analytical method and determine the numerical value for A.

Solution
The first step towards solving this problem is to classify all results into ascending order (from minimum to maximum value). For the determination of numerical values of t_{e_i}, the following two possibilities are available:

(a) Use the information given in Appendix Table 2, for $n = 14$.
(b) Calculate $M'(t_i)$ using equation 22.2 and then calculate t_{e_i} using equation 22.16.

Regardless of the method used, the results should be as shown in Table 22.1.

Table 22.1 Values for Example 22.3

i	$t_i = Y_{e_i}$	$Y_{e_i}^2$	$M'(t_i)$	t_{e_i}	$t_{e_i}^2$	$Y_{e_i} \times t_{e_i}$
1	14	196	0.04861	−0.0498	0.00248	−0.69720
2	27	729	0.11806	−0.1256	0.01578	−3.39120
3	32	1024	0.18750	−0.2076	0.04310	−6.64320
4	34	1156	0.25694	−0.2970	0.08821	−10.09800
5	54	2916	0.32639	−0.3951	0.15610	−22.52070
6	57	3249	0.39583	−0.5039	0.25392	−28.72230
7	61	3721	0.46528	−0.6260	0.39188	−38.18600
8	66	4356	0.53472	−0.7651	0.58538	−50.49660
9	67	4489	0.60417	−0.9268	0.85896	−62.09560
10	102	10404	0.67361	−1.1197	1.25373	−114.20940
11	134	17956	0.74306	−1.3589	1.84661	−182.09260
12	152	23104	0.81250	−1.6740	2.80228	−254.44800
13	209	43681	0.88194	−2.1366	4.56506	−446.54940
14	230	52900	0.95139	−3.0239	9.14397	−695.49700
	1239	169911	–	−13.21	22.00746	−1915.64720

The first step is to evaluate the coefficient of determination by applying equation 22.15, thus:

$$CD_e = \frac{(-1915.65 - (-13.21 \times 1239)/14)^2}{(22.007 - (174.5/14)(169911 - (1535121/14))} = 0.966$$

The high value of CD_e (maximum 1) means that the hypothesis about exponential distribution has not been proved incorrect. Hence, support time for the provision of spares for the aircraft can be modelled by the exponential distribution function.

Numerical values of the parameters a and b can be found by applying equations 22.13 and 22.14. Thus, $a = -78.26$ and $b = 14.66$. Consequently, by making use of equation 22.18, the value of the scale parameter $A = 92.9$, which is very close to the value obtained by using the graphical method in Example 22.4.

Graphical representation of this example is shown in Figure 22.12 where the straight line $Y_e = -78.26 \times t_e + 14.66$ is plotted through the empirical data.

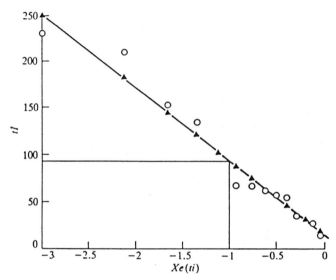

Figure 22.12 Plotted empirical data.

22.5.2 Graphical method

Probability paper for the exponential distribution is presented in Figure 22.13.

In order to determine the numerical value of the scale parameter, A, as the single parameter which defines the distribution, the following procedure is applied. From the point on the ordinate $M(x) = 0.632$ draw a horizontal line to the best fit line and then a vertical line down to the horizontal axis, where the numerical value for A can be read. This is shown in Figure 22.14 and is justified because the cumulative distribution function for the exponential distribution has a value of 0.632 for $x = A$, thus:

$$M(A) = 1 - \exp[-(A/A)] = 1 - \exp(-1) = 0.632$$

A particular member of the exponential distribution family is fully defined by fixing the value of A.

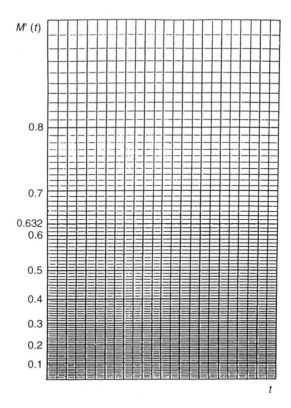

Figure 22.13 Probability paper for the exponential distribution.

Example 22.4

Using the empirical data given in Table 22.1 check the hypothesis of exponential distribution, determine its parameter and plot the functions $m(t)$ and $M(t)$, by applying the graphical method.

Solution
The first step towards solving this problem is to classify all results into ascending order (from minimum to maximum value). Then numerical values of $M'(t_i)$ should be determined. As the total number of results is less than 50, the median rank method must be used. The necessary values could be obtained either from Appendix Table 2, for $n = 14$, or calculated by applying equation 22.2. Values given below are extracted from Appendix Table 2.

The points with coordinates t_i, $M'(t_i)$, $i = 1, 14$, plotted on probability paper for the exponential distribution, are shown in Figure 22.15.

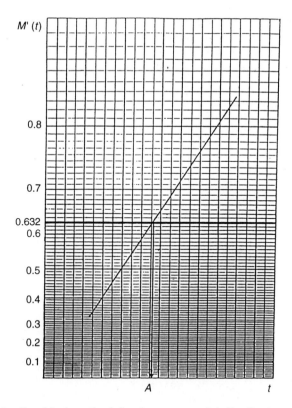

Figure 22.14 Graphical method for the exponential distribution.

Table 22.2 Values for Example 22.4

i	t_i	$M'(t_i)$
1	14.0	0.04861
2	27.0	0.11806
3	32.0	0.18750
4	34.0	0.25694
5	54.0	0.32639
6	57.0	0.39583
7	61.0	0.46528
8	66.0	0.53472
9	67.0	0.60417
10	102.0	0.67361
11	134.0	0.74306
12	152.0	0.81250
13	209.0	0.88194
14	230.0	0.95139

As the plotted points fall in a straight line there is no reason for the hypothesis to be rejected. After the best line is fitted through the points a numerical value of $A = 96$ hours is determined. Diagrams for $m(t)$ and $M(t)$ shown in Figure 22.16 are plotted by applying equations 4.7 and 4.8.

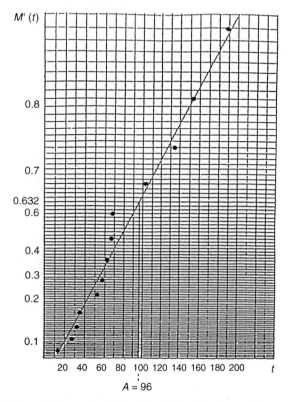

Figure 22.15 Graphical method for the determination of the parameter for the exponential distribution.

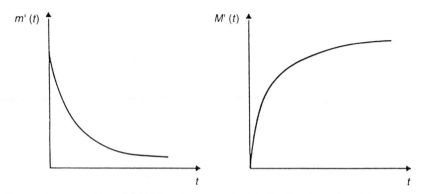

Figure 22.16 $m(t)$ and $M(t)$ for the exponential distribution, $A = 96$.

In conclusion, it can be said that:

(a) The hypothesis about exponential distribution has been proved correct.
(b) The maintenance time needed for the successful completion of the inspection for this particular type of aircraft could be represented by the exponential distribution with scale parameter $A = 96$ hours.
(c) It is possible to predict the duration of the maintenance time needed for the completion of a specified inspection task. For example, if we are interested to know what the probability is that this type of maintenance work will be accomplished within 59 hours, the following expression could be used:

$$P(DMT > 59) = M(59) = 1 - \exp(-(59/96)) = 0.54$$

This means that there is a probability of 0.54 (54%) that the required inspection will be accomplished within 59 hours of maintenance. However, this should not be accepted dogmatically, i.e. it should not be expected that exactly 54 inspection tasks out of 100 will be successfully completed within 59 operating hours.

22.6   NORMAL DISTRIBUTION

22.6.1   Analytical method

In the case of the analytical method, polynomial approximation for the determination of numerical values of t_{n_i} has been used by Abramovitz and Stegan (1964), thus:

$$t_{n_i} = \frac{p - c_0 + c_1 \times p + c_2 \times p^2}{1 + d_1 \times p + d_2 \times p^2 + d_3 \times p^3} \qquad (22.19)$$

where:

$$p = \sqrt{\ln[1/(1 - M'(t_i)^2]}$$

$$c_0 = 2.515517, \qquad c_1 = 0.802853, \qquad c_2 = 0.010328$$

$$d_1 = 1.432788, \qquad d_2 = 0.189269, \qquad d_3 = 0.001308$$

To facilitate calculations employing the above equation, column 4 of Appendix Table 2 presents numerical values for t_{n_i} where the total number of results is less than 50. (Appendix Table 3 should be consulted where there are more than 50 observed values.)

Coordinates on the vertical axis, Y_n, are fully defined by the numerical values of the empirical data, t_i, where $i = 1, n$. The specific member of the family can be found by making use of the following expressions:

$$A = b \qquad (22.20)$$

$$B = a \qquad (22.21)$$

Example 22.5

In a laboratory test of 19 washing machines, the maintenance time needed for the replacement of the drive belts was recorded. The data obtained (operating minutes) are presented below.

---

129 174 118 209 185 98 143 124 164 174 169 110 134 195 140 151 147 156 161

Assuming that we are dealing with a normal distribution, determine its parameters by applying the analytical method.

Solution
In order to determine numerical values for A and B it is necessary to calculate t_{n_i} and Y_{n_i}, $i = 1, 19$. Appendix Table 2 can be used for $n = 19$, or t_{n_i} can be calculated by using equation 22.19 where $M'(xi)$ should be determined by using the median rank method. The data obtained are presented in Table 22.3.

The coefficient of determination is then calculated as shown below.

$$CD_n = \frac{(523.92 - 0.22 \times 2886/19)^2}{(15.44 - 0.0484/19)(454114 - (2886)/19)} = 0.99$$

As the coefficient of determination is an extremely high value, there is no reason to reject the hypothesis of a normal distribution. Parameters A and B can be determined by using the above equations, thus:

$$B = a = \frac{523.92 - (0.22 \times (2886)/19)}{(15.44 - 0.0484/19)(454114 - (2886)/19} = 31.9$$

$$A = 2886/19 - 31.9 \times (0.22/19) = 151.9$$

Results obtained by the graphical method for the same set of empirical data (see Example 22.6) are very close to those calculated here.

A graphical representation of the empirical data analysed and plotted on an ordinary coordinate system is shown in Figure 22.17.

Table 22.3 Values for Example 22.5

i	$t_i = Y_{n_i}$	$Y^2_{n_i}$	$M'(t_i)$	t_{n_i}	$t^2_{n_i}$	$Y_{n_i} \times t_{n_i}$
1	98.0	9604	0.03608	−1.6784	2.82	−164.48
2	110.0	12100	0.08763	−1.3075	1.69	−143.83
3	118.0	13924	0.13918	−1.0590	1.12	−124.96
4	124.0	15376	0.19072	−0.8610	0.74	−106.76
5	129.0	16641	0.24227	−0.6907	0.48	−98.10
6	134.0	17956	0.29381	−0.5374	0.29	−72.01
7	139.0	19321	0.34536	−0.3952	0.16	−54.93
8	143.0	20449	0.39691	−0.2600	0.07	−37.18
9	147.0	21609	0.44845	−0.1291	0.02	−18.98
10	151.0	22801	0.50000	0.0000	0.00	0.00
11	156.0	24336	0.55155	0.1293	0.02	20.17
12	161.0	25921	0.60309	0.2609	0.07	42.00
13	164.0	26896	0.65464	0.3974	0.16	65.17
14	169.0	28561	0.70619	0.5419	0.29	91.58
15	174.0	30276	0.75773	0.6987	0.49	121.57
16	179.0	32041	0.80928	0.8751	0.77	162.77
17	186.0	34596	0.86082	1.0841	1.18	201.74
18	195.0	38025	0.91237	1.3557	1.84	264.36
19	209.0	43681	0.96392	1.7985	3.23	375.89
	2886.0	454114		0.2200	15.44	523.92

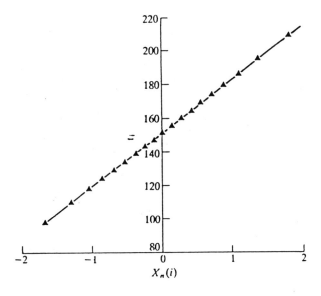

Figure 22.17 Plotted points for the normal distribution.

22.6.2 Graphical method

Probability paper for the normal distribution is shown in Figure 22.18.
   In the case of the normal distribution, the probability that the random variable will have a value less than or equal to scale parameter A is 0.5. Thus, $P(DMT \leq A) = M(A) = 0.5$. The numerical value of the scale parameter for the empirical data under consideration is the value of the point of intersection of the adopted best fit line and the horizontal line which corresponds to a cumulative probability of 0.5, denoted as $t_{0.5}$. The above statement can be mathematically expressed as follows:

$$A = t_{0.5} \quad \text{for which} \quad M(a) = 0.5 \tag{22.22}$$

   The numerical value of the shape parameter, B, is equal to the difference between the values of the abscissa whose ordinates have values 0.5 and 0.16 to 0.84 and 0.5. Thus:

$$B = DMT_{0.50} - DMT_{0.16} = DMT_{0.84} - DMT_{0.50} \tag{22.23}$$

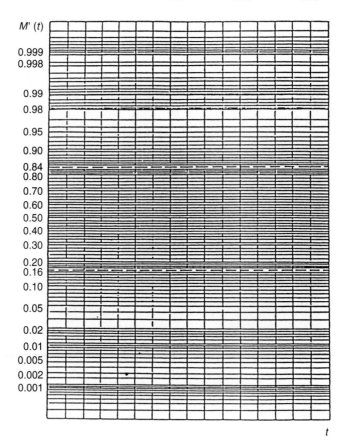

Figure 22.18 Probability paper for the normal distribution.

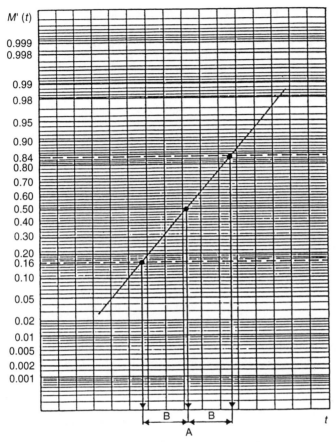

Figure 22.19 Graphical methods for determination of the parameters of the normal distribution.

The procedure for the determination of the numerical values for parameters A and B using the graphical method is presented in Figure 22.19.

Example 22.6

Assuming that the observed, empirical data given in Table 22.3 can be modelled by the normal distribution, determine the parameters which define it and calculate the probability that the belt will not be replaced in the first 125 minutes.

Solution
In order to solve the first part of this task the numerical values of the cumulative distribution function, $M'(t_i)$, for $i = 1, 19$ must be calculated.

Table 22.4 Values for Example 22.6

i	t_i	$M'(t_i)$
1	98.0	0.03608
2	110.0	0.08763
3	118.0	0.13918
4	124.0	0.19072
5	129.0	0.24227
6	134.0	0.29381
7	139.0	0.34536
8	143.0	0.39691
9	147.0	0.44845
10	151.0	0.50000
11	156.0	0.55155
12	161.0	0.60309
13	164.0	0.65464
14	169.0	0.70619
15	174.0	0.75773
16	179.0	0.80928
17	186.0	0.86082
18	195.0	0.91237
19	209.0	0.96392

As the total number of results is less than 50, equation 22.2 should be used. The data calculated are presented in Table 22.4.

This data, plotted on probability paper for the normal distribution, is shown in Figure 22.20. As the data form a straight line a normal distribution can be assumed. Based on the position of the best line of fit, parameters A and B are found to be $A = 152$, $B = 32$ minutes.

The cumulative distribution function can now be plotted by making use of equation 4.16 as shown in Figure 22.21.

The probability that the drive belts will not be replaced in the first 125 minutes of maintenance can be found by using the expression:

$$P(DMT > 125) = 1 - P(DMT < 125) = 1 - M(125) = 1 - \Phi\left(\frac{125 - 152}{32}\right)$$

$$= 0.8$$

The same result can be obtained using the function plotted in the Figure 22.21.

22.7 LOGNORMAL DISTRIBUTION

22.7.1 Analytical method

The process of determining numerical values of parameters which define the lognormal distribution is similar to the process for the normal

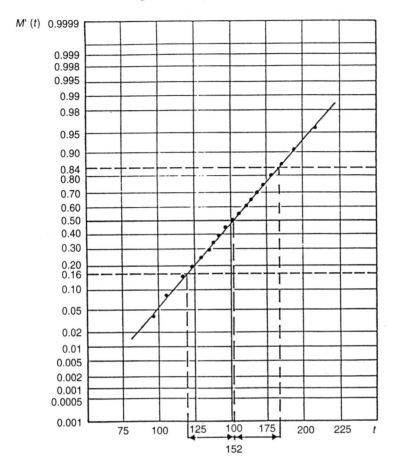

Figure 22.20 Graphical method for the determination of the normal distribution.

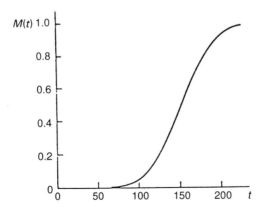

Figure 22.21 Graphical presentation of $M(t)$.

distribution, which has already been explained. The linearized values of t_l can be obtained in the same way as Xn, whereas for the values of Y_l the following expression should be used:

$$Y_l = \ln(t_i) \qquad (22.24)$$

The scale parameter, A, and shape parameter, B, can be determined by using the following relationships:

$$A = b \qquad (22.25)$$

$$B = a \qquad (22.26)$$

Example 22.7

Five copies of a newly designed engine were exposed to the disassembly and assembly process in order to determine the maintainability measures associated with this maintenance task. The following table represents the observed data, in seconds, from five different maintenance facilities.

i	1	2	3	4	5
t_i	2600	3800	5100	7700	10200

(a) Can the data be represented by the lognormal distribution?
(b) If so, estimate the expected operating time of that particular component.

Solution
The answer to the first task can be found by determining the numerical value for CD. Making use of equation 22.4 and the values established in Table 22.5, it can be found that $CD_l = 0.9918$.

Table 22.5 Values for Example 22.7

i	t_i	Y_{l_i}	$Y_{l_i}^2$	$M'(t_i)$	t_{l_i}	$t_{l_i}^2$	$Y_{l_i} \times t_{l_i}$
1	2600	7.8632	61.8299	0.12938	−1.1002	1.2104	−8.65109
2	3800	8.2427	67.9421	0.31541	−0.4784	0.2321	−3.94330
3	5100	8.5369	72.8786	0.50000	0.0000	0.0000	0.00000
4	7700	8.9489	80.0828	0.68519	0.4818	0.2321	4.31158
5	10200	9.2301	85.1947	0.87037	1.1282	1.2728	10.41339
	29400	34.8218	367.9281	–	−0.2234	2.9474	−2.13057

As the answer to part (a) is positive, parameters A and B must be evaluated in order to estimate the expected operating life. This can be done by making use of equations 22.25 and 22.26 and the values obtained are $A = 8.546$ and $B = 0.6254$.

The expected duration of the maintenance task can be obtained using equation 6.28, thus:

$$E(DMT) = MDMT = \exp(A + 0.5B^2) = \exp(8.546 + 0.157) = 6373.9$$

Figure 22.22 shows the empirical data plotted on the ordinary coordinate system.

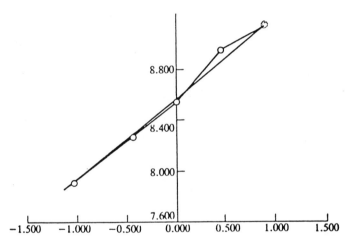

Figure 22.22 Graphical presentation of Example 22.7.

22.7.2 Graphical method

Figure 22.23 shows the probability paper for the lognormal distribution. The similarity with probability paper for the normal distribution is obvious.

According to the description of the lognormal distribution given in Chapter 4, it is reasonable to expect that the process of determining numerical values for its parameters is similar to the process already explained for the normal distribution.

The points of interest are the numerical values of the horizontal axis which correspond to the numerical values of the intersection of the best line and the cumulative probabilities of 0.16, 0.5 and 0.84. Once these

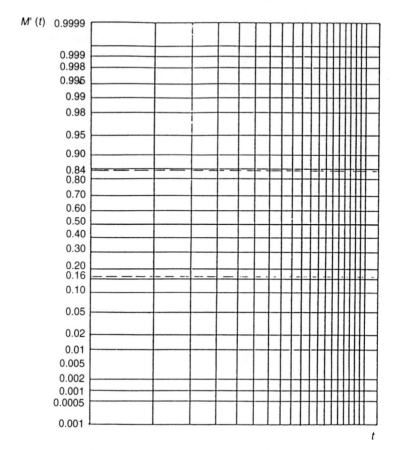

Figure 22.23 Probability paper for the lognormal distribution.

values are determined the distribution parameters can be obtained according to the following expressions:

$$A = \ln(DMT_{50}) \tag{22.27}$$

$$B = \ln(1/2(DMT_{50}/DMT_{16} + DMT_{84}/DMT_{50})) \tag{22.28}$$

where DMT_{50} is the value of t for which $M(t) = 0.5$, etc. The whole procedure is illustrated by Figure 22.24.

Example 22.8

Can the empirical data given in Example 22.5 be modelled by the lognormal distribution? If so determine the expected value of the duration of this particular maintenance task by applying the graphical method.

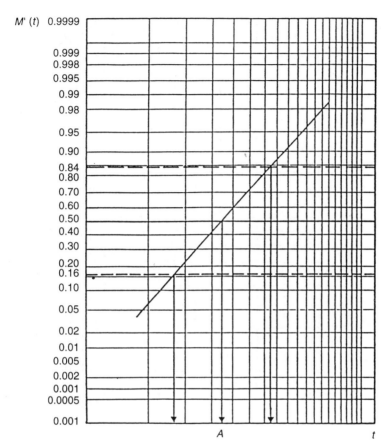

Figure 22.24 Graphical method for the determination of parameters for the lognormal distribution.

Solution

In order to check the above hypothesis, it is necessary to plot the points with coordinates t_i, $M'(t_i)$, $i = 1, 5$ on probability paper for the lognormal distribution. According to the distribution of the points shown in Figure 22.25 there is no reason for rejecting the lognormal distribution.

In order to calculate the mean value it is necessary to determine numerical values for parameters A and B. From Figure 22.25, the numerical values obtained are:

$$A = \ln(5150) = 8.546$$

$$B = \ln(1/2(5150/2970 + 9125/5150)) = 0.561$$

Thus, the expected value of the operating time to the failure of the component considered is:

$$E(DMT) = \exp(A+0.5B^2) = \exp(8.546+0.157) =$$
$$6025 \text{ seconds} = 100.4 \text{ hours}$$

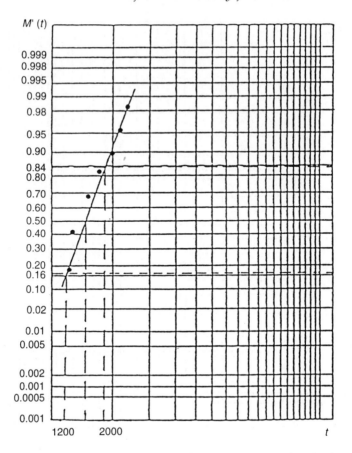

Figure 22.25 Graphical presentation of the method used for the lognormal distribution.

22.8   WEIBULL DISTRIBUTION

22.8.1   Analytical method

In order to determine the parameters which define a Weibull distribution it is necessary to linearize coordinates defined by numerical values of the random variable considered and the corresponding cumulative probability in the following way:

$$t_{w_i} = \ln[\ln(1/(1 - M'(t_i))] \qquad (22.29)$$

$$Y_{w_i} = \ln(t_i) \qquad (22.30)$$

Calculated values for t_{e_i} are to be found in column 5 of Appendix Table 2 for cases where the total number of empirical data is less than 50. Once the parameters a and b of the line of best fit are determined, it is possible to calculate the distribution parameters A and B by using the following equations:

$$A = \exp(b) \qquad (22.31)$$

$$B = 1/a \qquad (22.32)$$

Example 22.9

The results tabulated represent the duration of the adjustment of geometry of the front wheels of a specific motor vehicle, in minutes, observed during 11 trials as a part of a maintainability demonstration process:

Spring no.	1	2	3	4	5	6	7	8	9	10	11
No. of cycles	39	33	98	42.7	140	56.4	129	132	37	17	21

According to experience with this type of maintenance task of the previous design configuration, the maintenance time can be modelled by the Weibull probability distribution. Check this hypothesis.

Solution
Making use of Appendix Table 2 the data presented in Table 22.6 were obtained.

Table 22.6 Values for Example 22.9

i	t_i	$Y_{w_i} = \ln(t_i)$	$Y_{w_i}^2$	$F'(t_i)$	t_{w_i}	$t_{w_i}^2$	$t_{w_i} \times Y_{w_i}$
1	17	7.44	55.33	0.06140	−2.7588	7.61	−20.53
2	21	7.65	58.52	0.14912	−1.8233	3.32	−13.95
3	33	8.10	65.64	0.23684	−1.3083	1.71	−10.60
4	37	8.22	67.50	0.32456	−0.9355	0.88	−7.69
5	39	8.27	68.37	0.41228	−0.6320	0.40	−5.23
6	42.7	8.36	69.88	0.50000	−0.3665	0.13	−3.06
7	56.4	8.64	74.61	0.58772	0.1210	0.01	−1.05
8	98	9.19	84.46	0.67544	0.1180	0.01	−1.08
9	129	9.46	89.59	0.76316	0.3649	0.13	3.45
10	132	9.49	90.02	0.85088	0.6434	0.41	6.11
11	140	9.55	91.14	0.93860	1.0261	1.05	9.80
	−	94.37	815.06	−	−5.7978	15.66	−41.66

According to equation 22.15 the coefficient of determination in this case is $CD_w = 0.912$. Solving this problem by the analytical method the following values were obtained for parameters a and b:

$$a = \frac{-41.67 - (5.79 \times 94.37)/11}{15.66 - (33.52)/11} = 0.63$$

$$b = 94.37/11 - 0.63(-5.79)/11 = 8.91$$

Therefore, from the above expressions, the scale parameter, A, and the shape parameter, B, take the values $A = \exp(8.91) = 7700$ and $B = 1/0.63 = 1.59$, see Figure 22.26.

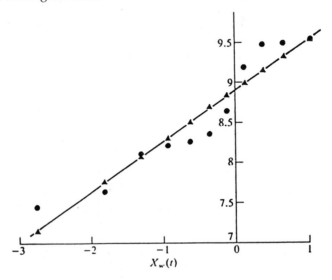

Figure 22.26 Graphical presentation of Example 22.9.

22.8.2 Graphical method

In spite of the fact that several different probability papers for the Weibull distribution exist, they are all based on the same principle and all give the same values for parameters A, B and C. Only one type of paper is shown in Figure 22.27.

After the line of best fit is plotted the parameters which define the distribution can be found by the following procedure:

(a) Draw a vertical line down from the point of intersection of the best fit line with ordinate $M(t) = 0.632$, and read the value for A on the horizontal axis, as shown in Figure 22.28. This holds true because:

$$M(A) = 1 - \exp[-(A/(A)^B] = 1 - \exp[-(1)^B] = 0.632$$

(b) Draw a line parallel to the line of best fit through the point O_1 and the value for B can be read on the scale above.

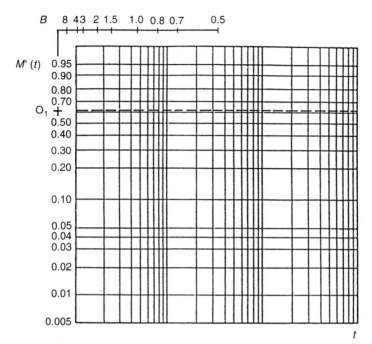

Figure 22.27 Probability paper for the Weibull distribution.

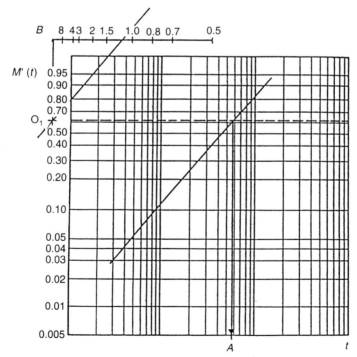

Figure 22.28 Graphical presentation of the graphical method.

Example 22.10

This example uses the same data relating to adjustment of the front wheels of a motor vehicle as Example 22.9. According to experience with this type of maintenance task of the previous design configuration, the maintenance time can be modelled by the Weibull probability distribution. Check this hypothesis, plot the maintainability function and determine the duration of the maintenance task which corresponds to the probability of adjustment completion of 0.85.

Solution

Applying equation 22.2, the numerical values for $M'(t_i)$ were calculated as shown in Table 22.7.

Table 22.7   Values for Example 22.10

i	t_i	$M'(t_i)$
1	17	0.06140
2	21	0.14912
3	33	0.23684
4	37	0.32456
5	39	0.41228
6	42.7	0.50000
7	56.4	0.58772
8	98	0.67544
9	129	0.76316
10	132	0.85088
11	140	0.93860

In order to apply the graphical method it is necessary to plot the empirical data on Weibull probability paper. Figure 22.9 clearly shows that the plotted points form a straight line, and there is no reason to reject the hypothesis made.

In order to determine the parameters A and B, the procedure described above should be applied. From Figure 22.29 the numerical values obtained are $A = 77$ and $B = 1.6$. The maintainability function shown in Figure 22.30 can be used to determine the length of the maintenance time up to which the adjustment of the geometry of the wheels will be completed with a probability of 0.85.

If the plotted points on Weibull probability paper form a curve this can mean that we are dealing with a three-parameter Weibull distribution, i.e. that the source parameter C is not zero.

The process to determine a numerical value of C is an iterative one which starts with an arbitrarily selected value for C, which obviously must be less than the minimum result obtained. The idea is to try different feasible values for C until something approximating to a

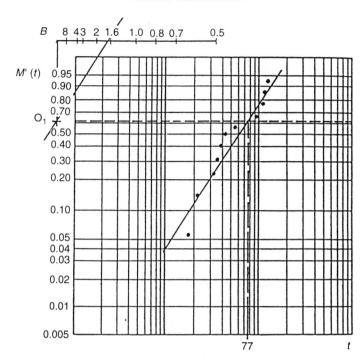

Figure 22.29 Graphical presentation of the example considered.

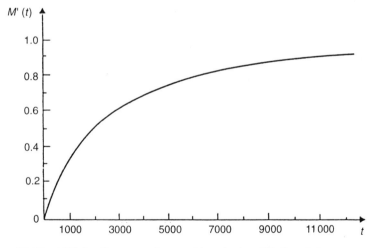

Figure 22.30 $M(t)$ for the example considered, $A = 77$, $B = 1.6$.

straight line is arrived at. If C is underestimated or overestimated there will be a tendency towards curvature in the directions indicated in Figure 22.31, and adjustments should be made in the appropriate direction.

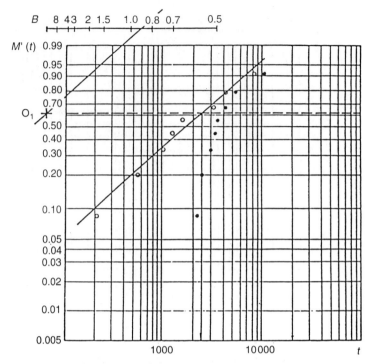

Figure 22.31 Graphical method for determination of parameters.

According to some authors the best arbitrary value for the source parameter is $C = 0.9(t_1)$. When the value for C is chosen it is necessary to replot all points again, using the following coordinates, $(t'\ M'(t'))$ where $t' = (t - C)$. If the data follow the Weibull distribution the points will lie in a straight line. As explained, this is an iterative process and sometimes it will be necessary to do a few corrections until the best value for C is found. The determination of the other parameters is identical to the case of the two-parameter distribution. In the case of the three-parameter Weibull distribution an explanation for the existence of the source parameter must be considered.

Example 22.11

The empirical data tabulated below represent the time for restoration of the clutch for eight tractors.

Tractor no.	1	2	3	4	5	6	7	8
Repair time (min.)	330	220	630	300	350	1040	255	520

Determine the probability of replacing the clutch in 300 minutes.

Solution
Numerical values of $M'(t_i)$ were determined using the method explained for the Weibull distribution (see Table 22.8) and plotted as shown in Figure 22.32.

Table 22.8 Values for Example 22.11

i	t_i	$M'(t_i)$
1	220.0	0.08333
2	255.0	0.20238
3	300.0	0.32143
4	330.0	0.44048
5	350.0	0.55952
6	520.0	0.67857
7	630.0	0.79762
8	1040.0	0.91667

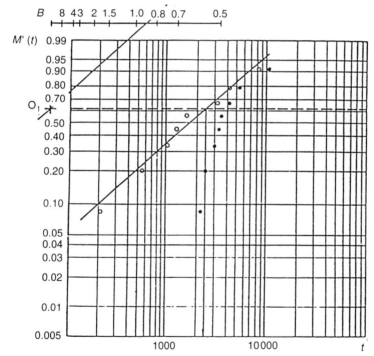

Figure 22.32 Graphical method for the three parameter Weibull distribution.

It is clear that the points do not follow a straight line, so the hypothesis of a two-parameter Weibull distribution has to be rejected.

Assuming that there is a basis for the existence of a third parameter, we will adopt a value of 200 = 0.9 (220) for the minimum value parameter. The recalculated values of t' are given in the table below:

i	1	2	3	4	5	6	7	8
t_i	220	255	300	330	350	520	630	1040
t'_i	20	55	100	110	250	320	430	840

Points with coordinates t'_i, $M'(t'_i)$ are replotted on the Weibull probability paper as shown in Figure 22.32. As the new distribution of the points, marked by circles, forms a straight line, the hypothesis of a three-parameter Weibull distribution can be accepted. The numerical values of the scale and shape parameters are found in the usual way, thus $A = 448$, $B = 0.9$.

The probability of replacing a clutch in 300 minutes can now be calculated, according to equation 3.35.

$$P(\text{replacement} < 300) = \exp(-[(300-200)/(448-200)]^{0.9}) = 0.643$$

Thus, there is a chance of 68% that the clutch will be replaced within 300 minutes of maintenance.

22.9 COMPLEX DISTRIBUTION

In situations where the plotted points on probability paper do not form a straight line, and a straight line cannot be obtained after several trials for three parameter distribution, it can mean that we are dealing with a complex distribution. This means that the empirical data consists of two or more subsets of data, related to different variables. This is related to the durations of maintenance task which are performed under different conditions (peace/war), with use of different resources (like tools, equipment, facilities and similar).

A typical example of complex distribution for a hypothetical probability is presented in Figure 22.33a. In this case all points related to the empirical data must be separated into two or more subsets. Each subset is defined by the points through which a straight line can be fitted (see Figure 22.33b).

After the separation the same procedure for determination of probability distribution explained above must be applied to each set of data.

Example 22.12

In order to determine the impact of the development of the test equipment on the duration of a specific maintenance task, the operator tested the execution of the task considered under identical conditions in

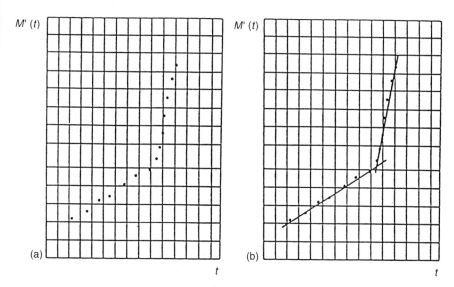

M' (t)   M' (t)

(a)   (b)

t   t

Figure 22.33 Complex distribution.

two different maintenance workshops. The data obtained are presented together in the table below.

196	198	199	214	219	226	239	251	263
266	269	273	283	288	292	293	299	315
389	397	412	423	444	451	458	462	466
469	475	479	483	485	488	492	495	499
512	516	527	538	542	551	562	567	580

Solution

Applying the method explained, the empirical data are plotted on Weibull paper as shown in Figure 22.34. It is clear that the distribution of data does not follow a straight line and that it is possible to divide them into two subsets, as shown in Figure 22.35.

In order to determine the probability distribution of the first subset of data, the points defined by the following data:

196	198	199	214	219	226	239	251	263	269	273	283	288	292	293	299	315

are replotted on separate probability paper and the values obtained are $A = 267.03$, $B = 2.5$ and $C = 166.6$.

The same procedure is then repeated for the second subset of data and the calculated values are $A = 506$, $B = 3.7$ and $C = 39.2$.

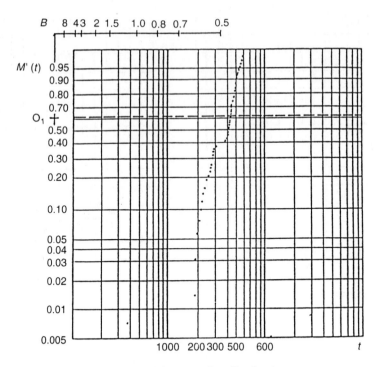

Figure 22.34 Graphical method for complex distribution.

Clearly, there is an impact of the duration of the maintenance task caused by the development of the new test equipment, as illustrated by the maintainability measures shown in Table 22.9.

22.10   HYPOTHESIS TESTING

According to the algorithm given in Figure 22.1, the last phase of this procedure is to test the hypothesis made. The main objective here is to determine how close the empirical probability distribution is to the postulated theoretical distribution. The hypothesis can be tested by graphical and analytical methods. Regardless of the method used for testing the hypothesis, one of two outputs is possible:

(a) Acceptance with certain confidence, which means that the test performed has not rejected the hypothesis made and the theoretical distribution function under consideration can be used as the interpretation of the behaviour of the population defined by the random variable.

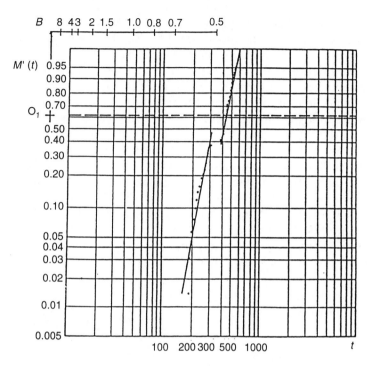

Figure 22.35 Graphical method for the first set of data.

Table 22.9

Test equipment	Original	New
MDMT	460.4	255.7
SDDMT	126.9	38.1
DMT_{10}	293.3	207.4
DMT_{50}	461.9	253.3
DMT_{90}	667.1	322.4

(b) Rejection with certain probability, which means that there are strong discrepancies between the pattern of distribution of the empirical data analysed and the theoretical distribution considered.

22.10.1 Graphical method

This method for testing the hypothesis is applicable to all cases where the graphical method has been used for determination of the parameters of the theoretical probability distribution.

As has already been mentioned, the rank distribution exists for any level of probability. As the median rank is used for the determination of the numerical values of $M'(t_i)$, very often the 5 and 95% rank are used to obtain confidence limits on probability paper (Appendix Table 2). By plotting the best line through the points defined by median rank we assume that this line represents the true probability distribution. Any deviation from this straight line can be regarded as an error of that particular data. Thus, any point which does not lie on a straight line can be projected horizontally to it and then the 5 and 95 ranks are related to the straight line. If the confidence limits are projected about the original sample points, the confidence limits will take an irregular rather than a smooth curve. The area between the 5 and 95% curves presents a 90% confidence band about the theoretical distribution function. Thus, if all the data points lie within the envelope formed by the confidence limits then there is no reason for rejecting the distribution as a reasonable model for the empirical data, Figure 22.36.

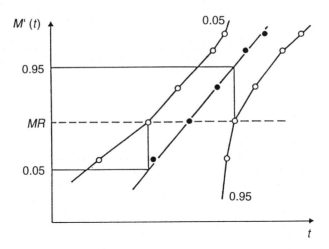

Figure 22.36 Confidence limits.

Example 22.13

Let us assume that empirical data of a random variable considered are as follows:

i	1	2	3	4	5	6	7
t_i	75	82	93	99	102	121	137

As the main objective of this example is to demonstrate the procedure for hypothesis testing, using the graphical method, the probability paper of some hypothetical probability distribution will be used.

For a sample size 7 values for 50 (median rank), 5 and 95% ranks are obtained from Appendix Table 2 as listed in Table 22.10 and shown diagrammatically in Figure 22.37.

Table 22.10 Rank values for sample size 7

i	1	2	3	4	5	6	7
t_i	75	82	93	99	102	121	137
5	0.00730	0.05337	0.12876	0.22532	0.34126	0.47830	0.65184
50	0.09428	0.22849	0.36412	0.50000	0.63588	0.77151	0.90572
95	0.34816	0.52070	0.65874	0.77468	0.87124	0.94662	0.99270

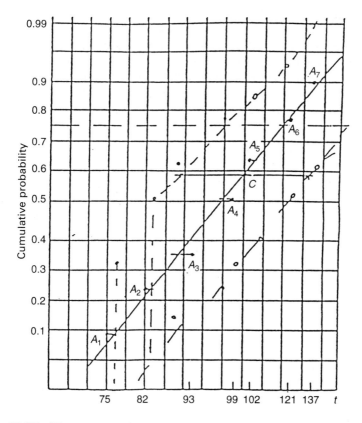

Figure 22.37 Diagrammatical representation of confidence limits.

Point A_1 denotes the plot obtained from the lowest value of random variable *DMT*, i.e. t_i, and corresponding value of *MR*. Thus, point A_1 is fully defined by coordinates 75 and 0.09428. The point A_2 is defined by 82 and 0.22849, and the other five points are defined by the values given in Table 22.9. According to the algorithm given in Figure 22.1, the straight line through the points Ai, $i = 1, 7$, should be fitted next, as shown in Figure 22.37. The plotted best line misses point A_1. Thus, to plot confidence limits about the theoretical distribution function, point A_1 must be projected horizontally to the best line. This projection defines point A_1 as shown in Figure 22.37.

It can be seen from Table 22.9 that the 5 and 95 ranks for the first point are 0.09428 and 0.34816 respectively. These values should be projected vertically to the best line at point B_1. If the same procedure is repeated for all seven points the confidence band would be obtained (see Figure 22.37). In order to clarify the obtained confidence interval let us consider point $C = 100$ on the abscissa. It can be said that we are 90% confident that between 34 and 81% of the population will have a duration of the maintenance task considered less than or equal to 100 minutes. The large width of this interval can be attributed to the small number of results.

The graph can be used as a source for other kinds of information. For example, if we are interested in the range of values within which, say, 60% of results will have a value with a confidence of 90%, we should look at the line defined by probability 0.60 on the ordinate. From Figure 22.37 it can be concluded that the interval in question is defined by 90.5 and 138.5.

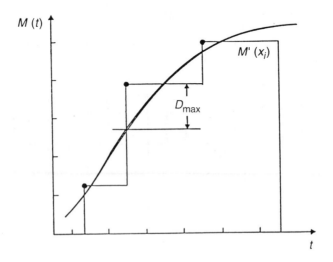

Figure 22.38   Illustration of the Kolmogorov–Smirnov test.

22.10.2 Analytical method

One of the most frequently used methods for hypothesis testing is the Kolmogorov–Smirnov test. This involves comparing the empirical cumulative distribution function, $M'(t_i)$, with the theoretical one defined by the distribution parameters. The latter should be calculated for each empirical data point $M(t_i)$, $i = 1$, n. The essence of this test is the determination of the absolute differences (d_i) between the following two maintainability functions:

$$d_i = \{| M(t_i) - M'(t_i) |, \ i = 1, n\}$$

as illustrated in Figure 22.38.

The difference with the highest numerical value should be identified, noted as D_{max}, and compared with the value assigned to the level of confidence in percentage terms given in Appendix Table 5. If the value obtained for the maximum difference is less than the recommended value p_r, then the hypothesis made can be accepted with the chosen confidence level.

$$D_{max} = \{d_i, \ i = 1, n\}_{max} < p_r$$

CHAPTER 23

Maintainability demonstration test

In some contracts, mainly defence based, the user of the system (Department of Defence, government agency, industrial customers and similar) specifies the requirements for demonstration of maintainability. Cases when demonstrations of maintainability have been required have usually involved functional items or functional assemblies. Where maintainability demonstration is actually a procurement requirement, the maintainability function takes a key part in negotiating the exact terms and conditions of the demonstration requirement. In addition, the maintainability personnel of both the buyer and seller usually participate in the actual maintainability demonstration testing. Since a substantial portion of profit may be at stake, maintainability demonstration testing usually receives a great deal of corporate attention.

Thus, during the design, acquisition and operational phase of a system/product a large number of maintainability tests and predictions are conducted by maintainability engineers and managers in order to collect data relative to the length of time needed for the successful completion of the maintenance task considered. Thus, the final product of this effort is a set of numbers, denoted as dmt_i where $i = 1, n$, each of which represents a length of time needed for the successful completion of the task analysed, when it is performed as specified. These data are the starting point for the statistical inference about maintainability measures.

The measures of maintainability addressed in Chapter 12 provide very useful information for design, operation and maintenance engineers relative to the planning of logistic support resources (personnel, tools, equipment, facilities and similar), provision of which has a great impact on the logistic delay time and consequently the operational availability of a product or system.

Thus, the main objective of this chapter is to present current maintainability demonstration methods used for the assessment of achieved maintainability measures, based on existing empirical maintainability data.

Demonstration of maintainability at the time of acceptance normally concentrates on repair (corrective maintenance tasks) rather than preventive maintenance tasks. It is usually easy to compare the contractor's recommended preventive maintenance with the requirement's constraints on downtime, frequency, maintenance personnel demands and so forth by measuring times or by using the contractor's estimates.

Demonstration testing of the durations of repair maintenance tasks consists of performing a number of typical repair tasks and measuring the time each one takes. The distribution of observed repair times is then compared with accept/reject criteria based on the repair times in the requirements specification.

It is necessary to point out that maintainability demonstrations require the following:

- The test must be on a sample of the fixed final build standard.
- The test applies to the same level of repair as the requirement's repair times (e.g. first line) and must therefore use the same repair as the requirement's repair times (e.g. first line) and must therefore use the same repair concept.
- The test conditions must be representative, and equipment/tools, maintenance manuals, tools, lighting and similar factors must be carefully considered.
- The repairs should be conducted by a mix of repairers representative in skills, training and experience of those who would do the actual repair in service.
- Maintenance tasks must be on a mix of failures representative of the proportions expected to occur in operation. The test personnel must have no advance knowledge of what they are required to undertake.
- The failures must be introduced in a safe way.

23.1 POSSIBLE APPROACHES TO ANALYSIS OF EXISTING DATA

Statistical inference is, generally speaking, a process of drawing conclusions about an entire population of similar objects, events or tasks based on a sample of a few. The two following approaches to statistical inference are mainly used (Knezevic, 1995a).

(a) Parametric, which is primarily concerned with inference about certain summary measures of distributions (mean, variance and similar). This approach is based on explicit assumptions about the normality of population distributions and parameters.
(b) Distribution, which is concerned with inference about entire probability distribution, free from the assumptions regarding the parameters of the population sampled.

Although both approaches are applicable to the analysis of obtained empirical data, currently in the defence-orientated applications the parametric method is used. Hence, this method will be addressed in this chapter, whereas the distribution approach was presented in Chapter 22 as the method which extracts more information from the existing data, and as such should be known to maintainability experts.

23.2 PARAMETRIC APPROACH TO MAINTAINABILITY DATA

Following the main statistical principles regarding the parametric approach which are based on the central limit theorem (Blanchard, 1991), in today's maintainability engineering practice, the numerical value of the mean duration of maintenance task, $MDMT^*$, of a particular sample size N, is computed according to the following expression:

$$MDMT^* = \sum_{i=1}^{N} \frac{dmt_i}{N} \qquad (23.1)$$

As the result obtained represents the mean value of this particular sample, which has been selected at random, it is necessary to determine the interval within which the mean of the entire population lies. Thus, if one is prepared to accept the chance of being wrong, say 10% of the time, which corresponds to the 90% confidence limit, then the upper limit of the mean time to repair, $MDMT^u$, should be determined according to the following equation (Blanchard, 1991):

$$MDMT^u = MDMT^* + z(SDDMT'/\sqrt{N}) \qquad (23.2)$$

where $SDDMT'$ represents the standard deviation of the obtained empirical data, $SDDMT'/\sqrt{N}$ is known as a standard error, and the value of z is selected from the table for the normal distribution based on the confidence level desired (see Appendix Table 1). In practice this means that, say for $z = 1.28$, there is a 90% chance that the $MDMT$ of the entire population is less than the value obtained for $MDMT^u$.

Once the numerical values for $MDMT^*$ and $SDDMT'$ have been calculated, according to this approach, the maximum maintenance time, M_{max} could be obtained according to the following expression (Patton, 1988; Blanchard, 1991):

$$M_{max} = \text{antilog}(\log MDMT^* + 1.65 \times SDDMT'_{\log(dmt_i)}) \qquad (23.3)$$

where $SDDMT'_{\log(dmt_i)}$ is the standard deviation of the logarithm of the initial values of dmt_i, for $i = 1, 2, \ldots, n$. The expression for the calculation of $SDDMT'_{\log(dmt_i)}$ could be found in Blanchard (1991). According to Knezevic (1994), analysis of specific maintainability data regarding M_{max} has shown its variations of 2.4 to 4 times of the $MDMT^*$, depending on the standard deviation.

However, being aware of the existing body of statistical knowledge available and its large number of applications in many different scientific disciplines, it is necessary to stress that:

(a) Equation 23.3, which is used in practical maintainability calculations, is lacking theoretical justification for universal application, and as such it has to be treated with extreme caution in daily engineering practice.
(b) Maintainability measures like maintainability function, restoration success and similar cannot be calculated at all if this approach is adopted.

In spite of these limitations the parametric approach, described above, has been: (a) fostered by existing military standards; (b) promoted by technical literature (Patton, 1988; Blanchard, 1991; Knezevic, 1994); and (c) adopted as the main contractual requirement regarding maintainability issues by many sophisticated customers/users in a large number of projects.

Example 23.1

The mean corrective replacement time, DMT^c, requirement for an equipment item is 55 minutes and the established risk factor is 10%. A maintainability demonstration is accomplished and yields the following results for the 50 tasks demonstrated.

34	82	36	63	30	32	52	48	86	36
30	67	71	96	45	58	82	32	56	58
39	57	37	51	44	33	31	42	33	36
42	43	54	35	47	40	53	32	50	30
31	91	75	74	67	73	49	62	64	62

Did the equipment item pass the maintainability demonstration?
If the maintainability demonstration test was successful, plot the corresponding maintainability function and graphically determine DMT_{10}, DMT_{50} and DMT_{90}.

Solution
The range of observations is 96 through 30. The number of intervals are:

$$K = 1 + 3.3\log_{10} N$$

where $N = 50$, $K = 6.6$.
The length of each interval (*LI*) is:

$$LI = (96 - 30)/6.6 = 9.69$$

The intervals are:

$$x_{min_1} = 32 + (1 - 1) \times 9.69 = 32$$
$$x_{max_1} = 32 + 9.69 = 41.69$$
$$x_{max_2} = 41.69 + 9.69 = 51.38$$
$$x_{max_3} = 51.38 + 9.69 = 61.07$$
$$x_{max_4} = 61.07 + 9.69 = 70.76$$
$$x_{max_5} = 70.76 + 9.69 = 80.45$$
$$x_{max_6} = 80.45 + 9.69 = 90.14$$
$$x_{max_7} = 90.14 + 9.69 = 99.83$$

The mean value is calculated as:

$$MDMTC = \sum_{i=1}^{50} \frac{dtm_i}{50} = \frac{2571}{50} = 51.42$$

The standard deviation (SD) is calculated as follows:

$$SD = \sqrt{\frac{\sum_{i=1}^{n} (dmt_i - MDMTC)^2}{N - 1}}$$

$$= \sqrt{\frac{15876}{49}} = 18.01$$

For a 10% risk factor, which equates to a 90% confidence factor, the standardized variable = 1.28.

Calculating the upper limit:

$$UL = 51.42 + 1.28 \times \frac{SD}{\sqrt{N}}$$

$$= 51.42 + (1.28) \times \left(\frac{18.01}{7.07}\right)$$

$$= 54.6$$

Given the requirement of 55 minutes, with an upper limit calculated from the data of 54.6 minutes, it can be concluded that the item will pass the maintainability demonstration.

23.3 COMPARISON BETWEEN PARAMETRIC AND DISTRIBUTION METHODS

The following example will be used in order to compare the possible methods for analysis of existing empirical data. Thus, let us assume that the empirical data related to the length of time needed for the successful completion of a certain maintenance task, related to two design alternatives of the same item, say A and B, are collected during the maintainability tests and are as presented in Table 23.1.

The main task here is to select the alternative which will meet the design specification of $MDMT = 250$.

According to the parametric approach the upper limits for both alternatives could be found by making use of equations 23.1 and 23.2, and are as given in Table 23.2.

It seems that both configurations are meeting specified design requirements with a confidence level of 85%.

The above information calculated from the existing empirical data is everything which could be extracted from this data. Thus, either alternative could be recommended for adoption.

However, if the distribution approach (described in Chapter 22) was applied to the existing empirical data, by using software 'PROBCHAR' (Knezevic, 1993), the results shown in Table 23.4 could be obtained.

Table 23.1 Empirical maintainability data in hours

i	1	2	3	4	5	6	7	8	9	10
dmt_{A_i}	206	167	232	193	128	181	218	249	151	275
dmt_{B_i}	189	92	273	158	35	121	221	360	64	486

Table 23.2 Maintainability measures based on *MDMT* approach

Alternative	MDMT	$MDMT^u$
A	200	214.8
B	200	245.7

Table 22.3 Maintainability measures based on distribution approach

	Alternative A	Alternative B
Distribution type	Normal	Weibull
Scale parameter	200	215
Shape parameter	50	1.25
MDMT	200	200
DMT_{10}	135.9	35.5
DMT_{50}	200	160.4
DMT_{90}	264.1	419
Standard deviation	45.4	139.5

Comparing the data listed in Tables 23.2 and 23.3, it can be found that the distribution method is able to extract much more information from empirical data. Among other things, it is possible to determine the length of restoration time up to which 10, 50, 90, or any other percentage of maintenance tasks attempted will be successfully completed. Maintainability functions for both alternatives are shown in Figure 23.1.

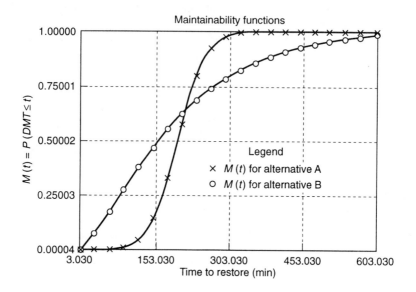

Figure 23.1   Maintainability characteristics for tasks considered.

Clearly, the amount of information extracted from the existing empirical maintainability data in the latter case is much higher and potentially more beneficial to the decision maker. Also some of the additional maintainability measures extracted from the data are providing a new light in maintainability studies.

Making use of the results given in Table 23.2, some limitations of the parameteric approach are discussed below:

(a) As both design alternatives have an identical value for *MDMT*, according to this maintainability measure, either alternative could be recommended for adoption.

(b) Assuming that the contractual requirement was that $MDMT \leqslant$ say 250 minutes with a chosen confidence level of 85%, based on the calculated values for $MDMT^u$ there is not a clear winner among competing alternatives, which in practice means that both designs have equal legal right, despite the fact that alternative A has some advantages ($MDMT_u^A < MDMT_u^B$).

(c) The information obtained is not sufficient enough for the determining and plotting of the maintainability function for either alternative.

In summary, it could be said that when identical empirical data have been exposed to both approaches, the effectiveness in extracting information from existing data by the distribution approach is by far superior.

Standardized normal variable

z	M(z)	m(z)	z	M(z)	m(z)
−4.00	0.00003	0.00013	−3.51	0.00022	0.00084
−3.99	0.00003	0.00014	−3.50	0.00023	0.00087
−3.98	0.00003	0.00014	−3.49	0.00024	0.00090
−3.97	0.00004	0.00015	−3.48	0.00025	0.00094
−3.96	0.00004	0.00016	−3.47	0.00026	0.00097
−3.95	0.00004	0.00016	−3.46	0.00027	0.00100
−3.94	0.00004	0.00017	−3.45	0.00028	0.00104
−3.93	0.00004	0.00018	−3.44	0.00029	0.00107
−3.92	0.00004	0.00018	−3.43	0.00030	0.00111
−3.91	0.00005	0.00019	−3.42	0.00031	0.00115
−3.90	0.00005	0.00020	−3.41	0.00032	0.00119
−3.89	0.00005	0.00021	−3.40	0.00034	0.00123
−3.88	0.00005	0.00021	−3.39	0.00035	0.00127
−3.87	0.00005	0.00022	−3.38	0.00036	0.00132
−3.86	0.00006	0.00023	−3.37	0.00038	0.00136
−3.85	0.00006	0.00024	−3.36	0.00039	0.00141
−3.84	0.00006	0.00025	−3.35	0.00040	0.00146
−3.83	0.00006	0.00026	−3.34	0.00042	0.00151
−3.82	0.00007	0.00027	−3.33	0.00043	0.00156
−3.81	0.00007	0.00028	−3.32	0.00045	0.00161
−3.80	0.00007	0.00029	−3.31	0.00047	0.00167
−3.79	0.00008	0.00030	−3.30	0.00048	0.00172
−3.78	0.00008	0.00031	−3.29	0.00050	0.00178
−3.77	0.00008	0.00033	−3.28	0.00052	0.00184
−3.76	0.00008	0.00034	−3.27	0.00054	0.00190
−3.75	0.00009	0.00035	−3.26	0.00056	0.00196
−3.74	0.00009	0.00037	−3.25	0.00058	0.00203
−3.73	0.00010	0.00038	−3.24	0.00060	0.00210
−3.72	0.00010	0.00039	−3.23	0.00062	0.00216
−3.71	0.00010	0.00041	−3.22	0.00064	0.00224
−3.70	0.00011	0.00042	−3.21	0.00066	0.00231
−3.69	0.00011	0.00044	−3.20	0.00069	0.00238
−3.68	0.00012	0.00046	−3.19	0.00071	0.00246
−3.67	0.00012	0.00047	−3.18	0.00074	0.00254
−3.66	0.00013	0.00049	−3.17	0.00076	0.00262
−3.65	0.00013	0.00051	−3.16	0.00079	0.00271
−3.64	0.00014	0.00053	−3.15	0.00082	0.00279
−3.63	0.00014	0.00055	−3.14	0.00084	0.00288
−3.62	0.00015	0.00057	−3.13	0.00087	0.00297
−3.61	0.00015	0.00059	−3.12	0.00090	0.00307
−3.60	0.00016	0.00061	−3.11	0.00094	0.00317
−3.59	0.00017	0.00063	−3.10	0.00097	0.00327
−3.58	0.00017	0.00066	−3.09	0.00100	0.00337
−3.57	0.00018	0.00068	−3.08	0.00103	0.00347
−3.56	0.00019	0.00071	−3.07	0.00107	0.00358
−3.55	0.00019	0.00073	−3.06	0.00111	0.00369
−3.54	0.00020	0.00076	−3.05	0.00114	0.00381
−3.53	0.00021	0.00079	−3.04	0.00118	0.00393
−3.52	0.00022	0.00081	−3.03	0.00122	0.00405

Table 1 *cont.*

z	M(z)	m(z)	z	M(z)	m(z)
−3.02	0.00126	0.00417	−2.53	0.00570	0.01625
−3.01	0.00131	0.00430	−2.52	0.00587	0.01667
−3.00	0.00135	0.00443	−2.51	0.00604	0.01709
−2.99	0.00139	0.00457	−2.50	0.00621	0.01752
−2.98	0.00144	0.00470	−2.49	0.00639	0.01797
−2.97	0.00149	0.00485	−2.48	0.00657	0.01842
−2.96	0.00154	0.00499	−2.47	0.00675	0.01888
−2.95	0.00159	0.00514	−2.46	0.00695	0.01935
−2.94	0.00164	0.00530	−2.45	0.00714	0.01983
−2.93	0.00169	0.00545	−2.44	0.00734	0.02032
−2.92	0.00175	0.00561	−2.43	0.00755	0.02082
−2.91	0.00181	0.00578	−2.42	0.00776	0.02134
−2.90	0.00187	0.00595	−2.41	0.00797	0.02186
−2.89	0.00193	0.00613	−2.40	0.00820	0.02239
−2.88	0.00199	0.00631	−2.39	0.00842	0.02293
−2.87	0.00205	0.00649	−2.38	0.00865	0.02349
−2.86	0.00212	0.00668	−2.37	0.00889	0.02405
−2.85	0.00219	0.00687	−2.36	0.00914	0.02463
−2.84	0.00226	0.00707	−2.35	0.00938	0.02521
−2.83	0.00233	0.00727	−2.34	0.00964	0.02581
−2.82	0.00240	0.00748	−2.33	0.00990	0.02642
−2.81	0.00248	0.00769	−2.32	0.01017	0.02704
−2.80	0.00255	0.00791	−2.31	0.01044	0.02768
−2.79	0.00264	0.00814	−2.30	0.01072	0.02832
−2.78	0.00272	0.00837	−2.29	0.01101	0.02898
−2.77	0.00280	0.00860	−2.28	0.01130	0.02965
−2.76	0.00289	0.00884	−2.27	0.01160	0.03033
−2.75	0.00298	0.00909	−2.26	0.01191	0.03103
−2.74	0.00307	0.00934	−2.25	0.01222	0.03173
−2.73	0.00317	0.00960	−2.24	0.01254	0.03245
−2.72	0.00326	0.00987	−2.23	0.01287	0.03319
−2.71	0.00336	0.01014	−2.22	0.01321	0.03393
−2.70	0.00347	0.01042	−2.21	0.01355	0.03469
−2.69	0.00357	0.01070	−2.20	0.01390	0.03547
−2.68	0.00368	0.01099	−2.19	0.01426	0.03625
−2.67	0.00379	0.01129	−2.18	0.01463	0.03705
−2.66	0.00391	0.01160	−2.17	0.01500	0.03787
−2.65	0.00402	0.01191	−2.16	0.01538	0.03870
−2.64	0.00414	0.01223	−2.15	0.01577	0.03954
−2.63	0.00427	0.01256	−2.14	0.01617	0.04040
−2.62	0.00440	0.01289	−2.13	0.01658	0.04127
−2.61	0.00453	0.01323	−2.12	0.01700	0.04216
−2.60	0.00466	0.01358	−2.11	0.01743	0.04306
−2.59	0.00480	0.01394	−2.10	0.01786	0.04397
−2.58	0.00494	0.01430	−2.09	0.01830	0.04490
−2.57	0.00508	0.01468	−2.08	0.01876	0.04585
−2.56	0.00523	0.01506	−2.07	0.01922	0.04681
−2.55	0.00539	0.01545	−2.06	0.01969	0.04779
−2.54	0.00554	0.01584	−2.05	0.02018	0.04878

Table 1 *cont.*

z	M(z)	m(z)	z	M(z)	m(z)
−2.04	0.02067	0.04979	−1.55	0.06056	0.11999
−2.03	0.02117	0.05081	−1.54	0.06177	0.12185
−2.02	0.02169	0.05185	−1.53	0.06299	0.12374
−2.01	0.02221	0.05291	−1.52	0.06424	0.12564
−2.00	0.02274	0.05398	−1.51	0.06551	0.12756
−1.99	0.02329	0.05507	−1.50	0.06679	0.12949
−1.98	0.02385	0.05617	−1.49	0.06810	0.13144
−1.97	0.02441	0.05729	−1.48	0.06942	0.13341
−1.96	0.02499	0.05843	−1.47	0.07077	0.13539
−1.95	0.02558	0.05958	−1.46	0.07213	0.13739
−1.94	0.02618	0.06075	−1.45	0.07351	0.13940
−1.93	0.02680	0.06194	−1.44	0.07492	0.14143
−1.92	0.02742	0.06314	−1.43	0.07634	0.14348
−1.91	0.02806	0.06436	−1.42	0.07779	0.14554
−1.90	0.02871	0.06560	−1.41	0.07925	0.14761
−1.89	0.02937	0.06686	−1.40	0.08074	0.14970
−1.88	0.03005	0.06813	−1.39	0.08225	0.15180
−1.87	0.03073	0.06942	−1.38	0.08378	0.15392
−1.86	0.03144	0.07073	−1.37	0.08533	0.15605
−1.85	0.03215	0.07205	−1.36	0.08690	0.15819
−1.84	0.03288	0.07339	−1.35	0.08849	0.16035
−1.83	0.03362	0.07475	−1.34	0.09010	0.16252
−1.82	0.03437	0.07613	−1.33	0.09174	0.16471
−1.81	0.03514	0.07752	−1.32	0.09340	0.16691
−1.80	0.03592	0.07893	−1.31	0.09508	0.16912
−1.79	0.03672	0.08036	−1.30	0.09678	0.17134
−1.78	0.03753	0.08181	−1.29	0.09850	0.17357
−1.77	0.03835	0.08328	−1.28	0.10025	0.17581
−1.76	0.03919	0.08476	−1.27	0.10202	0.17807
−1.75	0.04005	0.08626	−1.26	0.10381	0.18034
−1.74	0.04092	0.08778	−1.25	0.10563	0.18262
−1.73	0.04181	0.08931	−1.24	0.10747	0.18490
−1.72	0.04271	0.09087	−1.23	0.10933	0.18720
−1.71	0.04362	0.09244	−1.22	0.11121	0.18951
−1.70	0.04456	0.09403	−1.21	0.11312	0.19183
−1.69	0.04550	0.09564	−1.20	0.11505	0.19415
−1.68	0.04647	0.09726	−1.19	0.11700	0.19649
−1.67	0.04745	0.09891	−1.18	0.11898	0.19883
−1.66	0.04845	0.10057	−1.17	0.12098	0.20118
−1.65	0.04946	0.10224	−1.16	0.12300	0.20353
−1.64	0.05049	0.10394	−1.15	0.12505	0.20590
−1.63	0.05154	0.10565	−1.14	0.12712	0.20827
−1.62	0.05260	0.10738	−1.13	0.12921	0.21065
−1.61	0.05369	0.10913	−1.12	0.13133	0.21303
−1.60	0.05479	0.11090	−1.11	0.13347	0.21542
−1.59	0.05591	0.11268	−1.10	0.13564	0.21781
−1.58	0.05704	0.11448	−1.09	0.13783	0.22021
−1.57	0.05819	0.11630	−1.08	0.14004	0.22261
−1.56	0.05937	0.11813	−1.07	0.14228	0.22502

Table 1 *cont.*

z	M(z)	m(z)	z	M(z)	m(z)
−1.06	0.14454	0.22743	−0.57	0.28429	0.33907
−1.05	0.14683	0.22984	−0.56	0.28769	0.34099
−1.04	0.14914	0.23226	−0.55	0.29111	0.34289
−1.03	0.15147	0.23467	−0.54	0.29455	0.34477
−1.02	0.15383	0.23709	−0.53	0.29800	0.34662
−1.01	0.15622	0.23951	−0.52	0.30148	0.34844
−1.00	0.15862	0.24193	−0.51	0.30497	0.35024
−0.99	0.16106	0.24435	−0.50	0.30848	0.35201
−0.98	0.16351	0.24677	−0.49	0.31201	0.35376
−0.97	0.16599	0.24919	−0.48	0.31556	0.35548
−0.96	0.16849	0.25160	−0.47	0.31912	0.35717
−0.95	0.17102	0.25402	−0.46	0.32270	0.35884
−0.94	0.17357	0.25643	−0.45	0.32630	0.36048
−0.93	0.17615	0.25884	−0.44	0.32991	0.36208
−0.92	0.17875	0.26124	−0.43	0.33354	0.36366
−0.91	0.18138	0.26364	−0.42	0.33718	0.36521
−0.90	0.18402	0.26604	−0.41	0.34084	0.36673
−0.89	0.18670	0.26843	−0.40	0.34452	0.36822
−0.88	0.18939	0.27082	−0.39	0.34821	0.36968
−0.87	0.19211	0.27320	−0.38	0.35191	0.37110
−0.86	0.19486	0.27557	−0.37	0.35563	0.37250
−0.85	0.19762	0.27794	−0.36	0.35936	0.37386
−0.84	0.20042	0.28030	−0.35	0.36311	0.37519
−0.83	0.20323	0.28265	−0.34	0.36686	0.37649
−0.82	0.20607	0.28499	−0.33	0.37064	0.37775
−0.81	0.20893	0.28732	−0.32	0.37442	0.37898
−0.80	0.21182	0.28964	−0.31	0.37822	0.38017
−0.79	0.21472	0.29196	−0.30	0.38202	0.38134
−0.78	0.21765	0.29426	−0.29	0.38584	0.38246
−0.77	0.22061	0.29655	−0.28	0.38967	0.38355
−0.76	0.22359	0.29882	−0.27	0.39351	0.38461
−0.75	0.22658	0.30109	−0.26	0.39736	0.38563
−0.74	0.22961	0.30334	−0.25	0.40123	0.38662
−0.73	0.23265	0.30558	−0.24	0.40510	0.38757
−0.72	0.23572	0.30780	−0.23	0.40898	0.38848
−0.71	0.23881	0.31001	−0.22	0.41287	0.38935
−0.70	0.24192	0.31221	−0.21	0.41676	0.39019
−0.69	0.24505	0.31438	−0.20	0.42067	0.39099
−0.68	0.24821	0.31654	−0.19	0.42458	0.39175
−0.67	0.25138	0.31869	−0.18	0.42850	0.39248
−0.66	0.25458	0.32081	−0.17	0.43243	0.39317
−0.65	0.25780	0.32292	−0.16	0.43637	0.39382
−0.64	0.26104	0.32501	−0.15	0.44031	0.39443
−0.63	0.26430	0.32708	−0.14	0.44426	0.39500
−0.62	0.26758	0.32913	−0.13	0.44821	0.39554
−0.61	0.27088	0.33116	−0.12	0.45217	0.39603
−0.60	0.27420	0.33317	−0.11	0.45613	0.39649
−0.59	0.27754	0.33516	−0.10	0.46010	0.39690
−0.58	0.28091	0.33713	−0.09	0.46407	0.39728

Table 1 *cont.*

z	M(z)	m(z)	z	M(z)	m(z)
−0.08	0.46804	0.39762	0.41	0.65912	0.36675
−0.07	0.47202	0.39792	0.42	0.66278	0.36523
−0.06	0.47600	0.39818	0.43	0.66642	0.36368
−0.05	0.47998	0.39839	0.44	0.67005	0.36210
−0.04	0.48397	0.39857	0.45	0.67367	0.36049
−0.03	0.48795	0.39871	0.46	0.67726	0.35886
−0.02	0.49194	0.39881	0.47	0.68084	0.35719
−0.01	0.49593	0.39887	0.48	0.68441	0.35550
0.00	0.49992	0.39889	0.49	0.68795	0.35378
0.01	0.50403	0.39887	0.50	0.69148	0.35203
0.02	0.50802	0.39881	0.51	0.69499	0.35026
0.03	0.51200	0.39872	0.52	0.69849	0.34846
0.04	0.51599	0.39858	0.53	0.70196	0.34664
0.05	0.51998	0.39840	0.54	0.70542	0.34479
0.06	0.52396	0.39818	0.55	0.70886	0.34291
0.07	0.52794	0.39792	0.56	0.71228	0.34101
0.08	0.53192	0.39762	0.57	0.71568	0.33909
0.09	0.53589	0.39728	0.58	0.71906	0.33715
0.10	0.53986	0.39691	0.59	0.72242	0.33518
0.11	0.54383	0.39649	0.60	0.72576	0.33319
0.12	0.54779	0.39603	0.61	0.72908	0.33119
0.13	0.55175	0.39554	0.62	0.73239	0.32915
0.14	0.55570	0.39501	0.63	0.73567	0.32710
0.15	0.55965	0.39443	0.64	0.73893	0.32503
0.16	0.56359	0.39382	0.65	0.74217	0.32294
0.17	0.56753	0.39318	0.66	0.74539	0.32084
0.18	0.57146	0.39249	0.67	0.74858	0.31871
0.19	0.57538	0.39176	0.68	0.75176	0.31657
0.20	0.57929	0.39100	0.69	0.75492	0.31441
0.21	0.58320	0.39020	0.70	0.75805	0.31223
0.22	0.58709	0.38936	0.71	0.76116	0.31003
0.23	0.59098	0.38849	0.72	0.76425	0.30783
0.24	0.59486	0.38757	0.73	0.76732	0.30560
0.25	0.59873	0.38663	0.74	0.77036	0.30336
0.26	0.60260	0.38564	0.75	0.77338	0.30111
0.27	0.60645	0.38462	0.76	0.77638	0.29885
0.28	0.61029	0.38357	0.77	0.77936	0.29657
0.29	0.61412	0.38247	0.78	0.78232	0.29428
0.30	0.61794	0.38135	0.79	0.78525	0.29198
0.31	0.62175	0.38019	0.80	0.78816	0.28967
0.32	0.62554	0.37899	0.81	0.79104	0.28735
0.33	0.62933	0.37776	0.82	0.79390	0.28501
0.34	0.63310	0.37650	0.83	0.79674	0.28267
0.35	0.63686	0.37520	0.84	0.79956	0.28032
0.36	0.64060	0.37387	0.85	0.80235	0.27796
0.37	0.64433	0.37251	0.86	0.80511	0.27560
0.38	0.64805	0.37112	0.87	0.80786	0.27322
0.39	0.65175	0.36969	0.88	0.81058	0.27084
0.40	0.65544	0.36823	0.89	0.81328	0.26846

Table 1 *cont.*

z	M(z)	m(z)	z	M(z)	m(z)
0.90	0.81595	0.26607	1.39	0.91774	0.15182
0.91	0.81860	0.26367	1.40	0.91925	0.14972
0.92	0.82122	0.26127	1.41	0.92073	0.14763
0.93	0.82382	0.25886	1.42	0.92220	0.14556
0.94	0.82640	0.25645	1.43	0.92364	0.14350
0.95	0.82895	0.25404	1.44	0.92507	0.14145
0.96	0.83148	0.25163	1.45	0.92647	0.13942
0.97	0.83398	0.24921	1.46	0.92786	0.13741
0.98	0.83646	0.24679	1.47	0.92922	0.13541
0.99	0.83892	0.24437	1.48	0.93056	0.13343
1.00	0.84135	0.24195	1.49	0.93189	0.13146
1.01	0.84376	0.23953	1.50	0.93319	0.12951
1.02	0.84614	0.23712	1.51	0.93448	0.12758
1.03	0.84850	0.23470	1.52	0.93575	0.12566
1.04	0.85084	0.23228	1.53	0.93699	0.12376
1.05	0.85315	0.22987	1.54	0.93822	0.12187
1.06	0.85543	0.22745	1.55	0.93943	0.12000
1.07	0.85770	0.22505	1.56	0.94062	0.11815
1.08	0.85993	0.22264	1.57	0.94179	0.11632
1.09	0.86215	0.22024	1.58	0.94295	0.11450
1.10	0.86434	0.21784	1.59	0.94408	0.11270
1.11	0.86651	0.21544	1.60	0.94520	0.11092
1.12	0.86865	0.21306	1.61	0.94630	0.10915
1.13	0.87077	0.21067	1.62	0.94738	0.10740
1.14	0.87286	0.20830	1.63	0.94845	0.10567
1.15	0.87493	0.20592	1.64	0.94950	0.10396
1.16	0.87698	0.20356	1.65	0.95053	0.10226
1.17	0.87900	0.20120	1.66	0.95154	0.10058
1.18	0.88100	0.19885	1.67	0.95254	0.09892
1.19	0.88298	0.19651	1.68	0.95352	0.09728
1.20	0.88493	0.19417	1.69	0.95449	0.09565
1.21	0.88686	0.19185	1.70	0.95543	0.09405
1.22	0.88877	0.18953	1.71	0.95637	0.09246
1.23	0.89065	0.18722	1.72	0.95728	0.09088
1.24	0.89252	0.18493	1.73	0.95819	0.08933
1.25	0.89435	0.18264	1.74	0.95907	0.08779
1.26	0.89617	0.18036	1.75	0.95994	0.08627
1.27	0.89796	0.17809	1.76	0.96080	0.08477
1.28	0.89973	0.17584	1.77	0.96164	0.08329
1.29	0.90148	0.17359	1.78	0.96246	0.08183
1.30	0.90320	0.17136	1.79	0.96327	0.08038
1.31	0.90490	0.16914	1.80	0.96407	0.07895
1.32	0.90658	0.16693	1.81	0.96485	0.07754
1.33	0.90824	0.16473	1.82	0.96562	0.07614
1.34	0.90988	0.16255	1.83	0.96638	0.07476
1.35	0.91149	0.16038	1.84	0.96712	0.07341
1.36	0.91309	0.15822	1.85	0.96784	0.07206
1.37	0.91466	0.15607	1.86	0.96856	0.07074
1.38	0.91621	0.15394	1.87	0.96926	0.06943

Table 1 *cont.*

z	M(z)	m(z)	z	M(z)	m(z)
1.88	0.96995	0.06814	2.37	0.99111	0.02406
1.89	0.97062	0.06687	2.38	0.99134	0.02349
1.90	0.97128	0.06561	2.39	0.99158	0.02294
1.91	0.97193	0.06438	2.40	0.99180	0.02240
1.92	0.97257	0.06316	2.41	0.99202	0.02186
1.93	0.97320	0.06195	2.42	0.99224	0.02134
1.94	0.97381	0.06076	2.43	0.99245	0.02083
1.95	0.97441	0.05959	2.44	0.99266	0.02033
1.96	0.97500	0.05844	2.45	0.99286	0.01984
1.97	0.97558	0.05730	2.46	0.99305	0.01936
1.98	0.97615	0.05618	2.47	0.99324	0.01889
1.99	0.97670	0.05508	2.48	0.99343	0.01842
2.00	0.97725	0.05399	2.49	0.99361	0.01797
2.01	0.97778	0.05292	2.50	0.99379	0.01753
2.02	0.97831	0.05186	2.51	0.99396	0.01710
2.03	0.97882	0.05082	2.52	0.99413	0.01667
2.04	0.97932	0.04980	2.53	0.99430	0.01626
2.05	0.97982	0.04879	2.54	0.99446	0.01585
2.06	0.98030	0.04780	2.55	0.99461	0.01545
2.07	0.98077	0.04682	2.56	0.99477	0.01506
2.08	0.98124	0.04586	2.57	0.99491	0.01468
2.09	0.98169	0.04491	2.58	0.99506	0.01431
2.10	0.98214	0.04398	2.59	0.99520	0.01394
2.11	0.98257	0.04307	2.60	0.99534	0.01358
2.12	0.98300	0.04217	2.61	0.99547	0.01323
2.13	0.98341	0.04128	2.62	0.99560	0.01289
2.14	0.98382	0.04041	2.63	0.99573	0.01256
2.15	0.98422	0.03955	2.64	0.99585	0.01223
2.16	0.98461	0.03871	2.65	0.99598	0.01191
2.17	0.98500	0.03788	2.66	0.99609	0.01160
2.18	0.98537	0.03706	2.67	0.99621	0.01130
2.19	0.98574	0.03626	2.68	0.99632	0.01100
2.20	0.98610	0.03547	2.69	0.99643	0.01071
2.21	0.98645	0.03470	2.70	0.99653	0.01042
2.22	0.98679	0.03394	2.71	0.99664	0.01014
2.23	0.98713	0.03319	2.72	0.99674	0.00987
2.24	0.98745	0.03246	2.73	0.99683	0.00961
2.25	0.98778	0.03174	2.74	0.99693	0.00935
2.26	0.98809	0.03103	2.75	0.99702	0.00909
2.27	0.98840	0.03034	2.76	0.99711	0.00885
2.28	0.98870	0.02966	2.77	0.99720	0.00861
2.29	0.98899	0.02899	2.78	0.99728	0.00837
2.30	0.98928	0.02833	2.79	0.99736	0.00814
2.31	0.98956	0.02768	2.80	0.99744	0.00792
2.32	0.98983	0.02705	2.81	0.99752	0.00770
2.33	0.99010	0.02643	2.82	0.99760	0.00748
2.34	0.99036	0.02582	2.83	0.99767	0.00728
2.35	0.99061	0.02522	2.84	0.99774	0.00707
2.36	0.99086	0.02463	2.85	0.99781	0.00687

Table 1 *cont.*

z	M(z)	m(z)	z	M(z)	m(z)
2.86	0.99788	0.00668	3.35	0.99960	0.00146
2.87	0.99795	0.00649	3.36	0.99961	0.00141
2.88	0.99801	0.00631	3.37	0.99962	0.00136
2.89	0.99807	0.00613	3.38	0.99964	0.00132
2.90	0.99813	0.00595	3.39	0.99965	0.00127
2.91	0.99819	0.00578	3.40	0.99966	0.00123
2.92	0.99825	0.00562	3.41	0.99968	0.00119
2.93	0.99830	0.00545	3.42	0.99969	0.00115
2.94	0.99836	0.00530	3.43	0.99970	0.00111
2.95	0.99841	0.00514	3.44	0.99971	0.00107
2.96	0.99846	0.00499	3.45	0.99972	0.00104
2.97	0.99851	0.00485	3.46	0.99973	0.00100
2.98	0.99856	0.00471	3.47	0.99974	0.00097
2.99	0.99860	0.00457	3.48	0.99975	0.00094
3.00	0.99865	0.00443	3.49	0.99976	0.00090
3.01	0.99869	0.00430	3.50	0.99977	0.00087
3.02	0.99874	0.00417	3.51	0.99978	0.00084
3.03	0.99878	0.00405	3.52	0.99978	0.00081
3.04	0.99882	0.00393	3.53	0.99979	0.00079
3.05	0.99886	0.00381	3.54	0.99980	0.00076
3.06	0.99889	0.00370	3.55	0.99981	0.00073
3.07	0.99893	0.00358	3.56	0.99981	0.00071
3.08	0.99896	0.00348	3.57	0.99982	0.00068
3.09	0.99900	0.00337	3.58	0.99983	0.00066
3.10	0.99903	0.00327	3.59	0.99983	0.00063
3.11	0.99906	0.00317	3.60	0.99984	0.00061
3.12	0.99910	0.00307	3.61	0.99985	0.00059
3.13	0.99913	0.00298	3.62	0.99985	0.00057
3.14	0.99916	0.00288	3.63	0.99986	0.00055
3.15	0.99918	0.00279	3.64	0.99986	0.00053
3.16	0.99921	0.00271	3.65	0.99987	0.00051
3.17	0.99924	0.00262	3.66	0.99987	0.00049
3.18	0.99926	0.00254	3.67	0.99988	0.00047
3.19	0.99929	0.00246	3.68	0.99988	0.00046
3.20	0.99931	0.00238	3.69	0.99989	0.00044
3.21	0.99934	0.00231	3.70	0.99989	0.00042
3.22	0.99936	0.00224	3.71	0.99990	0.00041
3.23	0.99938	0.00217	3.72	0.99990	0.00039
3.24	0.99940	0.00210	3.73	0.99990	0.00038
3.25	0.99942	0.00203	3.74	0.99991	0.00037
3.26	0.99944	0.00196	3.75	0.99991	0.00035
3.27	0.99946	0.00190	3.76	0.99991	0.00034
3.28	0.99948	0.00184	3.77	0.99992	0.00033
3.29	0.99950	0.00178	3.78	0.99992	0.00032
3.30	0.99952	0.00172	3.79	0.99992	0.00030
3.31	0.99953	0.00167	3.80	0.99993	0.00029
3.32	0.99955	0.00161	3.81	0.99993	0.00028
3.33	0.99957	0.00156	3.82	0.99993	0.00027
3.34	0.99958	0.00151	3.83	0.99994	0.00026

Table 1 *cont.*

z	M(z)	m(z)		z	M(z)	m(z)
3.84	0.99994	0.00025		3.93	0.99996	0.00018
3.85	0.99994	0.00024		3.94	0.99996	0.00017
3.86	0.99994	0.00023		3.95	0.99996	0.00016
3.87	0.99995	0.00022		3.96	0.99996	0.00016
3.88	0.99995	0.00021		3.97	0.99996	0.00015
3.89	0.99995	0.00021		3.98	0.99997	0.00014
3.90	0.99995	0.00020		3.99	0.99997	0.00014
3.91	0.99995	0.00019		4.00	0.99997	0.00013
3.92	0.99996	0.00018				

Median, 5 and 95% rank

i	M'(i	1 − M'(i)		Xn(i)	Xe(i)	Xw(i)	p0.05	p0.95
Total number of results: 1								
1	0.50000	0.50000		0.0000	−0.6931	−0.3665	0.05000	0.95000
			S3:	0.0000	−0.6931	−0.3665		
			S4:	0.0000	0.4804	0.1343		
Total number of results: 2								
1	0.29167	0.70833		−0.5436	−0.3448	−1.0647	0.02532	0.77639
2	0.70833	0.29167		0.5481	−1.2321	0.2088	0.22361	0.97468
			S3:	0.0045	−1.5769	−0.8559		
			S4:	0.5959	1.6371	1.1771		
Total number of results: 3								
1	0.20588	0.79412		−0.8087	−0.2305	−1.4674	0.01695	0.63160
2	0.50000	0.50000		0.0000	−0.6931	−0.3665	0.13535	0.86465
3	0.79412	0.20588		0.8206	−1.5805	0.4577	0.36840	0.98305
			S3:	0.0119	−2.5041	−1.3762		
			S4:	1.3273	3.0314	2.4971		
Total number of results: 4								
1	0.15909	0.84091		−0.9782	−0.1733	−1.7529	0.01274	0.52713
2	0.38636	0.61364		−0.2872	−0.4884	−0.7167	0.09761	0.75139
3	0.61364	0.38636		0.2884	−0.9510	−0.0503	0.24860	0.90239
4	0.84091	0.15909		0.9982	−1.8383	0.6088	0.47237	0.98726
			S3:	0.0211	−3.4508	−1.9110		
			S4:	2.1189	4.5521	3.9595		
Total number of results: 5								
1	0.12963	0.87037		−1.1002	−0.1388	−1.9745	0.01021	0.45072
2	0.31481	0.68519		−0.4784	−0.3781	−0.9727	0.07644	0.65741
3	0.50000	0.50000		0.0000	−0.6931	−0.3665	0.18925	0.81075
4	0.68519	0.31481		0.4818	−1.1558	0.1448	0.34259	0.92356
5	0.87037	0.12963		1.1282	−2.0431	0.7145	0.54928	0.98979
			S3:	0.0314	−4.4088	−2.4544		
			S4:	2.9443	6.1526	5.5103		
Total number of results: 6								
1	0.10938	0.89063		−1.1940	−0.1158	−2.1556	0.00851	0.39304
2	0.26563	0.73438		−0.6196	−0.3087	−1.1753	0.06285	0.58180
3	0.42188	0.57813		−0.1962	−0.5480	−0.6015	0.15316	0.72866
4	0.57813	0.42188		0.1967	−0.8630	−0.1473	0.27134	0.84684
5	0.73438	0.26563		0.6258	−1.3257	0.2819	0.41820	0.93715
6	0.89063	0.10938		1.2300	−2.2130	0.7943	0.60696	0.99149
			S3:	0.0427	−5.3742	−3.0034		
			S4:	3.7911	7.8085	7.1219		
Total number of results: 7								
1	0.09459	0.90541		−1.2694	−0.0994	−2.3089	0.00730	0.34816
2	0.22973	0.77027		−0.7303	−0.2610	−1.3432	0.05337	0.52070
3	0.36486	0.63514		−0.3434	−0.4539	−0.7898	0.12876	0.65874
4	0.50000	0.50000		0.0000	−0.6931	−0.3665	0.22532	0.77468
5	0.63514	0.36486		0.3450	−1.0082	0.0082	0.34126	0.87124
6	0.77027	0.22973		0.7395	−1.4709	0.3858	0.47930	0.94662

Table 2 *cont.*

i	M'(i	1 − M'(i)		Xn(i)	Xe(i)	Xw(i)	p0.05	p0.95
7	0.90541	0.09459		1.3132	−2.3582	0.8579	0.65184	0.99270
			S3:	0.0546	−6.3446	−3.5565		
			S4:	4.6528	9.5053	8.7781		
Total number of results: 8								
1	0.08333	0.91667		−1.3319	−0.0870	−2.4417	0.00639	0.31234
2	0.20238	0.79762		−0.8206	−0.2261	−1.4867	0.04639	0.47068
3	0.32143	0.67857		−0.4601	−0.3878	−0.9474	0.11111	0.59969
4	0.44048	0.55952		−0.1491	−0.5807	−0.5436	0.19290	0.71076
5	0.55952	0.44048		0.1494	−0.8199	−0.1986	0.28924	0.80710
6	0.67857	0.32143		0.4633	−1.1350	0.1266	0.40031	0.88889
7	0.79762	0.20238		0.8330	−1.5976	0.4685	0.52932	0.95361
8	0.91667	0.08333		1.3832	−2.4849	0.9102	0.68766	0.99361
			S3:	0.0671	−7.3189	−4.1125		
			S4:	5.5253	11.2338	10.4686		
Total number of results: 9								
1	0.07447	0.92553		−1.3849	−0.0774	−2.5589	0.00568	0.28313
2	0.18085	0.81915		−0.8964	−0.1995	−1.6120	0.04102	0.42914
3	0.28723	0.71277		−0.5563	−0.3386	−1.0829	0.09775	0.54964
4	0.39362	0.60638		−0.2685	−0.5002	−0.6927	0.16875	0.65506
5	0.50000	0.50000		0.0000	−0.6931	−0.3665	0.25137	0.74863
6	0.60638	0.39362		0.2695	−0.9324	−0.0700	0.34494	0.83125
7	0.71277	0.28723		0.5611	−1.2475	0.2211	0.45036	0.90225
8	0.81915	0.18085		0.9120	−1.7101	0.5365	0.57086	0.95898
9	0.92553	0.07447		1.4436	−2.5974	0.9545	0.71687	0.99432
			S3:	0.0813	−8.2961	−4.6709		
			S4:	6.4606	12.9874	12.1863		
Total number of results: 10								
1	0.06731	0.93269		−1.4307	−0.0697	−2.6638	0.00512	0.25887
2	0.16346	0.83654		−0.9613	−0.1785	−1.7233	0.03677	0.39416
3	0.25962	0.74038		−0.6376	−0.3006	−1.2020	0.08726	0.50690
4	0.35577	0.64423		−0.3674	−0.4397	−0.8217	0.15003	0.60662
5	0.45192	0.54808		−0.1203	−0.6013	−0.5086	0.22244	0.69646
6	0.54808	0.45192		0.1205	−0.7942	−0.2304	0.30354	0.77756
7	0.64423	0.35577		0.3693	−1.0335	0.0329	0.39338	0.84997
8	0.74038	0.25962		0.6442	−1.3486	0.2990	0.49310	0.91274
9	0.83654	0.16346		0.9803	−1.8112	0.5940	0.60584	0.96323
10	0.93269	0.06731		1.4964	−2.6985	0.9927	0.74113	0.99488
			S3:	0.0934	−9.2757	−5.2311		
			S4:	7.2933	14.7617	13.9262		
Total number of results: 11								
1	0.06140	0.93860		−1.4709	−0.0634	−2.7588	0.00465	0.23840
2	0.14912	0.85088		−1.0179	−0.1615	−1.8233	0.03332	0.36436
3	0.23684	0.76316		−0.7077	−0.2703	−1.3083	0.07882	0.47009
4	0.32456	0.67544		−0.4515	−0.3924	−0.9355	0.13507	0.56437
5	0.41228	0.58772		−0.2206	−0.5315	−0.6320	0.19958	0.65019
6	0.50000	0.50000		0.0000	−0.6931	−0.3665	0.27125	0.72875
7	0.58772	0.41228		0.2213	−0.8861	−0.1210	0.34981	0.80042

Table 2 *cont.*

i	M'(i	1 − M'(i)		Xn(i)	Xe(i)	Xw(i)	p0.05	p0.95
8	0.67544	0.32456		0.4546	− 1.1253	0.1180	0.43563	0.86492
9	0.76316	0.23684		0.7162	− 1.4404	0.3649	0.52991	0.92118
10	0.85088	0.14912		1.0402	− 1.9030	0.6434	0.63564	0.96668
11	0.93860	0.06140		1.5434	− 2.7903	1.0261	0.76160	0.99535
			S3:	0.1070	− 10.2572	− 5.7928		
			S4:	8.1856	16.5531	15.6845		
Total number of results: 12								
1	0.05645	0.94355		− 1.5065	− 0.0581	− 2.8455	0.00426	0.22092
2	0.13710	0.86290		− 1.0678	− 0.1475	− 1.9142	0.03046	0.33868
3	0.21774	0.78226		− 0.7691	− 0.2456	− 1.4042	0.07187	0.43811
4	0.29839	0.70161		− 0.5244	− 0.3544	− 1.0374	0.12285	0.52733
5	0.37903	0.62097		− 0.3063	− 0.4765	− 0.7413	0.18102	0.60914
6	0.45968	0.54032		− 0.1009	− 0.6156	− 0.4852	0.24530	0.68476
7	0.54032	0.45968		0.1010	− 0.7772	− 0.2520	0.31524	0.75470
8	0.62097	0.37903		0.3076	− 0.9701	− 0.0303	0.39086	0.81898
9	0.70161	0.29839		0.5286	− 1.2094	0.1901	0.47267	0.87715
10	0.78226	0.21774		0.7796	− 1.5244	0.4216	0.56189	0.92813
11	0.86290	0.13710		1.0935	− 1.9871	0.6867	0.66132	0.96954
12	0.94355	0.05645		1.5856	− 2.8744	1.0558	0.77908	0.99573
			S3:	0.1209	− 11.2402	− 6.3559		
			S4:	9.0822	18.3592	17.4585		
Total number of results: 13								
1	0.05224	0.94776		− 1.5383	− 0.0537	− 2.9252	0.00394	0.20582
2	0.12687	0.87313		− 1.1124	− 0.1357	− 1.9976	0.02805	0.31634
3	0.20149	0.79851		− 0.8236	− 0.2250	− 1.4916	0.07187	0.41010
4	0.27612	0.72388		− 0.5886	− 0.3231	− 1.1297	0.11267	0.49465
5	0.35075	0.64925		− 0.3808	− 0.4319	− 0.8395	0.16566	0.57262
6	0.42537	0.57463		− 0.1873	− 0.5540	− 0.5905	0.22395	0.64520
7	0.50000	0.50000		0.0000	− 0.6931	− 0.3665	0.28705	0.71295
8	0.57463	0.42537		0.1878	− 0.8548	− 0.1569	0.35480	0.77604
9	0.64925	0.35075		0.3829	− 1.0477	0.0466	0.42738	0.83434
10	0.72388	0.27612		0.5941	− 1.2869	0.2523	0.50535	0.88733
11	0.79851	0.20149		0.8361	− 1.6020	0.4713	0.58990	0.93395
12	0.87313	0.12687		1.1414	− 2.0646	0.7249	0.68366	0.97195
13	0.94776	0.05224		1.6239	− 2.9519	1.0825	0.79418	0.99606
			S3:	0.1351	− 12.2245	− 6.9200		
			S4:	9.9824	20.1778	19.2459		
Total number of results: 14								
1	0.04861	0.95139		− 1.5671	− 0.0498	− 2.9991	0.00366	0.19264
2	0.11806	0.88194		− 1.1526	− 0.1256	− 2.0744	0.02600	0.29673
3	0.18750	0.81250		− 0.8724	− 0.2076	− 1.5720	0.06110	0.38539
4	0.25694	0.74306		− 0.6457	− 0.2970	− 1.2141	0.10405	0.46566
5	0.32639	0.67361		− 0.4465	− 0.3951	− 0.9286	0.15272	0.54000
6	0.39583	0.60417		− 0.2628	− 0.5039	− 0.6854	0.20607	0.60928
7	0.46528	0.53472		− 0.0868	− 0.6260	− 0.4684	0.26358	0.67497
8	0.53472	0.46528		0.0869	− 0.7651	− 0.2677	0.32503	0.73641
9	0.60417	0.39583		0.2637	− 0.9268	− 0.0761	0.39041	0.79393
10	0.67361	0.32639		0.4495	− 1.1197	0.1130	0.45999	0.84728

Table 2 *cont.*

i	M'(i	1 − M'(i)	Xn(i)	Xe(i)	Xw(i)	p0.05	p0.95
11	0.74306	0.25694	0.6525	−1.3589	0.3067	0.53434	0.89595
12	0.81250	0.18750	0.8870	−1.6740	0.5152	0.61461	0.93890
13	0.88194	0.11806	1.1849	−2.1366	0.7592	0.70327	0.97400
14	0.95139	0.04861	1.6588	−3.0239	1.1065	0.80736	0.99634
			S3: 0.1494	−13.2100	−7.4850		
			S4: 10.8856	22.0073	21.0449		

Total number of results: 15

i	M'(i	1 − M'(i)	Xn(i)	Xe(i)	Xw(i)	p0.05	p0.95
1	0.04545	0.95455	−1.5932	−0.0465	−3.0679	0.00341	0.18104
2	0.11039	0.88961	−1.1891	−0.1170	−2.1458	0.02423	0.27940
3	0.17532	0.82468	−0.9166	−0.1928	−1.6463	0.05685	0.36344
4	0.24026	0.75974	−0.6970	−0.2748	−1.2918	0.09666	0.43978
5	0.30519	0.69481	−0.5052	−0.3641	−1.0103	0.14166	0.51075
6	0.37013	0.62987	−0.3295	−0.4622	−0.7717	0.19086	0.51075
7	0.43506	0.56494	−0.1628	−0.5710	−0.5603	0.24373	0.57744
8	0.50000	0.50000	0.0000	−0.6931	−0.3665	0.29999	0.64043
9	0.56494	0.43506	0.1632	−0.8323	−0.1836	0.35956	0.70001
10	0.62987	0.37013	0.3311	−0.9939	−0.0061	0.42256	0.75627
11	0.69481	0.30519	0.5091	−1.1868	0.1713	0.48925	0.80913
12	0.75974	0.24026	0.7052	−1.4260	0.3549	0.56022	0.85834
13	0.82468	0.17532	0.9332	−1.7411	0.5545	0.63656	0.90334
14	0.88961	0.11039	1.2246	−2.2037	0.7902	0.72060	0.94315
15	0.95455	0.04545	1.6910	−3.0910	1.1285	0.81896	0.99659
			S3: 0.1639	−14.1965	−8.0587		
			S4: 11.7913	23.8464	22.8541		

Total number of results: 16

i	M'(i	1 − M'(i)	Xn(i)	Xe(i)	Xw(i)	p0.05	p0.95
1	0.04268	0.95732	−1.6172	−0.0436	−3.1322	0.00320	0.17075
2	0.10366	0.89634	−1.2224	−0.1094	−2.2124	0.02268	0.26396
3	0.16463	0.83537	−0.9568	−0.1799	−1.7154	0.05315	0.34383
4	0.22561	0.77439	−0.7435	−0.2557	−1.3638	0.09025	0.41657
5	0.28659	0.71341	−0.5582	−0.3377	−1.0856	0.13211	0.48440
6	0.34756	0.65244	−0.3893	−0.4270	−0.8509	0.17777	0.54835
7	0.40854	0.59146	−0.2302	−0.5252	−0.6441	0.22669	0.60899
8	0.46951	0.53049	−0.0762	−0.6340	−0.4558	0.27860	0.66663
9	0.53049	0.46951	0.0763	−0.7561	−0.2796	0.33337	0.72140
10	0.59146	0.40854	0.2309	−0.8952	−0.1107	0.39101	0.77331
11	0.65244	0.34756	0.3915	−1.0568	0.0553	0.45165	0.82223
12	0.71341	0.28659	0.5630	−1.2497	0.2229	0.51560	0.86789
13	0.77439	0.22561	0.7531	−1.4889	0.3981	0.58343	0.90975
14	0.83537	0.16463	0.9755	−1.8040	0.5900	0.65617	0.94685
15	0.89634	0.10366	1.2611	−2.2667	0.8183	0.73604	0.97732
16	0.95732	0.04268	1.7207	−3.1540	1.1487	0.82925	0.99680
			S3: 0.1785	−15.1838	−8.6174		
			S4: 12.6993	25.6940	24.6724		

Total number of results: 17

i	M'(i	1 − M'(i)	Xn(i)	Xe(i)	Xw(i)	p0.05	p0.95
1	0.04023	0.95977	−1.6392	−0.0411	−3.1927	0.00301	0.16157
2	0.09770	0.90230	−1.2530	−0.1028	−2.2749	0.02132	0.25012
3	0.15517	0.84483	−0.9936	−0.1686	−1.7801	0.04990	0.32619
4	0.21264	0.78736	−0.7860	−0.2391	−1.4310	0.08464	0.39564

Table 2 *cont.*

i	M'(i	1 − M'(i)		Xn(i)	Xe(i)	Xw(i)	p0.05	p0.95
5	0.27011	0.72989		−0.6062	−0.3149	−1.1556	0.12377	0.46055
6	0.32759	0.67241		−0.4433	−0.3969	−0.9241	0.16636	0.52192
7	0.38506	0.61494		−0.2906	−0.4862	−0.7211	0.21191	0.58029
8	0.44253	0.55747		−0.1440	−0.5843	−0.5373	0.26011	0.63599
9	0.50000	0.50000		0.0000	−0.6931	−0.3665	0.31083	0.68917
10	0.55747	0.44253		0.1443	−0.8152	−0.2043	0.36401	0.73989
11	0.61494	0.38506		0.2918	−0.9544	−0.0467	0.41970	0.78809
12	0.67241	0.32759		0.4462	−1.1160	0.1098	0.47808	0.83364
13	0.72989	0.27011		0.6121	−1.3089	0.2692	0.53945	0.87623
14	0.78736	0.21264		0.7971	−1.5481	0.4371	0.60436	0.91535
15	0.84483	0.15517		1.0145	−1.8632	0.6223	0.67381	0.95010
16	0.90230	0.09770		1.2949	−2.3258	0.8441	0.74988	0.97869
17	0.95977	0.04023		1.7484	−3.2131	1.1673	0.83843	0.99699
			S3:	0.1933	−16.1719	−9.1845		
			S4:	13.6092	27.5491	26.4987		

Total number of results: 18

i	M'(i	1 − M'(i)		Xn(i)	Xe(i)	Xw(i)	p0.05	p0.95
1	0.03804	0.96196		−1.6595	−0.0388	−3.2497	0.00285	0.15332
2	0.09239	0.90761		−1.2812	−0.0969	−2.3336	0.02011	0.23766
3	0.14674	0.85326		−1.0276	−0.1587	−1.8408	0.04702	0.31026
4	0.20109	0.79891		−0.8250	−0.2245	−1.4939	0.07696	0.37668
5	0.25543	0.74457		−0.6502	−0.2950	−1.2209	0.11643	0.43888
6	0.30978	0.69022		−0.4924	−0.3707	−0.9922	0.15634	0.49783
7	0.36413	0.63587		−0.3453	−0.4528	−0.7924	0.19895	0.55404
8	0.41848	0.58152		−0.2048	−0.5421	−0.6123	0.24396	0.69784
9	0.47283	0.52717		−0.0679	−0.6402	−0.4459	0.29120	0.65940
10	0.52717	0.47283		0.0680	−0.7490	−0.2890	0.34060	0.70880
11	0.58152	0.41848		0.2054	−0.8711	−0.1380	0.39215	0.75604
12	0.63587	0.36413		0.3470	−1.0102	0.0102	0.44595	0.80105
13	0.69022	0.30978		0.4961	−1.1719	0.1586	0.50217	0.84366
14	0.74457	0.25543		0.6572	−1.3648	0.3110	0.56112	0.88357
15	0.79891	0.20109		0.8376	−1.6040	0.4725	0.62332	0.92030
16	0.85326	0.14674		1.0505	−1.9191	0.6519	0.68974	0.95297
17	0.90761	0.09239		1.3264	−2.3817	0.8678	0.76234	0.97989
18	0.96196	0.03804		1.7743	−3.2690	1.1845	0.84668	0.99715
			S3:	0.2083	−17.1607	−9.7522		
			S4:	14.5208	29.4111	28.3323		

Total number of results: 19

i	M'(i	1 − M'(i)		Xn(i)	Xe(i)	Xw(i)	p0.05	p0.95
1	0.03608	0.96392		−1.6784	−0.0367	−3.3036	0.00270	0.14587
2	0.08763	0.91237		−1.3075	−0.0917	−2.3891	0.01903	0.22637
3	0.13918	0.86082		−1.0590	−0.1499	−1.8980	0.04446	0.29580
4	0.19072	0.80928		−0.8610	−0.2116	−1.5530	0.07529	0.35943
5	0.24227	0.75773		−0.6907	−0.2774	−1.2822	0.10991	0.41912
6	0.29381	0.70619		−0.5374	−0.3479	−1.0559	0.14747	0.47580
7	0.34536	0.65464		−0.3952	−0.4237	−0.8588	0.18750	0.52997
8	0.39691	0.60309		−0.2600	−0.5057	−0.6818	0.22972	0.58194
9	0.44845	0.55155		−0.1291	−0.5950	−0.5191	0.27395	0.63188
10	0.50000	0.50000		0.0000	−0.6931	−0.3665	0.32009	0.67991
11	0.55155	0.44845		0.1293	−0.8019	−0.2207	0.36811	0.72605

Table 2 *cont.*

i	M'(i	1 − M'(i)		Xn(i)	Xe(i)	Xw(i)	p0.05	p0.95
12	0.60309	0.39691		0.2609	−0.9241	−0.0790	0.41806	0.77028
13	0.65464	0.34536		0.3974	−1.0632	0.0613	0.47003	0.81250
14	0.70619	0.29381		0.5419	−1.2248	0.2028	0.52420	0.85253
15	0.75773	0.24227		0.6987	−1.4177	0.3490	0.58088	0.89009
16	0.80928	0.19072		0.8751	−1.6569	0.5050	0.64057	0.92471
17	0.86082	0.13918		1.0841	−1.9720	0.6791	0.70420	0.95553
18	0.91237	0.08763		1.3557	−2.4346	0.8898	0.77363	0.98097
19	0.96392	0.03608		1.7985	−3.3219	1.2006	0.85413	0.99730
			S3:	0.2233	−18.1500	−10.3204		
			S4:	15.4339	31.2792	30.1724		

Total number of results: 20

i	M'(i	1 − M'(i)		Xn(i)	Xe(i)	Xw(i)	p0.05	p0.95
1	0.03431	0.96569		−1.6960	−0.0349	−3.3548	0.00256	0.13911
2	0.08333	0.91667		−1.3319	−0.0870	−2.4417	0.01806	0.21611
3	0.13235	0.86765		−1.0883	−0.1420	−1.9521	0.04217	0.28262
4	0.18137	0.81863		−0.8945	−0.2001	−1.6088	0.07135	0.34366
5	0.23039	0.76961		−0.7282	−0.2619	−1.3399	0.10408	0.40103
6	0.27941	0.72059		−0.5789	−0.3277	−1.1157	0.13955	0.45558
7	0.32843	0.67157		−0.4410	−0.3981	−0.9210	0.17731	0.50782
8	0.37745	0.62255		−0.3104	−0.4739	−0.7467	0.21707	0.55803
9	0.42647	0.57353		−0.1845	−0.5559	−0.5871	0.25865	0.60641
10	0.47549	0.52451		−0.0613	−0.6453	−0.4381	0.30195	0.65307
11	0.52451	0.47549		0.0613	−0.7434	−0.2965	0.34693	0.69805
12	0.57353	0.42647		0.1850	−0.8522	−0.1599	0.39358	0.74135
13	0.62255	0.37745		0.3117	−0.9743	−0.0260	0.44197	0.78293
14	0.67157	0.32843		0.4438	−1.1134	0.1074	0.49218	0.82269
15	0.72059	0.27941		0.5842	−1.2751	0.2430	0.54442	0.86045
16	0.76961	0.23039		0.7373	−1.4680	0.3839	0.59897	0.89592
17	0.81863	0.18137		0.9100	−1.7072	0.5349	0.65634	0.92865
18	0.86765	0.13235		1.1154	−2.0223	0.7042	0.71738	0.95783
19	0.91667	0.08333		1.3832	−2.4849	0.9102	0.78389	0.98193
20	0.96569	0.03431		1.8213	−3.3722	1.2156	0.86089	0.99744
			S3:	0.2384	−19.1399	−10.8891		
			S4:	16.3484	33.1529	32.0184		

Total number of results: 21

i	M'(i	1 − M'(i)	Xn(i)	Xe(i)	Xw(i)	p0.05	p0.95
1	0.03271	0.96729	−1.7124	−0.0333	−3.4035	0.00244	0.13295
2	0.07944	0.92056	−1.3547	−0.0828	−2.4917	0.01719	0.20673
3	0.12617	0.87383	−1.1156	−0.1349	−2.0035	0.04010	0.27055
4	0.17290	0.82710	−0.9256	−0.1898	−1.6616	0.06781	0.32921
5	0.21963	0.78037	−0.7629	−0.2480	−1.3944	0.09884	0.38441
6	0.26636	0.73364	−0.6174	−0.3097	−1.1721	0.13245	0.43698
7	0.31308	0.68692	−0.4832	−0.3755	−0.9794	0.16818	0.48739
8	0.35981	0.64019	−0.3567	−0.4460	−0.8074	0.20575	0.53594
9	0.40654	0.59346	−0.2353	−0.5218	−0.6505	0.24499	0.58280
10	0.45327	0.54673	−0.1170	−0.6038	−0.5045	0.28580	0.62810
11	0.50000	0.50000	0.0000	−0.6931	−0.3665	0.32811	0.67189
12	0.54673	0.45327	0.1171	−0.7913	−0.2341	0.37190	0.71420
13	0.59346	0.40654	0.2360	−0.9001	−0.1053	0.41720	0.75501
14	0.64019	0.35981	0.3585	−1.0222	0.0219	0.46406	0.79425

Table 2 *cont.*

i	M'(i	1 − M'(i)		Xn(i)	Xe(i)	Xw(i)	p0.05	p0.95
15	0.68692	0.31308		0.4867	− 1.1613	0.1495	0.51261	0.83182
16	0.73364	0.26636		0.6235	− 1.3229	0.2798	0.56302	0.86755
17	0.78037	0.21963		0.7732	− 1.5158	0.4160	0.61559	0.90116
18	0.82710	0.17290		0.9427	− 1.7551	0.5625	0.67079	0.93219
19	0.87383	0.12617		1.1448	− 2.0701	0.7276	0.72945	0.95990
20	0.92056	0.07944		1.4091	− 2.5328	0.9293	0.79327	0.98281
21	0.96729	0.03271		1.8428	− 3.4201	1.2297	0.86705	0.99756
			S3:	0.2537	− 20.1303	− 11.4581		
			S4:	17.2641	35.0318	33.8698		

Total number of results: 22

1	0.03125	0.96875		− 1.7279	− 0.0317	− 3.4499	0.00233	0.12731
2	0.07589	0.92411		− 1.3761	− 0.0789	− 2.5392	0.01640	0.19812
3	0.12054	0.87946		− 1.1411	− 0.1284	− 2.0523	0.03822	0.25947
4	0.16518	0.83482		− 0.9547	− 0.1805	− 1.7118	0.06460	0.31591
5	0.20982	0.79018		− 0.7954	− 0.2355	− 1.4461	0.09411	0.36909
6	0.25446	0.74554		− 0.6532	− 0.2937	− 1.2254	0.12603	0.41980
7	0.29911	0.70089		− 0.5224	− 0.3554	− 1.0345	0.15994	0.46849
8	0.34375	0.65625		− 0.3995	− 0.4212	− 0.8646	0.19556	0.51546
9	0.38839	0.61161		− 0.2820	− 0.4917	− 0.7100	0.23272	0.56087
10	0.43304	0.56696		− 0.1679	− 0.5675	− 0.5666	0.27131	0.60484
11	0.47768	0.52232		− 0.0558	− 0.6495	− 0.4316	0.31126	0.64746
12	0.52232	0.47768		0.0558	− 0.7388	− 0.3027	0.35254	0.68874
13	0.56696	0.43304		0.1683	− 0.8369	− 0.1780	0.39516	0.72869
14	0.61161	0.38839		0.2831	− 0.9457	− 0.0558	0.43913	0.76728
15	0.65625	0.34375		0.4018	− 1.0678	0.0656	0.48454	0.80444
16	0.70089	0.29911		0.5266	− 1.2070	0.1881	0.53151	0.84006
17	0.74554	0.25446		0.6602	− 1.3686	0.3138	0.58020	0.87397
18	0.79018	0.20982		0.8068	− 1.5615	0.4456	0.63091	0.90589
19	0.83482	0.16518		0.9733	− 1.8007	0.5882	0.68409	0.93540
20	0.87946	0.12054		1.1724	− 2.1158	0.7494	0.74053	0.96178
21	0.92411	0.07589		1.4335	− 2.5784	0.9472	0.80188	0.98360
22	0.96875	0.03125		1.8631	− 3.4657	1.2429	0.87269	0.99767
			S3:	0.2690	− 21.1211	− 12.0275		
			S4:	18.1809	36.9153	35.7262		

Total number of results: 23

1	0.02991	0.97009		− 1.7424	− 0.0304	− 3.4943	0.00223	0.12212
2	0.07265	0.92735		− 1.3963	− 0.0754	− 2.5846	0.01567	0.19020
3	0.11538	0.88462		− 1.1651	− 0.1226	− 2.0988	0.03651	0.24925
4	0.15812	0.84188		− 0.9820	− 0.1721	− 1.7596	0.06167	0.30364
5	0.20085	0.79915		− 0.8258	− 0.2242	− 1.4952	0.08981	0.35193
6	0.24359	0.75641		− 0.6866	− 0.2792	− 1.2759	0.12021	0.40390
7	0.28632	0.71368		− 0.5589	− 0.3373	− 1.0867	0.15248	0.45097
8	0.32906	0.67094		− 0.4392	− 0.3991	− 0.9186	0.18634	0.49643
9	0.37179	0.62821		− 0.3252	− 0.4649	− 0.7660	0.22164	0.54046
10	0.41453	0.58547		− 0.2149	− 0.5353	− 0.6249	0.25824	0.58315
11	0.45727	0.54274		− 0.1069	− 0.6111	− 0.4924	0.29609	0.62461
12	0.50000	0.50000		0.0000	− 0.6931	− 0.3665	0.33515	0.66485
13	0.54274	0.45727		0.1071	− 0.7825	− 0.2453	0.37539	0.70391

Table 2 *cont.*

i	M'(i	1 − M'(i)		Xn(i)	Xe(i)	Xw(i)	p0.05	p0.95
14	0.58547	0.41453		0.2155	−0.8806	−0.1271	0.41684	0.74176
15	0.62821	0.37179		0.3267	−0.9894	−0.0106	0.45954	0.77836
16	0.67094	0.32906		0.4421	−1.1115	0.1057	0.50356	0.81366
17	0.71368	0.28632		0.5638	−1.2506	0.2236	0.54902	0.84752
18	0.75641	0.24359		0.6945	−1.4123	0.3452	0.59610	0.87978
19	0.79915	0.20085		0.8384	−1.6052	0.4732	0.64507	0.91019
20	0.84188	0.15812		1.0022	−1.8444	0.6122	0.69636	0.93832
21	0.88462	0.11538		1.1985	−2.1595	0.7699	0.75075	0.96348
22	0.92735	0.07265		1.4566	−2.6221	0.9640	0.80980	0.98433
23	0.97009	0.02991		1.8825	−3.5094	1.2554	0.87788	0.99777
			S3:	0.2844	−22.1123	−12.5973		
			S4:	19.0987	38.8031	37.5871		

Total number of results: 24

i	M'(i	1 − M'(i)		Xn(i)	Xe(i)	Xw(i)	p0.05	p0.95
1	0.02869	0.97131		−1.7560	−0.0291	−3.5367	0.00213	0.11735
2	0.06967	0.93033		−1.4153	−0.0722	−2.6281	0.01501	0.18289
3	0.11066	0.88934		−1.1878	−0.1173	−2.1433	0.03495	0.23980
4	0.15164	0.84836		−1.0077	−0.1644	−1.8052	0.05901	0.29227
5	0.19262	0.80738		−0.8543	−0.2140	−1.5419	0.08588	0.34181
6	0.23361	0.76639		−0.7179	−0.2661	−1.3240	0.11491	0.38914
7	0.27459	0.72541		−0.5931	−0.3210	−1.1363	0.14569	0.43469
8	0.31557	0.68443		−0.4763	−0.3792	−0.9698	0.17796	0.47853
9	0.35656	0.64344		−0.3653	−0.4409	−0.8189	0.21157	0.52142
10	0.39754	0.60246		−0.2584	−0.5067	−0.6798	0.24639	0.56289
11	0.43852	0.56148		−0.1541	−0.5772	−0.5496	0.28236	0.60321
12	0.47951	0.52049		−0.0512	−0.6530	−0.4262	0.31942	0.64244
13	0.52049	0.47951		0.0513	−0.7350	−0.3079	0.35756	0.68058
14	0.56148	0.43852		0.1544	−0.8243	−0.1932	0.39678	0.71764
15	0.60246	0.39754		0.2593	−0.9225	−0.0807	0.43711	0.75361
16	0.64344	0.35656		0.3672	−1.0313	0.0308	0.47858	0.78843
17	0.68443	0.31557		0.4797	−1.1534	0.1427	0.52127	0.82204
18	0.72541	0.27459		0.5986	−1.2925	0.2566	0.56531	0.85431
19	0.76639	0.23361		0.7268	−1.4541	0.3744	0.61086	0.88509
20	0.80738	0.19262		0.8681	−1.6470	0.4990	0.65819	0.91411
21	0.84836	0.15164		1.0294	−1.8863	0.6346	0.70773	0.94099
22	0.88934	0.11066		1.2232	−2.2013	0.7891	0.76020	0.96505
23	0.93033	0.06967		1.4785	−2.6640	0.9798	0.81711	0.98499
24	0.97131	0.02869		1.9008	−3.5513	1.2673	0.88265	0.99786
			S3:	0.2999	−23.1039	−13.1673		
			S4:	20.0174	40.6949	39.4522		

Total number of results: 25

i	M'(i	1 − M'(i)		Xn(i)	Xe(i)	Xw(i)	p0.05	p0.95
1	0.02756	0.97244		−1.7690	−0.0279	−3.5775	0.00205	0.11293
2	0.06693	0.93307		−1.4332	−0.0693	−2.6697	0.01440	0.17612
3	0.10630	0.89370		−1.2091	−0.1124	−2.1858	0.03352	0.23104
4	0.14567	0.85433		−1.0320	−0.1574	−1.8487	0.05656	0.28172
5	0.18504	0.81496		−0.8812	−0.2046	−1.5866	0.08229	0.32961
6	0.22441	0.77559		−0.7474	−0.2541	−1.3699	0.11006	0.37541
7	0.26378	0.73622		−0.6251	−0.3062	−1.1834	0.13947	0.41952
8	0.30315	0.69685		−0.5110	−0.3612	−1.0184	0.17030	0.46221

Table 2 *cont.*

i	$M'(i$	$1 - M'(i)$	$Xn(i)$	$Xe(i)$	$Xw(i)$	$p0.05$	$p0.95$
9	0.34252	0.65748	−0.4028	−0.4193	−0.8691	0.20238	0.50364
10	0.38189	0.61811	−0.2988	−0.4811	−0.7317	0.23559	0.54393
11	0.42126	0.57874	−0.1978	−0.5469	−0.6035	0.26985	0.58316
12	0.46063	0.53937	−0.0985	−0.6174	−0.4823	0.30513	0.62138
13	0.50000	0.50000	0.0000	−0.6931	−0.3665	0.34139	0.65861
14	0.53937	0.46063	0.0986	−0.7752	−0.2547	0.37862	0.69487
15	0.57874	0.42126	0.1983	−0.8645	−0.1456	0.41684	0.73015
16	0.61811	0.38189	0.3001	−0.9626	−0.0381	0.45607	0.76441
17	0.65748	0.34252	0.4052	−1.0714	0.0690	0.49636	0.79762
18	0.69685	0.30315	0.5150	−1.1935	0.1769	0.53779	0.82970
19	0.73622	0.26378	0.6314	−1.3326	0.2872	0.58048	0.86052
20	0.77559	0.22441	0.7572	−1.4943	0.4016	0.62459	0.88994
21	0.81496	0.18504	0.8962	−1.6872	0.5231	0.67039	0.91771
22	0.85433	0.14567	1.0552	−1.9264	0.6557	0.71828	0.94344
23	0.89370	0.10630	1.2466	−2.2415	0.8071	0.76896	0.96648
24	0.93307	0.06693	1.4994	−2.7041	0.9948	0.82388	0.98560
25	0.97244	0.02756	1.9184	−3.5914	1.2785	0.88707	0.99795
			S3: 0.3155	−24.0958	−13.7376		
			S4: 20.9370	42.5904	41.3212		

Total number of results: 26

i	$M'(i$	$1 - M'(i)$	$Xn(i)$	$Xe(i)$	$Xw(i)$	$p0.05$	$p0.95$
1	0.02652	0.97348	−1.7812	−0.0269	−3.6166	0.00197	0.10883
2	0.06439	0.93561	−1.4502	−0.0666	−2.7096	0.01384	0.16983
3	0.10227	0.89773	−1.2294	−0.1079	−2.2267	0.03220	0.22289
4	0.14015	0.85985	−1.0549	−0.1510	−1.8905	0.05431	0.27190
5	0.17803	0.82197	−0.9066	−0.1961	−1.6294	0.07899	0.31824
6	0.21591	0.78409	−0.7752	−0.2432	−1.4137	0.10560	0.36260
7	0.25379	0.74621	−0.6552	−0.2927	−1.2285	0.13377	0.40535
8	0.29167	0.70833	−0.5436	−0.3448	−1.0647	0.16328	0.44677
9	0.32955	0.67045	−0.4379	−0.3998	−0.9168	0.19396	0.48700
10	0.36742	0.63258	−0.3366	−0.4580	−0.7810	0.22570	0.52616
11	0.40530	0.59470	−0.2385	−0.5197	−0.6545	0.25842	0.56434
12	0.44318	0.55682	−0.1423	−0.5855	−0.5353	0.29508	0.60158
13	0.48106	0.51894	−0.0473	−0.6560	−0.4216	0.32664	0.63791
14	0.51894	0.48106	0.0474	−0.7318	−0.3123	0.36209	0.67336
15	0.55682	0.44318	0.1426	−0.8138	−0.2061	0.39842	0.70792
16	0.59470	0.40530	0.2392	−0.9031	−0.1019	0.43566	0.74158
17	0.63258	0.36742	0.3382	−1.0012	0.0012	0.47384	0.77430
18	0.67045	0.32955	0.4407	−1.1100	0.1044	0.51300	0.80604
19	0.70833	0.29167	0.5481	−1.2321	0.2088	0.55323	0.83672
20	0.74621	0.25379	0.6623	−1.3713	0.3157	0.59465	0.86623
21	0.78409	0.21591	0.7859	−1.5329	0.4272	0.63740	0.89440
22	0.82197	0.17803	0.9228	−1.7258	0.5457	0.68176	0.92101
23	0.85985	0.14015	1.0797	−1.9650	0.6755	0.72810	0.94569
24	0.89773	0.10227	1.2689	−2.2801	0.8242	0.77711	0.96780
25	0.93561	0.06439	1.5192	−2.7427	1.0090	0.83017	0.98616
26	0.97348	0.02652	1.9351	−3.6300	1.2892	0.89117	0.99803
			S3: 0.3311	−25.0881	−14.3082		
			S4: 21.8573	44.4893	43.1939		

Table 2 *cont.*

i	M'(i	1 − M'(i)	Xn(i)	Xe(i)	Xw(i)	p0.05	p0.95
Total number of results: 27							
1	0.02555	0.97445	−1.7928	−0.0259	−3.6543	0.00190	0.10502
2	0.06204	0.93796	−1.4664	−0.0641	−2.7481	0.01332	0.16397
3	0.09854	0.90146	−1.2486	−0.1037	−2.2659	0.03098	0.21530
4	0.13504	0.86496	−1.0767	−0.1451	−1.9306	0.05223	0.26274
5	0.17153	0.82847	−0.9307	−0.1882	−1.6704	0.07594	0.30763
6	0.20803	0.79197	−0.8014	−0.2332	−1.4557	0.10148	0.35062
7	0.24453	0.75547	−0.6837	−0.2804	−1.2715	0.12852	0.39210
8	0.28102	0.71898	−0.5743	−0.3299	−1.1089	0.15682	0.43230
9	0.31752	0.68248	−0.4709	−0.3820	−0.9623	0.18622	0.47139
10	0.35401	0.64599	−0.3721	−0.4370	−0.8279	0.21662	0.50489
11	0.39051	0.60949	−0.2765	−0.4951	−0.7029	0.24793	0.54664
12	0.42701	0.57299	−0.1832	−0.5569	−0.5854	0.28012	0.58293
13	0.46350	0.53650	−0.0913	−0.6227	−0.4737	0.31314	0.61839
14	0.50000	0.50000	0.0000	−0.6931	−0.3665	0.34697	0.65303
15	0.53650	0.46350	0.0914	−0.7689	−0.2627	0.38161	0.68686
16	0.57299	0.42701	0.1836	−0.8510	−0.1614	0.41707	0.71988
17	0.60949	0.39051	0.2776	−0.9403	−0.0616	0.45336	0.75207
18	0.64599	0.35401	0.3741	−1.0384	0.0377	0.49052	0.78338
19	0.68248	0.31752	0.4742	−1.1472	0.1373	0.52861	0.81378
20	0.71898	0.28102	0.5794	−1.2693	0.2385	0.56770	0.84318
21	0.75547	0.24453	0.6915	−1.4084	0.3425	0.60790	0.87148
22	0.79197	0.20803	0.8131	−1.5701	0.4511	0.64936	0.89851
23	0.82847	0.17153	0.9481	−1.7630	0.5670	0.69237	0.92406
24	0.86496	0.13504	1.1029	−2.0022	0.6943	0.73726	0.94777
25	0.90146	0.09854	1.2901	−2.3173	0.8404	0.78470	0.96902
26	0.93796	0.06204	1.5382	−2.7799	1.0224	0.83603	0.98668
27	0.97445	0.02555	1.9511	−3.6672	1.2994	0.89498	0.99810
		S3:	0.3467	−26.0806	−14.8790		
		S4:	22.7783	46.3914	45.0699		
Total number of results: 28							
1	0.02465	0.97535	−1.8039	−0.0250	−3.6906	0.00183	0.10147
2	0.05986	0.94014	−1.4818	−0.0617	−2.7851	0.01284	0.15851
3	0.09507	0.90493	−1.2668	−0.0999	−2.3036	0.02985	0.20821
4	0.13028	0.86972	−1.0973	−0.1396	−1.9691	0.05031	0.25417
5	0.16549	0.83451	−0.9535	−0.1809	−1.7097	0.07311	0.29769
6	0.20070	0.79930	−0.8263	−0.2240	−1.4960	0.09768	0.33940
7	0.23592	0.76408	−0.7106	−0.2691	−1.3128	0.12367	0.37967
8	0.27113	0.72887	−0.6033	−0.3163	−1.1512	0.15085	0.41873
9	0.30634	0.69366	−0.5020	−0.3658	−1.0057	0.17908	0.45673
10	0.34155	0.65845	−0.4054	−0.4179	−0.8726	0.20824	0.49379
11	0.37676	0.62324	−0.3122	−0.4728	−0.7490	0.23827	0.52998
12	0.41197	0.58803	−0.2214	−0.5310	−0.6330	0.26911	0.56536
13	0.44718	0.55282	−0.1323	−0.5927	−0.5230	0.30072	0.59996
14	0.48239	0.51761	−0.0440	−0.6585	−0.4177	0.33309	0.63380
15	0.51761	0.48239	0.0440	−0.7290	−0.3161	0.36620	0.66691
16	0.55282	0.44718	0.1325	−0.8048	−0.2172	0.40004	0.69927
17	0.58803	0.41197	0.2221	−0.8868	−0.1201	0.43464	0.73089

Table 2 *cont.*

i	M'(i	1 − M'(i)	Xn(i)	Xe(i)	Xw(i)	p0.05	p0.95
18	0.62324	0.37676	0.3136	−0.9761	−0.0241	0.47002	0.46173
19	0.65845	0.34155	0.4078	−1.0743	0.0716	0.50621	0.79176
20	0.69366	0.30634	0.5058	−1.1831	0.1681	0.54327	0.82092
21	0.72887	0.27113	0.6091	−1.3052	0.2663	0.58127	0.84915
22	0.76408	0.23592	0.7192	−1.4443	0.3676	0.62033	0.87633
23	0.79930	0.20070	0.8389	−1.6059	0.4737	0.66060	0.90232
24	0.83451	0.16549	0.9721	−1.7988	0.5871	0.70231	0.92689
25	0.86972	0.13028	1.1251	−2.0381	0.7120	0.74583	0.94696
26	0.90493	0.09507	1.3104	−2.3531	0.8557	0.79179	0.97015
27	0.94014	0.05986	1.5563	−2.8158	1.0352	0.84149	0.98716
28	0.97535	0.02465	1.9665	−3.7031	1.3092	0.89853	0.99817
		S3:	0.3625	−27.0734	−15.4501		
		S4:	23.7000	48.2965	46.9491		

Total number of results: 29

i	M'(i	1 − M'(i)	Xn(i)	Xe(i)	Xw(i)	p0.05	p0.95
1	0.02381	0.97619	−1.8144	−0.0241	−3.7256	0.00177	0.09814
2	0.05782	0.94218	−1.4964	−0.0596	−2.8207	0.01239	0.15339
3	0.09184	0.90816	−1.2842	−0.0963	−2.3400	0.02879	0.20156
4	0.12585	0.87415	−1.1170	−0.1345	−2.0062	0.04852	0.24614
5	0.15986	0.84014	−0.9752	−0.1742	−1.7476	0.07049	0.28837
6	0.19388	0.80612	−0.8500	−0.2155	−1.5347	0.09415	0.32887
7	0.22789	0.77211	−0.7362	−0.2586	−1.3524	0.11917	0.36800
8	0.26190	0.73810	−0.6307	−0.3037	−1.1918	0.14532	0.40597
9	0.29592	0.70408	−0.5314	−0.3509	−1.0474	0.17246	0.44294
10	0.32993	0.67007	−0.4369	−0.4004	−0.9154	0.20050	0.47901
11	0.36395	0.63605	−0.3458	−0.4525	−0.7930	0.22934	0.51427
12	0.39796	0.60204	−0.2573	−0.5074	−0.6784	0.25894	0.54877
13	0.43197	0.56803	−0.1706	−0.5656	−0.5699	0.28927	0.58254
14	0.46599	0.53401	−0.0851	−0.6273	−0.4663	0.32030	0.61561
15	0.50000	0.50000	0.0000	−0.6931	−0.3665	0.35200	0.64799
16	0.53401	0.46599	0.0852	−0.7636	−0.2697	0.38439	0.67970
17	0.56803	0.43197	0.1710	−0.8394	−0.1751	0.41746	0.71073
18	0.60204	0.39796	0.2582	−0.9214	−0.0819	0.45123	0.74106
19	0.63605	0.36395	0.3475	−1.0108	0.0107	0.48573	0.77066
20	0.67007	0.32993	0.4397	−1.1089	0.1033	0.52099	0.79950
21	0.70408	0.29592	0.5358	−1.2177	0.1969	0.55076	0.82753
22	0.73810	0.26190	0.6372	−1.3398	0.2925	0.59403	0.85468
23	0.77211	0.22789	0.7456	−1.4789	0.3913	0.63200	0.88083
24	0.80612	0.19388	0.8635	−1.6405	0.4950	0.67113	0.90584
25	0.84014	0.15986	0.9950	−1.8334	0.6062	0.71168	0.92951
26	0.87415	0.12585	1.1463	−2.0727	0.7288	0.75386	0.95148
27	0.90816	0.09184	1.3297	−2.3877	0.8703	0.79844	0.97120
28	0.94218	0.05782	1.5736	−2.8504	1.0474	0.84661	0.98761
29	0.97619	0.02381	1.9812	−3.7377	1.3185	0.90185	0.99823
		S3:	0.3783	−28.0665	−16.0214		
		S4:	24.6223	50.2044	48.8312		

Total number of results: 30

i	M'(i	1 − M'(i)	Xn(i)	Xe(i)	Xw(i)	p0.05	p0.95
1	0.02303	0.97697	−1.8244	−0.0233	−3.7595	0.00171	0.09503
2	0.05592	0.94408	−1.5104	−0.0575	−2.8552	0.01189	0.14860

Table 2 *cont.*

i	M'(i	1 − M'(i)	Xn(i)	Xe(i)	Xw(i)	p0.05	p0.95
3	0.08882	0.91118	− 1.3008	− 0.0930	− 2.3750	0.02781	0.19533
4	0.12171	0.87829	− 1.1357	− 0.1298	− 2.0419	0.04685	0.23860
5	0.15461	0.84539	− 0.9959	− 0.1680	− 1.7841	0.06806	0.27962
6	0.18750	0.81250	− 0.8724	− 0.2076	− 1.5720	0.09087	0.31897
7	0.22039	0.77961	− 0.7604	− 0.2490	− 1.3904	0.11499	0.35701
8	0.25329	0.74671	− 0.6568	− 0.2921	− 1.2307	0.14018	0.39395
9	0.28618	0.71382	− 0.5593	− 0.3371	− 1.0873	0.16633	0.42993
10	0.31908	0.68092	− 0.4666	− 0.3843	− 0.9563	0.19331	0.46507
11	0.35197	0.64803	− 0.3775	− 0.4338	− 0.8351	0.22106	0.49944
12	0.38487	0.61513	− 0.2911	− 0.4859	− 0.7217	0.24953	0.53309
13	0.41776	0.58224	− 0.2067	− 0.5409	− 0.6146	0.27867	0.56605
14	0.45066	0.54934	− 0.1235	− 0.5990	− 0.5124	0.30846	0.59837
15	0.48355	0.51645	− 0.0411	− 0.6608	− 0.4143	0.33889	0.63005
16	0.51645	0.48355	0.0411	− 0.7266	− 0.3194	0.36995	0.66111
17	0.54934	0.45066	0.1237	− 0.7970	− 0.2268	0.40163	0.69154
18	0.58224	0.41776	0.2072	− 0.8728	− 0.1360	0.43394	0.72133
19	0.61513	0.38487	0.2923	− 0.9549	− 0.0462	0.46691	0.75047
20	0.64803	0.35197	0.3796	− 1.0442	0.0432	0.50056	0.77894
21	0.68092	0.31908	0.4699	− 1.1423	0.1331	0.53493	0.80669
22	0.71382	0.28618	0.5642	− 1.2511	0.2240	0.57007	0.83367
23	0.74671	0.25329	0.6639	− 1.3732	0.3172	0.60605	0.85981
24	0.77961	0.22039	0.7706	− 1.5123	0.4137	0.64299	0.88501
25	0.81250	0.18750	0.8870	− 1.6740	0.5152	0.68103	0.90913
26	0.84539	0.15461	1.0169	− 1.8669	0.6243	0.72038	0.93194
27	0.87829	0.12171	1.1666	− 2.1061	0.7448	0.76140	0.95314
28	0.91118	0.08882	1.3483	− 2.4212	0.8843	0.80467	0.97218
29	0.94408	0.05592	1.5903	− 2.8838	1.0591	0.85140	0.98802
30	0.97697	0.02303	1.9953	− 3.7711	1.3274	0.90497	0.99829
			S3: 0.3941	− 29.0598	− 16.5928		
			S4: 25.5451	52.1149	50.7161		

Total number of results: 31

i	M'(i	1 − M'(i)	Xn(i)	Xe(i)	Xw(i)	p0.05	p0.95
1	0.02229	0.97771	− 1.8340	− 0.0225	− 3.7922	0.00165	0.09211
2	0.05414	0.94586	− 1.5237	− 0.0557	− 2.8885	0.01158	0.14409
3	0.08599	0.91401	− 1.3167	− 0.0899	− 2.4089	0.02690	0.18946
4	0.11783	0.88217	− 1.1536	− 0.1254	− 2.0764	0.04530	0.23150
5	0.14968	0.85032	− 1.0156	− 0.1621	− 1.8193	0.06578	0.27137
6	0.18153	0.81847	− 0.8939	− 0.2003	− 1.6079	0.08781	0.30964
7	0.21338	0.78662	− 0.7836	− 0.2400	− 1.4271	0.11109	0.34665
8	0.24522	0.75478	− 0.6815	− 0.2813	− 1.2682	0.13540	0.38261
9	0.27707	0.72293	− 0.5858	− 0.3244	− 1.1256	0.16061	0.41766
10	0.30892	0.69108	− 0.4948	− 0.3695	− 0.9956	0.18662	0.45190
11	0.34076	0.65924	− 0.4075	− 0.4167	− 0.8755	0.21336	0.48542
12	0.37261	0.62739	− 0.3230	− 0.4662	− 0.7632	0.24077	0.51825
13	0.40446	0.59554	− 0.2406	− 0.5183	− 0.6572	0.26883	0.55044
14	0.43631	0.56369	− 0.1597	− 0.5732	− 0.5564	0.29749	0.58203
15	0.46815	0.53185	− 0.0796	− 0.6314	− 0.4598	0.32674	0.61302
16	0.50000	0.50000	0.0000	− 0.6931	− 0.3665	0.35657	0.64343
17	0.53185	0.46815	0.0797	− 0.7590	− 0.2758	0.38698	0.67326

Table 2 *cont.*

i	M'(i	1 − M'(i)	Xn(i)	Xe(i)	Xw(i)	p0.05	p0.95
18	0.56369	0.43631	0.1600	−0.8294	−0.1870	0.41797	0.70251
19	0.59554	0.40446	0.2414	−0.9052	−0.0996	0.44956	0.73117
20	0.62739	0.37261	0.3245	−0.9872	−0.0129	0.48175	0.75922
21	0.65924	0.34076	0.4099	−1.0766	0.0738	0.51458	0.78664
22	0.69108	0.30892	0.4985	−1.1747	0.1610	0.54810	0.81338
23	0.72293	0.27707	0.5912	−1.2835	0.2496	0.58234	0.83939
24	0.75478	0.24522	0.6893	−1.4056	0.3405	0.64739	0.86460
25	0.78662	0.21338	0.7946	−1.5447	0.4348	0.65336	0.88891
26	0.81847	0.18153	0.9094	−1.7063	0.5344	0.69036	0.91219
27	0.85032	0.14968	1.0378	−1.8992	0.6415	0.72563	0.93422
28	0.88217	0.11783	1.1860	−2.1385	0.7601	0.76650	0.95470
29	0.91401	0.08599	1.3661	−2.4536	0.8975	0.81054	0.97310
30	0.94586	0.05414	1.6063	−2.9162	1.0703	0.85591	0.98841
31	0.97771	0.02229	2.0090	−3.8035	1.3359	0.90789	0.99835
			S3: 0.4100	−30.0533	−17.1644		
			S4: 26.4685	54.0279	52.6035		

Total number of results: 32

i	M'(i	1 − M'(i)	Xn(i)	Xe(i)	Xw(i)	p0.05	p0.95
1	0.02160	0.97840	−1.8432	−0.0218	−3.8239	0.00160	0.08937
2	0.05247	0.94753	−1.5365	−0.0539	−2.9207	0.01122	0.13985
3	0.08333	0.91667	−1.3319	−0.0870	−2.4417	0.02604	0.18394
4	0.11420	0.88580	−1.1708	−0.1213	−2.1098	0.04384	0.22482
5	0.14506	0.85494	−1.0345	−0.1567	−1.8533	0.06365	0.26360
6	0.17593	0.82407	−0.9144	−0.1935	−1.6425	0.08495	0.30084
7	0.20679	0.79321	−0.8056	−0.2317	−1.4625	0.10745	0.33687
8	0.23765	0.76235	−0.7051	−0.2714	−1.3043	0.13093	0.37190
9	0.26852	0.73148	−0.6110	−0.3127	−1.1626	0.15528	0.40606
10	0.29938	0.70062	−0.5216	−0.3558	−1.0334	0.18038	0.43945
11	0.33025	0.66975	−0.4360	−0.4008	−0.9142	0.20618	0.47214
12	0.36111	0.63889	−0.3533	−0.4480	−0.8029	0.23262	0.50419
13	0.39198	0.60802	−0.2727	−0.4975	−0.6981	0.25966	0.53564
14	0.42284	0.57716	−0.1938	−0.5496	−0.5985	0.28727	0.56651
15	0.45370	0.54630	−0.1159	−0.6046	−0.5032	0.31544	0.59683
16	0.48457	0.51543	−0.0386	−0.6627	−0.4114	0.34415	0.62661
17	0.51543	0.48457	0.0386	−0.7245	−0.3223	0.37339	0.65585
18	0.54630	0.45370	0.1160	−0.7903	−0.2353	0.40317	0.68456
19	0.57716	0.42284	0.1943	−0.8608	−0.1499	0.43349	0.71272
20	0.60802	0.39198	0.2737	−0.9366	−0.0655	0.46436	0.74034
21	0.63889	0.36111	0.3550	−1.0186	0.0184	0.49581	0.76738
22	0.66975	0.33025	0.4388	−1.1079	0.1025	0.52786	0.79382
23	0.70062	0.29938	0.5258	−1.2060	0.1873	0.56055	0.81961
24	0.73148	0.26852	0.6170	−1.3148	0.2737	0.59314	0.84472
25	0.76235	0.23765	0.7136	−1.4369	0.3625	0.62810	0.86907
26	0.79321	0.20679	0.8174	−1.5761	0.4549	0.66313	0.89255
27	0.82407	0.17593	0.9309	−1.7377	0.5526	0.69916	0.91505
28	0.85494	0.14506	1.0579	−1.9306	0.6578	0.73640	0.93635
29	0.88580	0.11420	1.2046	−2.1698	0.7746	0.77518	0.95615
30	0.91667	0.08333	1.3832	−2.4849	0.9102	0.81606	0.97396
31	0.94753	0.05247	1.6217	−2.9475	1.0810	0.86015	0.98878

Table 2 *cont.*

i	M'(i	1 − M'(i)		Xn(i)	Xe(i)	Xw(i)	p0.05	p0.95
32	0.97840	0.02160		2.0221	−3.8348	1.3441	0.91063	0.99840
			S3:	0.4259	−31.0470	−17.7363		
			S4:	27.3923	55.9432	54.4935		
Total number of results: 33								
1	0.02096	0.97904		−1.8520	−0.0212	−3.8547	0.00155	0.08678
2	0.05090	0.94910		−1.5488	−0.0522	−2.9519	0.01086	0.13585
3	0.08084	0.91916		−1.3464	−0.0843	−2.4735	0.02524	0.17873
4	0.11078	0.88922		−1.1872	−0.1174	−2.1421	0.04246	0.21850
5	0.14072	0.85928		−1.0525	−0.1517	−1.8861	0.06166	0.25625
6	0.17066	0.82934		−0.9340	−0.1871	−1.6760	0.08227	0.29252
7	0.20060	0.79940		−0.8267	−0.2239	−1.4966	0.10404	0.32763
8	0.23054	0.76946		−0.7277	−0.2621	−1.3392	0.12675	0.36176
9	0.26048	0.73952		−0.6350	−0.3018	−1.1981	0.15029	0.39507
10	0.29042	0.70958		−0.5471	−0.3431	−1.0698	0.17455	0.42765
11	0.32036	0.67964		−0.4631	−0.3862	−0.9514	0.19948	0.45956
12	0.35030	0.64970		−0.3820	−0.4312	−0.8411	0.22501	0.49086
13	0.38024	0.61976		−0.3031	−0.4784	−0.7373	0.25111	0.52159
14	0.41018	0.58982		−0.2260	−0.5279	−0.6388	0.27775	0.55177
15	0.44012	0.55988		−0.1500	−0.5800	−0.5447	0.30491	0.58144
16	0.47006	0.52994		−0.0749	−0.6350	−0.4541	0.33258	0.61060
17	0.50000	0.50000		0.0000	−0.6931	−0.3665	0.36074	0.63926
18	0.52994	0.47006		0.0749	−0.7549	−0.2812	0.38940	0.66742
19	0.55988	0.44012		0.1503	−0.8207	−0.1976	0.41656	0.69509
20	0.58982	0.41018		0.2267	−0.8912	−0.1152	0.44823	0.72225
21	0.61976	0.38024		0.3044	−0.9670	−0.0336	0.47841	0.74889
22	0.64970	0.35030		0.3841	−1.0490	0.0478	0.50914	0.77499
23	0.67964	0.32036		0.4663	−1.1383	0.1295	0.54344	0.80052
24	0.70958	0.29042		0.5518	−1.2364	0.2122	0.57235	0.82545
25	0.73952	0.26048		0.6415	−1.3452	0.2966	0.60493	0.84971
26	0.76946	0.23054		0.7368	−1.4673	0.3834	0.63824	0.87325
27	0.79940	0.20060		0.8393	−1.6064	0.4740	0.67237	0.89596
28	0.82934	0.17066		0.9515	−1.7681	0.5699	0.70748	0.91772
29	0.85928	0.14072		1.0771	−1.9610	0.6735	0.74375	0.93834
30	0.88922	0.11078		1.2225	−2.2002	0.7886	0.76150	0.95752
31	0.91916	0.08084		1.3997	−2.5153	0.9224	0.82127	0.97476
32	0.94910	0.05090		1.6366	−2.9779	1.0912	0.86415	0.98912
33	0.97904	0.02096		2.0348	−3.8652	1.3520	0.91322	0.99845
			S3:	0.4441	−32.0409	−18.3082		
			S4:	28.3166	57.8607	56.3858		
Total number of results: 34								
1	0.02035	0.97965		−1.8605	−0.0206	−3.8845	0.00151	0.08434
2	0.04942	0.95058		−1.5606	−0.0507	−2.9822	0.01055	0.13207
3	0.07849	0.92151		−1.3604	−0.0817	−2.5042	0.02448	0.17381
4	0.10756	0.89244		−1.2029	−0.1138	−2.1734	0.04120	0.21253
5	0.13663	0.86337		−1.0698	−0.1469	−1.9179	0.05978	0.24931
6	0.16570	0.83430		−0.9527	−0.1812	−1.7084	0.07976	0.28465
7	0.19477	0.80523		−0.8468	−0.2166	−1.5296	0.10084	0.31887
8	0.22384	0.77616		−0.7492	−0.2534	−1.3728	0.12283	0.35216

Table 2 *cont.*

i	M'(i	1 − M'(i)	Xn(i)	Xe(i)	Xw(i)	p0.05	p0.95
9	0.25291	0.74709	−0.6579	−0.2916	−1.2325	0.14561	0.38466
10	0.28198	0.71802	−0.5715	−0.3313	−1.1049	0.16909	0.41645
11	0.31105	0.68895	−0.4889	−0.3726	−0.9873	0.19319	0.44761
12	0.34012	0.65988	−0.4093	−0.4157	−0.8778	0.21788	0.47819
13	0.36919	0.63081	−0.3320	−0.4607	−0.7749	0.24310	0.50823
14	0.39826	0.60174	−0.2565	−0.5079	−0.6774	0.26884	0.53775
15	0.42733	0.57267	−0.1824	−0.5574	−0.5844	0.29507	0.56678
16	0.45640	0.54360	−0.1091	−0.6095	−0.4951	0.32177	0.59534
17	0.48547	0.51453	−0.0363	−0.6645	−0.4087	0.34894	0.62343
18	0.51453	0.48547	0.0363	−0.7226	−0.3248	0.37657	0.65106
19	0.54360	0.45640	0.1093	−0.7844	−0.2428	0.40466	0.67823
20	0.57267	0.42733	0.1828	−0.8502	−0.1623	0.43321	0.70493
21	0.60174	0.39826	0.2574	−0.9207	−0.0827	0.46225	0.73116
22	0.63081	0.36919	0.3336	−0.9965	−0.0036	0.49177	0.75689
23	0.65988	0.34012	0.4117	−1.0785	0.0755	0.52181	0.78212
24	0.68895	0.31105	0.4925	−1.1678	0.1551	0.55239	0.80680
25	0.71802	0.28198	0.5766	−1.2659	0.2358	0.58355	0.83091
26	0.74709	0.25291	0.6651	−1.3747	0.3183	0.61534	0.85439
27	0.77616	0.22384	0.7591	−1.4968	0.4034	0.64754	0.87717
28	0.80523	0.19477	0.8603	−1.6359	0.4922	0.68113	0.89916
29	0.83430	0.16570	0.9713	−1.7976	0.5864	0.71535	0.92024
30	0.86337	0.13663	1.0956	−1.9905	0.6884	0.75069	0.94021
31	0.89244	0.10756	1.2398	−2.2297	0.8019	0.78747	0.95880
32	0.92151	0.07849	1.4156	−2.5448	0.9341	0.82619	0.97552
33	0.95058	0.04942	1.6509	−3.0074	1.1011	0.86793	0.98945
34	0.97965	0.02035	2.0470	−3.8947	1.3596	0.91566	0.99849
			S3: 0.4578	−33.0349	−18.8804		
			S4: 29.2413	59.7803	58.2802		

Total number of results: 35

i	M'(i	1 − M'(i)	Xn(i)	Xe(i)	Xw(i)	p0.05	p0.95
1	0.01977	0.98023	−1.8686	−0.0200	−3.9134	0.00146	0.08203
2	0.04802	0.95198	−1.5719	−0.0492	−3.0116	0.01025	0.12850
3	0.07627	0.92373	−1.3738	−0.0793	−2.5341	0.02377	0.16915
4	0.10452	0.89548	−1.2180	−0.1104	−2.2037	0.03999	0.20688
5	0.13277	0.86723	−1.0865	−0.1424	−1.9488	0.05802	0.24272
6	0.16102	0.83898	−0.9707	−0.1756	−1.7397	0.07739	0.27718
7	0.18927	0.81073	−0.8662	−0.2098	−1.5615	0.09783	0.31056
8	0.21751	0.78249	−0.7699	−0.2453	−1.4054	0.11914	0.34305
9	0.24576	0.75424	−0.6799	−0.2820	−1.2657	0.14122	0.37477
10	0.27401	0.72599	−0.5948	−0.3202	−1.1387	0.16396	0.40582
11	0.30226	0.69774	−0.5135	−0.3599	−1.0219	0.18730	0.43626
12	0.33051	0.66949	−0.4353	−0.4012	−0.9132	0.21119	0.46615
13	0.35876	0.64124	−0.3595	−0.4443	−0.8112	0.23560	0.49552
14	0.38701	0.61299	−0.2856	−0.4894	−0.7146	0.26049	0.52440
15	0.41525	0.58475	−0.2130	−0.5366	−0.6225	0.28585	0.55282
16	0.44350	0.55650	−0.1415	−0.5861	−0.5343	0.31165	0.58080
17	0.47175	0.52825	−0.0706	−0.6382	−0.4491	0.33789	0.60833
18	0.50000	0.50000	0.0000	−0.6931	−0.3665	0.36457	0.63543
19	0.52825	0.47175	0.0707	−0.7513	−0.2859	0.39167	0.66210

Table 2 *cont.*

i	M'(i	1 − M'(i)	Xn(i)	Xe(i)	Xw(i)	p0.05	p0.95
20	0.55650	0.44350	0.1418	−0.8131	−0.2070	0.41920	0.68835
21	0.58475	0.41525	0.2137	−0.8789	−0.1291	0.44717	0.71415
22	0.61299	0.38701	0.2867	−0.9493	−0.0520	0.47560	0.73951
23	0.64124	0.35876	0.3613	−1.0251	0.0248	0.50448	0.76440
24	0.66949	0.33051	0.4381	−1.1071	0.1018	0.53385	0.78881
25	0.69774	0.30226	0.5175	−1.1965	0.1794	0.56374	0.81270
26	0.72599	0.27401	0.6004	−1.2946	0.2582	0.59416	0.83604
27	0.75424	0.24576	0.6876	−1.4034	0.3389	0.62523	0.85878
28	0.78249	0.21751	0.7804	−1.5255	0.4223	0.65695	0.88086
29	0.81073	0.18927	0.8805	−1.6646	0.5096	0.68944	0.90217
30	0.83898	0.16102	0.9902	−1.8262	0.6023	0.72282	0.92261
31	0.86723	0.13277	1.1135	−2.0191	0.7027	0.75728	0.94198
32	0.89548	0.10452	1.2564	−2.2584	0.8146	0.79312	0.96001
33	0.92373	0.07627	1.4309	−2.5735	0.9453	0.83085	0.97623
34	0.95198	0.04802	1.6647	−3.0361	1.1106	0.87150	0.98975
35	0.98023	0.01977	2.0589	−3.9234	1.3670	0.91797	0.99854
			S3: 0.4739	−34.0292	−19.4526		
			S4: 30.1665	61.7019	60.1767		

Total number of results: 36

i	M'(i	1 − M'(i)	Xn(i)	Xe(i)	Xw(i)	p0.05	p0.95
1	0.01923	0.98077	−1.8764	−0.0194	−3.9416	0.00142	0.07985
2	0.04670	0.95330	−1.5828	−0.0478	−3.0401	0.00996	0.12512
3	0.07418	0.92582	−1.3867	−0.0771	−2.5630	0.02310	0.16474
4	0.10165	0.89835	−1.2326	−0.1072	−2.2331	0.03885	0.20152
5	0.12912	0.87088	−1.1024	−0.1383	−1.9787	0.05636	0.23648
6	0.15659	0.84341	−0.9880	−0.1703	−1.7702	0.07516	0.27010
7	0.18407	0.81593	−0.8847	−0.2034	−1.5925	0.09499	0.30268
8	0.21154	0.78846	−0.7897	−0.2377	−1.4369	0.11567	0.33439
9	0.23901	0.76099	−0.7009	−0.2731	−1.2978	0.13708	0.36537
10	0.26648	0.73352	−0.6170	−0.3099	−1.1715	0.15913	0.39571
11	0.29396	0.70604	−0.5370	−0.3481	−1.0553	0.18175	0.42546
12	0.32143	0.67857	−0.4601	−0.3878	−0.9474	0.20491	0.45468
13	0.34890	0.65110	−0.3857	−0.4291	−0.8461	0.22855	0.48341
14	0.37637	0.62363	−0.3132	−0.4722	−0.7503	0.25265	0.51168
15	0.40385	0.59615	−0.2422	−0.5173	−0.6592	0.27719	0.53951
16	0.43132	0.56868	−0.1723	−0.5644	−0.5719	0.30216	0.56691
17	0.45879	0.54121	−0.1031	−0.6139	−0.4878	0.32754	0.59391
18	0.48626	0.51374	−0.0343	−0.6660	−0.4064	0.35332	0.62049
19	0.51374	0.48626	0.0343	−0.7210	−0.3271	0.37951	0.64668
20	0.54121	0.45879	0.1032	−0.7792	−0.2495	0.40609	0.67246
21	0.56868	0.43132	0.1727	−0.8409	−0.1733	0.43309	0.69784
22	0.59615	0.40385	0.2430	−0.9067	−0.0979	0.46049	0.72280
23	0.62363	0.37637	0.3146	−0.9772	−0.0231	0.48832	0.74735
24	0.65110	0.34890	0.3878	−1.0530	0.0516	0.51658	0.77145
25	0.67857	0.32143	0.4633	−1.1350	0.1266	0.54532	0.79509
26	0.70604	0.29396	0.5415	−1.2243	0.2024	0.57454	0.81825
27	0.73352	0.26648	0.6231	−1.3224	0.2795	0.60429	0.84087
28	0.76099	0.23901	0.7092	−1.4312	0.3585	0.63483	0.86292
29	0.78846	0.21154	0.8009	−1.5533	0.4404	0.66561	0.88433

Table 2 *cont.*

i	M'(i	1 − M'(i)	Xn(i)	Xe(i)	Xw(i)	p0.05	p0.95
30	0.81593	0.18407	0.8999	−1.6925	0.5262	0.69732	0.90501
31	0.84341	0.15659	1.0085	−1.8541	0.6174	0.72990	0.92483
32	0.87088	0.12912	1.1306	−2.0470	0.7164	0.76352	0.94364
33	0.89835	0.10165	1.2724	−2.2862	0.8269	0.79848	0.96114
34	0.92582	0.07418	1.4457	−2.6013	0.9560	0.83526	0.97690
35	0.95330	0.04670	1.6781	−3.0639	1.1197	0.87488	0.99004
36	0.98077	0.01923	2.0703	−3.9512	1.3740	0.92015	0.99858
			S3: 0.4899	−35.0236	−20.0250		
			S4: 31.0920	63.6253	62.0751		

Total number of results: 37

i	M'(i	1 − M'(i)	Xn(i)	Xe(i)	Xw(i)	p0.05	p0.95
1	0.01872	0.98128	−1.8839	−0.0189	−3.9689	0.00138	0.07778
2	0.04545	0.95455	−1.5932	−0.0465	−3.0679	0.00969	0.12191
3	0.07219	0.92781	−1.3992	−0.0749	−2.5912	0.02246	0.16054
4	0.09893	0.90107	−1.2466	−0.1042	−2.2617	0.03778	0.19643
5	0.12567	0.87433	−1.1178	−0.1343	−2.0077	0.05479	0.23054
6	0.15241	0.84759	−1.0047	−0.1654	−1.7997	0.07306	0.26337
7	0.17914	0.82086	−0.9026	−0.1974	−1.6225	0.09232	0.29518
8	0.20588	0.79412	−0.8087	−0.2305	−1.4674	0.11240	0.32616
9	0.23262	0.76738	−0.7211	−0.2648	−1.3289	0.13318	0.35641
10	0.25936	0.74064	−0.6384	−0.3002	−1.2032	0.15458	0.38608
11	0.28610	0.71390	−0.5596	−0.3370	−1.0877	0.17653	0.41517
12	0.31283	0.68717	−0.4839	−0.3752	−0.9804	0.19898	0.44376
13	0.33957	0.66043	−0.4107	−0.4149	−0.8798	0.22191	0.47187
14	0.36631	0.63369	−0.3396	−0.4562	−0.7848	0.24527	0.49955
15	0.39305	0.60695	−0.2700	−0.4993	−0.6945	0.26905	0.52680
16	0.41979	0.58021	−0.2015	−0.5444	−0.6081	0.29324	0.55366
17	0.44652	0.55348	−0.1339	−0.5915	−0.5250	0.31781	0.58012
18	0.47326	0.52674	−0.0668	−0.6411	−0.4446	0.34276	0.60620
19	0.50000	0.50000	0.0000	−0.6931	−0.3665	0.36809	0.63190
20	0.52674	0.47326	0.0669	−0.7481	−0.2902	0.39380	0.65723
21	0.55348	0.44652	0.1342	−0.8063	−0.2153	0.41988	0.68219
22	0.58021	0.41979	0.2021	−0.8680	−0.1416	0.44634	0.70676
23	0.60695	0.39305	0.2710	−0.9338	−0.0685	0.47320	0.73094
24	0.63369	0.36631	0.3412	−1.0043	0.0043	0.50045	0.75473
25	0.66043	0.33957	0.4132	−1.0801	0.0770	0.52812	0.77809
26	0.68717	0.31283	0.4874	−1.1621	0.1502	0.55624	0.80101
27	0.71390	0.28610	0.5644	−1.2514	0.2243	0.58483	0.82347
28	0.74064	0.25936	0.6450	−1.3495	0.2998	0.61392	0.84542
29	0.76738	0.23262	0.7300	−1.4583	0.3773	0.64357	0.86682
30	0.79412	0.20588	0.8206	−1.5805	0.4577	0.67384	0.88760
31	0.82086	0.17914	0.9185	−1.7196	0.5421	0.70482	0.90768
32	0.84759	0.15241	1.0262	−1.8812	0.6319	0.73663	0.92694
33	0.87433	0.12567	1.1472	−2.0741	0.7295	0.76946	0.94521
34	0.90107	0.09893	1.2879	−2.3133	0.8387	0.80357	0.96222
35	0.92781	0.07219	1.4599	−2.6284	0.9664	0.83946	0.97754
36	0.95455	0.04545	1.6910	−3.0910	1.1285	0.87809	0.99031
37	0.98128	0.01872	2.0814	−3.9783	1.3809	0.92222	0.99861
			S3: 0.5061	−36.0182	−20.5976		
			S4: 32.0178	65.5505	63.9755		

Table 2 *cont.*

i	M'(i	1 − M'(i)	Xn(i)	Xe(i)	Xw(i)	p0.05	p0.95
Total number of results: 38							
1	0.01823	0.98177	−1.8911	−0.0184	−3.9956	0.00135	0.07581
2	0.04427	0.95573	−1.6033	−0.0453	−3.0949	0.00943	0.11885
3	0.07031	0.92969	−1.4111	−0.0729	−2.6186	0.02186	0.15656
4	0.09635	0.90365	−1.2600	−0.1013	−2.2895	0.03676	0.19159
5	0.12240	0.87760	−1.1326	−0.1306	−2.0359	0.05331	0.22490
6	0.14844	0.85156	−1.0207	−0.1607	−1.8283	0.07107	0.25696
7	0.17448	0.82552	−0.9197	−0.1917	−1.6516	0.08979	0.28804
8	0.20052	0.79948	−0.8269	−0.2238	−1.4970	0.10931	0.31832
9	0.22656	0.77344	−0.7404	−0.2569	−1.3590	0.12950	0.34791
10	0.25260	0.74740	−0.6588	−0.2912	−1.2339	0.15028	0.37691
11	0.27865	0.72135	−0.5812	−0.3266	−1.1189	0.17160	0.40537
12	0.30469	0.69531	−0.5067	−0.3634	−1.0123	0.19340	0.43334
13	0.33073	0.66927	−0.4347	−0.4016	−0.9124	0.21565	0.46086
14	0.35677	0.64323	−0.3648	−0.4413	−0.8181	0.23832	0.48796
15	0.38281	0.61719	−0.2964	−0.4826	−0.7286	0.26178	0.51466
16	0.40885	0.59115	−0.2294	−0.5257	−0.6430	0.28483	0.54098
17	0.43490	0.56510	−0.1632	−0.5707	−0.5608	0.30865	0.56693
18	0.46094	0.53906	−0.0977	−0.6179	−0.4814	0.33283	0.59252
19	0.48698	0.51302	−0.0325	−0.6674	−0.4043	0.35736	0.61776
20	0.51302	0.48698	0.0326	−0.7195	−0.3292	0.38224	0.64264
21	0.53906	0.46094	0.0978	−0.7745	−0.2555	0.40748	0.66717
22	0.56510	0.43490	0.1636	−0.8326	−0.1831	0.43307	0.69135
23	0.59115	0.40885	0.2301	−0.8944	−0.1116	0.45902	0.71517
24	0.61719	0.38281	0.2977	−0.9602	−0.0406	0.48534	0.73862
25	0.64323	0.35677	0.3667	−1.0307	0.0302	0.51204	0.76168
26	0.66927	0.33073	0.4375	−1.1065	0.1012	0.53914	0.78435
27	0.69531	0.30469	0.5106	−1.1885	0.1727	0.56666	0.80660
28	0.72135	0.27865	0.5865	−1.2778	0.2452	0.59463	0.82840
29	0.74740	0.25260	0.6660	−1.3759	0.3191	0.62309	0.84972
30	0.77344	0.22656	0.7500	−1.4847	0.3952	0.65209	0.87050
31	0.79948	0.20052	0.8396	−1.6068	0.4743	0.68168	0.89069
32	0.82552	0.17448	0.9365	−1.7459	0.5573	0.71196	0.91021
33	0.85156	0.14844	1.0432	−1.9076	0.6458	0.74304	0.92893
34	0.87760	0.12240	1.1632	−2.1005	0.7422	0.77510	0.94669
35	0.90365	0.09635	1.3028	−2.3397	0.8500	0.80841	0.96324
36	0.92969	0.07031	1.4738	−2.6548	0.9764	0.84344	0.97814
37	0.95573	0.04427	1.7035	−3.1174	1.1370	0.88115	0.99057
38	0.98177	0.01823	2.0922	−4.0047	1.3875	0.92419	0.99865
		S3:	0.5222	−37.0129	−21.1702		
		S4:	32.9440	67.4774	65.8776		
Total number of results: 39							
1	0.01777	0.98223	−1.8981	−0.0179	−4.0215	0.00131	0.07394
2	0.04315	0.95685	−1.6131	−0.0441	−3.1212	0.00919	0.11595
3	0.06853	0.93147	−1.4227	−0.0710	−2.6452	0.02129	0.15277
4	0.09391	0.90609	−1.2731	−0.0986	−2.3165	0.03580	0.18698
5	0.11929	0.88071	−1.1469	−0.1270	−2.0634	0.05190	0.21952
6	0.14467	0.85533	−1.0361	−0.1563	−1.8562	0.06919	0.25085

Table 2 *cont.*

i	M'(i	1 − M'(i)	Xn(i)	Xe(i)	Xw(i)	p0.05	p0.95
7	0.17005	0.82995	−0.9363	−0.1864	−1.6799	0.08740	0.28124
8	0.19543	0.80457	−0.8445	−0.2174	−1.5258	0.10638	0.31084
9	0.22081	0.77919	−0.7591	−0.2495	−1.3883	0.12601	0.33979
10	0.24619	0.75381	−0.6785	−0.2826	−1.2637	0.14622	0.36815
11	0.27157	0.72843	−0.6019	−0.3169	−1.1493	0.16694	0.39601
12	0.29695	0.70305	−0.5285	−0.3523	−1.0432	0.18812	0.42339
13	0.32234	0.67767	−0.4576	−0.3891	−0.9439	0.20973	0.45034
14	0.34772	0.65228	−0.3889	−0.4273	−0.8503	0.23175	0.47689
15	0.37310	0.62690	−0.3218	−0.4670	−0.7615	0.25414	0.50305
16	0.39848	0.60152	−0.2560	−0.5083	−0.6767	0.27690	0.52886
17	0.42386	0.57614	−0.1912	−0.5514	−0.5953	0.30001	0.55431
18	0.44924	0.55076	−0.1271	−0.5965	−0.5168	0.32346	0.57942
19	0.47462	0.52538	−0.0634	−0.6436	−0.4406	0.34725	0.60419
20	0.50000	0.50000	0.0000	−0.6931	−0.3665	0.37136	0.62864
21	0.52538	0.47462	0.0635	−0.7452	−0.2940	0.39581	0.65275
22	0.55076	0.44924	0.1273	−0.8002	−0.2229	0.42058	0.67654
23	0.57614	0.42386	0.1917	−0.8584	−0.1527	0.44569	0.69999
24	0.60152	0.39848	0.2569	−0.9201	−0.0833	0.47114	0.72310
25	0.62690	0.37310	0.3232	−0.9859	−0.0142	0.49694	0.74586
26	0.65228	0.34772	0.3911	−1.0564	0.0548	0.52311	0.76825
27	0.67767	0.32234	0.4608	−1.1322	0.1241	0.54966	0.79027
28	0.70305	0.29695	0.5328	−1.2142	0.1941	0.57661	0.81188
29	0.72843	0.27157	0.6077	−1.3035	0.2651	0.60399	0.83306
30	0.75381	0.24619	0.6862	−1.4016	0.3376	0.63185	0.85378
31	0.77919	0.22081	0.7692	−1.5104	0.4124	0.66021	0.87399
32	0.80457	0.19543	0.8579	−1.6325	0.4901	0.68916	0.89362
33	0.82995	0.17005	0.9539	−1.7717	0.5719	0.71876	0.91260
34	0.85533	0.14467	1.0596	−1.9333	0.6592	0.74915	0.93081
35	0.88071	0.11929	1.1787	−2.1262	0.7543	0.78048	0.94810
36	0.90609	0.09391	1.3173	−2.3654	0.8610	0.81302	0.96420
37	0.93147	0.06853	1.4871	−2.6805	0.9860	0.84723	0.97871
38	0.95685	0.04315	1.7157	−3.1431	1.1452	0.88405	0.99081
39	0.98223	0.01777	2.1027	−4.0304	1.3939	0.92606	0.99869
			S3: 0.5384	−38.0077	−21.7430		
			S4: 33.8705	69.4059	67.7814		

Total number of results: 40

i	M'(i	1 − M'(i)	Xn(i)	Xe(i)	Xw(i)	p0.05	p0.95
1	0.01733	0.98267	−1.9048	−0.0175	−4.0468	0.00128	0.07216
2	0.04208	0.95792	−1.6225	−0.0430	−3.1468	0.00896	0.11319
3	0.06683	0.93317	−1.4339	−0.0692	−2.6712	0.02079	0.14915
4	0.09158	0.90842	−1.2856	−0.0961	−2.3429	0.03488	0.18259
5	0.11634	0.88366	−1.1606	−0.1237	−2.0901	0.05057	0.21440
6	0.14109	0.85891	−1.0510	−0.1521	−1.8833	0.06740	0.24503
7	0.16584	0.83416	−0.9522	−0.1813	−1.7074	0.08513	0.27475
8	0.19059	0.80941	−0.8615	−0.2115	−1.5537	0.10361	0.30371
9	0.21535	0.78465	−0.7770	−0.2425	−1.4167	0.12271	0.33203
10	0.24010	0.75990	−0.6975	−0.2746	−1.2926	0.14237	0.35979
11	0.26485	0.73515	−0.6219	−0.3077	−1.1787	0.16252	0.38706
12	0.28960	0.71040	−0.5495	−0.3419	−1.0731	0.18312	0.41388

Table 2 *cont.*

i	M'(i	1 − M'(i)	Xn(i)	Xe(i)	Xw(i)	p0.05	p0.95
13	0.31436	0.68564	−0.4797	−0.3774	−0.9745	0.20413	0.44028
14	0.33911	0.66089	−0.4120	−0.4142	−0.8815	0.22553	0.46630
15	0.36386	0.63614	−0.3460	−0.4523	−0.7933	0.24729	0.49195
16	0.38861	0.61139	−0.2814	−0.4920	−0.7092	0.26941	0.51725
17	0.41337	0.58663	−0.2179	−0.5334	−0.6286	0.29189	0.54222
18	0.43812	0.56188	−0.1551	−0.5765	−0.5508	0.31461	0.56686
19	0.46287	0.53713	−0.0929	−0.6215	−0.4756	0.33770	0.59119
20	0.48762	0.51238	−0.0309	−0.6687	−0.4024	0.36105	0.61520
21	0.51238	0.48762	0.0309	−0.7182	−0.3310	0.38480	0.63891
22	0.53713	0.46287	0.0930	−0.7703	−0.2610	0.40881	0.66230
23	0.56188	0.43812	0.1554	−0.8253	−0.1921	0.43314	0.68539
24	0.58663	0.41337	0.2185	−0.8834	−0.1240	0.45778	0.70815
25	0.61139	0.38861	0.2825	−0.9452	−0.0564	0.48275	0.73060
26	0.63614	0.36386	0.3477	−1.0110	0.0109	0.50805	0.75270
27	0.66089	0.33911	0.4145	−1.0814	0.0783	0.53370	0.77447
28	0.68564	0.31436	0.4831	−1.1572	0.1460	0.55972	0.79587
29	0.71040	0.28960	0.5542	−1.2392	0.2145	0.58612	0.81688
30	0.73515	0.26485	0.6281	−1.3286	0.2841	0.61294	0.83746
31	0.75990	0.24010	0.7057	−1.4267	0.3554	0.64021	0.85763
32	0.78465	0.21535	0.7878	−1.5355	0.4289	0.66797	0.87729
33	0.80941	0.19059	0.8756	−1.6576	0.5054	0.69629	0.89639
34	0.83416	0.16584	0.9707	−1.7967	0.5860	0.72525	0.91487
35	0.85891	0.14109	1.0755	−1.9584	0.6721	0.75497	0.93260
36	0.88366	0.11634	1.1936	−2.1513	0.7661	0.78560	0.94943
37	0.90842	0.09158	1.3313	−2.3905	0.8715	0.81741	0.96511
38	0.93317	0.06683	1.5001	−2.7056	0.9953	0.85085	0.97925
39	0.95792	0.04208	1.7274	−3.1682	1.1532	0.88681	0.99104
40	0.98267	0.01733	2.1128	−4.0555	1.4001	0.92784	0.99872
			S3: 0.5545	−39.0027	−22.3159		
			S4: 34.7973	71.3360	69.6868		

Total number of results: 41

i	M'(i	1 − M'(i)	Xn(i)	Xe(i)	Xw(i)	p0.05	p0.95
1	0.01691	0.98309	−1.9113	−0.0171	−4.0715	0.00125	0.07046
2	0.04106	0.95894	−1.6316	−0.0419	−3.1718	0.00874	0.11055
3	0.06522	0.93478	−1.4447	−0.0674	−2.6965	0.02024	0.14571
4	0.08937	0.91063	−1.2978	−0.0936	−2.3685	0.03402	0.17840
5	0.11353	0.88647	−1.1740	−0.1205	−2.1161	0.04930	0.20951
6	0.13768	0.86232	−1.0654	−0.1481	−1.9097	0.06570	0.23947
7	0.16184	0.83816	−0.9676	−0.1765	−1.7342	0.08298	0.26854
8	0.18599	0.81401	−0.8778	−0.2058	−1.5809	0.10097	0.29689
9	0.21014	0.78986	−0.7943	−0.2359	−1.4443	0.11958	0.32461
10	0.23430	0.76570	−0.7157	−0.2670	−1.3206	0.13872	0.35180
11	0.25845	0.74155	−0.6411	−0.2990	−1.2073	0.15833	0.37851
12	0.28261	0.71739	−0.5697	−0.3321	−1.1022	0.17838	0.40478
13	0.30676	0.69324	−0.5008	−0.3664	−1.0041	0.19883	0.43065
14	0.33092	0.66908	−0.4342	−0.4018	−0.9117	0.21964	0.45615
15	0.35507	0.64493	−0.3693	−0.4386	−0.8241	0.24081	0.48131
16	0.37923	0.62077	−0.3058	−0.4768	−0.7407	0.26230	0.50612
17	0.40338	0.59662	−0.2434	−0.5165	−0.6607	0.28412	0.53062

Table 2 *cont.*

i	M'(i	1 − M'(i)	Xn(i)	Xe(i)	Xw(i)	p0.05	p0.95
18	0.42754	0.57246	−0.1818	−0.5578	−0.5837	0.30624	0.55482
19	0.45169	0.54831	−0.1209	−0.6009	−0.5093	0.32867	0.57871
20	0.47585	0.52415	−0.0604	−0.6460	−0.4370	0.35138	0.60230
21	0.50000	0.50000	0.0000	−0.6931	−0.3665	0.37440	0.62560
22	0.52415	0.47585	0.0604	−0.7427	−0.2975	0.39770	0.64861
23	0.54831	0.45169	0.1211	−0.7948	−0.2297	0.42129	0.67133
24	0.57246	0.42754	0.1823	−0.8497	−0.1629	0.44518	0.69376
25	0.59662	0.40338	0.2442	−0.9079	−0.0967	0.46937	0.71588
26	0.62077	0.37923	0.3071	−0.9696	−0.0309	0.49388	0.73769
27	0.64493	0.35507	0.3712	−1.0354	0.0348	0.51869	0.75919
28	0.66908	0.33092	0.4369	−1.1059	0.1006	0.54385	0.78035
29	0.69324	0.30676	0.5046	−1.1817	0.1669	0.56935	0.80117
30	0.71739	0.28261	0.5747	−1.2637	0.2340	0.59522	0.82162
31	0.74155	0.25845	0.6478	−1.3530	0.3024	0.62149	0.84166
32	0.76570	0.23430	0.7245	−1.4512	0.3724	0.64820	0.86128
33	0.78986	0.21014	0.8057	−1.5600	0.4447	0.67539	0.88042
34	0.81401	0.18599	0.8926	−1.6821	0.5200	0.70311	0.89903
35	0.83816	0.16184	0.9869	−1.8212	0.5995	0.73146	0.91702
36	0.86232	0.13768	1.0908	−1.9828	0.6845	0.76053	0.93430
37	0.88647	0.11353	1.2081	−2.1757	0.7774	0.79049	0.95070
38	0.91063	0.08937	1.3448	−2.4149	0.8817	0.82160	0.96598
39	0.93478	0.06522	1.5127	−2.7300	1.0043	0.85429	0.97976
40	0.95894	0.04106	1.7389	−3.1927	1.1609	0.88945	0.99126
41	0.98309	0.01691	2.1227	−4.0800	1.4061	0.92954	0.99875
			S3: 0.5708	−39.9978	−22.8888		
			S4: 35.7245	73.2675	71.5937		

Total number of results: 42

i	M'(i	1 − M'(i)	Xn(i)	Xe(i)	Xw(i)	p0.05	p0.95
1	0.01651	0.98349	−1.9176	−0.0166	−4.0955	0.00122	0.06884
2	0.04009	0.95991	−1.6404	−0.0409	−3.1961	0.00853	0.10804
3	0.06368	0.93632	−1.4551	−0.0658	−2.7212	0.01975	0.14241
4	0.08726	0.91274	−1.3095	−0.0913	−2.3935	0.03319	0.17439
5	0.11085	0.88915	−1.1868	−0.1175	−2.1414	0.04810	0.20483
6	0.13443	0.86557	−1.0793	−0.1444	−1.9354	0.06409	0.23416
7	0.15802	0.84198	−0.9824	−0.1720	−1.7603	0.08093	0.26262
8	0.18160	0.81840	−0.8936	−0.2004	−1.6074	0.09847	0.29037
9	0.20519	0.79481	−0.8110	−0.2297	−1.4712	0.11660	0.31752
10	0.22877	0.77123	−0.7333	−0.2598	−1.3479	0.13525	0.34415
11	0.25236	0.74764	−0.6596	−0.2908	−1.2350	0.15436	0.37032
12	0.27594	0.72406	−0.5891	−0.3229	−1.1305	0.17389	0.39607
13	0.29953	0.70047	−0.5212	−0.3560	−1.0328	0.19379	0.42143
14	0.32311	0.67689	−0.4555	−0.3903	−0.9410	0.21406	0.44644
15	0.34670	0.65330	−0.3916	−0.4257	−0.8540	0.23466	0.47110
16	0.37028	0.62972	−0.3291	−0.4625	−0.7711	0.25557	0.49546
17	0.39387	0.60613	−0.2678	−0.5007	−0.6918	0.27679	0.51950
18	0.41745	0.58255	−0.2074	−0.5403	−0.6155	0.29831	0.54326
19	0.44104	0.55896	−0.1477	−0.5817	−0.5418	0.32011	0.56672
20	0.46462	0.53538	−0.0885	−0.6248	−0.4704	0.34219	0.58991
21	0.48821	0.51179	−0.0295	−0.6698	−0.4007	0.36455	0.61281

Table 2 *cont.*

i	M'(i	1 − M'(i)	Xn(i)	Xe(i)	Xw(i)	p0.05	p0.95
22	0.51179	0.48821	0.0295	−0.7170	−0.3327	0.38719	0.63545
23	0.53538	0.46462	0.0886	−0.7665	−0.2659	0.41009	0.65781
24	0.55896	0.44104	0.1480	−0.8186	−0.2001	0.43328	0.67989
25	0.58255	0.41745	0.2080	−0.8736	−0.1352	0.45674	0.70169
26	0.60613	0.39387	0.2688	−0.9317	−0.0707	0.48050	0.72320
27	0.62972	0.37028	0.3307	−0.9935	−0.0065	0.50454	0.74443
28	0.65330	0.34670	0.3938	−1.0593	0.0576	0.52889	0.76534
29	0.67689	0.32311	0.4586	−1.1298	0.1220	0.55356	0.78594
30	0.70047	0.29953	0.5254	−1.2055	0.1869	0.57857	0.80621
31	0.72406	0.27594	0.5946	−1.2876	0.2527	0.60393	0.82611
32	0.74764	0.25236	0.6668	−1.3769	0.3198	0.62968	0.84564
33	0.77123	0.22877	0.7426	−1.4750	0.3887	0.56585	0.86475
34	0.79481	0.20519	0.8230	−1.5838	0.4598	0.68248	0.88340
35	0.81840	0.18160	0.9092	−1.7059	0.5341	0.70963	0.90153
36	0.84198	0.15802	1.0026	−1.8450	0.6125	0.73738	0.91907
37	0.86557	0.13443	1.1057	−2.0067	0.6965	0.76584	0.93591
38	0.88915	0.11085	1.2222	−2.1996	0.7883	0.79517	0.95190
39	0.91274	0.08726	1.3580	−2.4388	0.8915	0.82561	0.96681
40	0.93632	0.06368	1.5249	−2.7539	1.0130	0.85759	0.98025
41	0.95991	0.04009	1.7500	−3.2165	1.1683	0.89196	0.99147
42	0.98349	0.01651	2.1323	−4.1038	1.4119	0.93116	0.99878
			S3: 0.5870	−40.9930	−23.4619		
			S4: 36.6518	75.2004	73.5021		

Total number of results: 43

i	M'(i	1 − M'(i)	Xn(i)	Xe(i)	Xw(i)	p0.05	p0.95
1	0.01613	0.98387	−1.9237	−0.0163	−4.1190	0.00119	0.06730
2	0.03917	0.96083	−1.6489	−0.0400	−3.2199	0.00833	0.10563
3	0.06221	0.93779	−1.4652	−0.0642	−2.7453	0.01928	0.13927
4	0.08525	0.91475	−1.3209	−0.0891	−2.4179	0.03240	0.17056
5	0.10829	0.89171	−1.1993	−0.1146	−2.1661	0.04695	0.20036
6	0.13134	0.86866	−1.0927	−0.1408	−1.9604	0.06256	0.22907
7	0.15438	0.84562	−0.9968	−0.1677	−1.7857	0.07898	0.25694
8	0.17742	0.82258	−0.9089	−0.1953	−1.6332	0.09609	0.28413
9	0.20046	0.79954	−0.8272	−0.2237	−1.4974	0.11377	0.31073
10	0.22350	0.77650	−0.7503	−0.2530	−1.3745	0.13195	0.33682
11	0.24654	0.75346	−0.6775	−0.2831	−1.2620	0.15058	0.36247
12	0.26959	0.73041	−0.6078	−0.3141	−1.1579	0.16961	0.38772
13	0.29263	0.70737	−0.5408	−0.3462	−1.0607	0.18901	0.41259
14	0.31567	0.68433	−0.4760	−0.3793	−0.9694	0.20875	0.43712
15	0.33871	0.66129	−0.4131	−0.4136	−0.8829	0.22881	0.46132
16	0.36175	0.63825	−0.3516	−0.4490	−0.8007	0.24918	0.48522
17	0.38479	0.61521	−0.2913	−0.4858	−0.7220	0.26984	0.50883
18	0.40783	0.59217	−0.2320	−0.5240	−0.6463	0.29078	0.53215
19	0.43088	0.56912	−0.1734	−0.5637	−0.5733	0.31200	0.55520
20	0.45392	0.54608	−0.1153	−0.6050	−0.5026	0.33348	0.57799
21	0.47696	0.52304	−0.0576	−0.6481	−0.4337	0.35522	0.60051
22	0.50000	0.50000	0.0000	−0.6931	−0.3665	0.37722	0.62273
23	0.52304	0.47696	0.0576	−0.7403	−0.3007	0.39949	0.64478
24	0.54608	0.45392	0.1155	−0.7898	−0.2359	0.42201	0.66652

Table 2 *cont.*

i	M'(i	1 − M'(i)	Xn(i)	Xe(i)	Xw(i)	p0.05	p0.95
25	0.56912	0.43088	0.1738	−0.8419	−0.1721	0.44480	0.68800
26	0.59217	0.40783	0.2327	−0.8969	−0.1088	0.46785	0.70922
27	0.61521	0.38479	0.2925	−0.9551	−0.0460	0.49117	0.73016
28	0.63825	0.36175	0.3533	−1.0168	0.0167	0.51478	0.75082
29	0.66129	0.33871	0.4155	−1.0826	0.0794	0.53868	0.77119
30	0.68433	0.31567	0.4794	−1.1531	0.1424	0.56288	0.79125
31	0.70737	0.29263	0.5453	−1.2289	0.2061	0.58741	0.81099
32	0.73041	0.26959	0.6137	−1.3109	0.2707	0.61228	0.83039
33	0.75346	0.24654	0.6851	−1.4002	0.3366	0.63753	0.84942
34	0.77650	0.22350	0.7602	−1.4983	0.4044	0.66318	0.86805
35	0.79954	0.20046	0.8398	−1.6071	0.4745	0.68927	0.88623
36	0.82258	0.17742	0.9252	−1.7292	0.5477	0.71587	0.90391
37	0.84562	0.15438	1.0178	−1.8684	0.6251	0.74306	0.92102
38	0.86866	0.13134	1.1202	−2.0300	0.7080	0.77093	0.93744
39	0.89171	0.10829	1.2358	−2.2229	0.7988	0.79964	0.95305
40	0.91475	0.08525	1.3708	−2.4621	0.9010	0.82944	0.96760
41	0.93779	0.06221	1.5368	−2.7772	1.0214	0.86073	0.98071
42	0.96083	0.03917	1.7608	−3.2398	1.1755	0.89437	0.99167
43	0.98387	0.01613	2.1416	−4.1271	1.4176	0.93270	0.99881
			S3: 0.6033	−41.9883	−24.0351		
			S4: 37.5795	77.1347	75.4119		

Total number of results: 44

i	M'(i	1 − M'(i)	Xn(i)	Xe(i)	Xw(i)	p0.05	p0.95
1	0.01577	0.98423	−1.9296	−0.0159	−4.1420	0.00116	0.06582
2	0.03829	0.96171	−1.6572	−0.0390	−3.2432	0.00814	0.10334
3	0.06081	0.93919	−1.4750	−0.0627	−2.7688	0.01884	0.13626
4	0.08333	0.91667	−1.3319	−0.0870	−2.4417	0.03165	0.16690
5	0.10586	0.89414	−1.2113	−0.1119	−2.1903	0.04586	0.19608
6	0.12838	0.87162	−1.1057	−0.1374	−1.9849	0.06109	0.22420
7	0.15090	0.84910	−1.0107	−0.1636	−1.8105	0.07713	0.25151
8	0.17342	0.82658	−0.9236	−0.1905	−1.6583	0.09382	0.27814
9	0.19595	0.80405	−0.8427	−0.2181	−1.5229	0.11107	0.30422
10	0.21847	0.78153	−0.7667	−0.2465	−1.4004	0.12881	0.32980
11	0.24099	0.75901	−0.6947	−0.2757	−1.2883	0.14698	0.35495
12	0.26351	0.73649	−0.6259	−0.3059	−1.1846	0.16554	0.37971
13	0.28604	0.71396	−0.5597	−0.3369	−1.0879	0.18445	0.40410
14	0.30856	0.69144	−0.4958	−0.3690	−0.9970	0.20370	0.42817
15	0.33108	0.66892	−0.4337	−0.4021	−0.9111	0.22326	0.45193
16	0.35360	0.64640	−0.3732	−0.4363	−0.8293	0.24311	0.47539
17	0.37613	0.62387	−0.3139	−0.4718	−0.7512	0.26323	0.49857
18	0.39865	0.60135	−0.2555	−0.5086	−0.6761	0.28363	0.52148
19	0.42117	0.57883	−0.1980	−0.5467	−0.6038	0.30429	0.54413
20	0.44369	0.55631	−0.1410	−0.5864	−0.5337	0.32520	0.56653
21	0.46622	0.53378	−0.0845	−0.6278	−0.4656	0.34636	0.58867
22	0.48874	0.51126	−0.0281	−0.6709	−0.3992	0.36777	0.61057
23	0.51126	0.48874	0.0282	−0.7159	−0.3342	0.38943	0.63223
24	0.53378	0.46622	0.0846	−0.7631	−0.2704	0.41133	0.65363
25	0.55631	0.44369	0.1413	−0.8126	−0.2075	0.43347	0.67480
26	0.57883	0.42117	0.1985	−0.8647	−0.1454	0.45587	0.69571

Table 2 *cont.*

i	M'(i	1 − M'(i)		Xn(i)	Xe(i)	Xw(i)	p0.05	p0.95
27	0.60135	0.39865		0.2564	−0.9197	−0.0837	0.47852	0.71637
28	0.62387	0.37613		0.3152	−0.9778	−0.0224	0.50143	0.73677
29	0.64640	0.35360		0.3752	−1.0396	0.0388	0.52461	0.75689
30	0.66892	0.33108		0.4365	−1.1054	0.1002	0.54807	0.77674
31	0.69144	0.30856		0.4995	−1.1758	0.1620	0.57183	0.79630
32	0.71396	0.28604		0.5646	−1.2516	0.2245	0.59590	0.81554
33	0.73649	0.26351		0.6322	−1.3337	0.2879	0.62029	0.83446
34	0.75901	0.24099		0.7028	−1.4230	0.3528	0.64505	0.85302
35	0.78153	0.21847		0.7772	−1.5211	0.4194	0.67020	0.87119
36	0.80405	0.19595		0.8560	−1.6299	0.4885	0.69578	0.88892
37	0.82658	0.17342		0.9406	−1.7520	0.5608	0.72185	0.90618
38	0.84910	0.15090		1.0326	−1.8911	0.6372	0.74849	0.92287
39	0.87162	0.12838		1.1342	−2.0528	0.7192	0.77580	0.93891
40	0.89414	0.10586		1.2490	−2.2457	0.8090	0.80392	0.95414
41	0.91667	0.08333		1.3832	−2.4849	0.9102	0.83310	0.96835
42	0.93919	0.06081		1.5483	−2.8000	1.0296	0.86374	0.98116
43	0.96171	0.03829		1.7713	−3.2626	1.1825	0.89666	0.99186
44	0.98423	0.01577		2.1507	−4.1499	1.4231	0.93418	0.99883
			S3:	0.6196	−42.9838	−24.6083		
			S4:	38.5073	79.0703	77.3231		

Total number of results: 45

i	M'(i	1 − M'(i)	Xn(i)	Xe(i)	Xw(i)	p0.05	p0.95
1	0.01542	0.98458	−1.9353	−0.0155	−4.1644	0.00114	0.06440
2	0.03744	0.96256	−1.6652	−0.0382	−3.2659	0.00795	0.10113
3	0.05947	0.94053	−1.4845	−0.0613	−2.7918	0.01842	0.13338
4	0.08150	0.91850	−1.3426	−0.0850	−2.4650	0.03093	0.16339
5	0.10352	0.89648	−1.2230	−0.1093	−2.2138	0.04481	0.19198
6	0.12555	0.87445	−1.1183	−0.1342	−2.0087	0.05969	0.21954
7	0.14758	0.85242	−1.0242	−0.1597	−1.8346	0.07536	0.24630
8	0.16960	0.83040	−0.9379	−0.1859	−1.6828	0.09166	0.27241
9	0.19163	0.80837	−0.8578	−0.2127	−1.5477	0.10850	0.29797
10	0.21366	0.78634	−0.7826	−0.2404	−1.4256	0.12582	0.32306
11	0.23568	0.76432	−0.7114	−0.2688	−1.3139	0.14355	0.34773
12	0.25771	0.74229	−0.6433	−0.2980	−1.2106	0.16166	0.37202
13	0.27974	0.72026	−0.5780	−0.3281	−1.1143	0.18012	0.39596
14	0.30176	0.69824	−0.5149	−0.3592	−1.0239	0.19889	0.41958
15	0.32379	0.67621	−0.4537	−0.3912	−0.9384	0.21796	0.44290
16	0.34582	0.65419	−0.3940	−0.4244	−0.8572	0.23732	0.46594
17	0.36784	0.63216	−0.3355	−0.4586	−0.7795	0.25694	0.48871
18	0.38987	0.61013	−0.2782	−0.4941	−0.7051	0.27683	0.51122
19	0.41189	0.58811	−0.2216	−0.5308	−0.6333	0.29696	0.53348
20	0.43392	0.56608	−0.1657	−0.5690	−0.5638	0.31733	0.55549
21	0.45595	0.54405	−0.1102	−0.6087	−0.4964	0.33794	0.57727
22	0.47797	0.52203	−0.0551	−0.6500	−0.4307	0.35879	0.59882
23	0.50000	0.50000	0.0000	−0.6931	−0.3665	0.37987	0.62013
24	0.52203	0.47797	0.0551	−0.7382	−0.3035	0.40118	0.64121
25	0.54405	0.45595	0.1104	−0.7854	−0.2416	0.42273	0.66205
26	0.56608	0.43392	0.1661	−0.8349	−0.1805	0.44451	0.68267
27	0.58811	0.41189	0.2223	−0.8870	−0.1199	0.46652	0.70304

Table 2 *cont.*

i	M'(i	1 − M'(i)	Xn(i)	Xe(i)	Xw(i)	p0.05	p0.95
28	0.61013	0.38987	0.2792	−0.9419	−0.0598	0.48878	0.72317
29	0.63216	0.36784	0.3371	−1.0001	0.0001	0.51129	0.74306
30	0.65419	0.34582	0.3962	−1.0619	0.0600	0.53406	0.76368
31	0.67621	0.32379	0.4567	−1.1277	0.1201	0.55710	0.78203
32	0.69824	0.30176	0.5189	−1.1981	0.1808	0.58042	0.80111
33	0.72026	0.27974	0.5833	−1.2739	0.2421	0.60404	0.81988
34	0.74229	0.25771	0.6501	−1.3559	0.3045	0.62798	0.83834
35	0.76432	0.23568	0.7200	−1.4453	0.3683	0.65227	0.85645
36	0.78634	0.21366	0.7936	−1.5434	0.4340	0.67694	0.87418
37	0.80837	0.19163	0.8718	−1.6522	0.5021	0.70203	0.89150
38	0.83040	0.16960	0.9557	−1.7743	0.5734	0.72759˙	0.90834
39	0.85242	0.14758	1.0469	−1.9134	0.6489	0.75370	0.92464
40	0.87445	0.12555	1.1478	−2.0750	0.7300	0.78046	0.94030
41	0.89648	0.10352	1.2619	−2.2679	0.8189	0.80802	0.95518
42	0.91850	0.08150	1.3953	−2.5072	0.9192	0.83661	0.96907
43	0.94053	0.05947	1.5596	−2.8223	1.0375	0.86662	0.98158
44	0.96256	0.03744	1.7815	−3.2849	1.1893	0.89887	0.99205
45	0.98458	0.01542	2.1596	−4.1722	1.4284	0.93560	0.99886
			S3: 0.6359	−43.9793	−25.1817		
			S4: 39.4355	81.0071	79.2355		

Total number of results: 46

i	M'(i	1 − M'(i)	Xn(i)	Xe(i)	Xw(i)	p0.05	p0.95
1	0.01509	0.98491	−1.9408	−0.0152	−4.1864	0.00111	0.06305
2	0.03664	0.96336	−1.6730	−0.0373	−3.2881	0.00778	0.09902
3	0.05819	0.94181	−1.4937	−0.0600	−2.8142	0.01801	0.13061
4	0.07974	0.92026	−1.3529	−0.0831	−2.4877	0.03025	0.16002
5	0.10129	0.89871	−1.2344	−0.1068	−2.2368	0.04382	0.18804
6	0.12284	0.87716	−1.1306	−0.1311	−2.0320	0.05836	0.21506
7	0.14440	0.85560	−1.0372	−0.1559	−1.8582	0.07366	0.24130
8	0.16595	0.83405	−0.9518	−0.1815	−1.7067	0.08959	0.26691
9	0.18750	0.81250	−0.8724	−0.2076	−1.5720	0.10605	0.29198
10	0.20905	0.79095	−0.7980	−0.2345	−1.4502	0.12296	0.31659
11	0.23060	0.76940	−0.7275	−0.2621	−1.3388	0.14028	0.34079
12	0.25216	0.74784	−0.6602	−0.2906	−1.2359	0.15796	0.36463
13	0.27371	0.72629	−0.5956	−0.3198	−1.1401	0.17598	0.38813
14	0.29526	0.70474	−0.5333	−0.3499	−1.0500	0.19430	0.41132
15	0.31681	0.68319	−0.4729	−0.3810	−0.9650	0.21292	0.43422
16	0.33836	0.66164	−0.4140	−0.4130	−0.8842	0.23180	0.45665
17	0.35991	0.64009	−0.3564	−0.4462	−0.8071	0.25085	0.47922
18	0.38147	0.61853	−0.2999	−0.4804	−0.7331	0.27034	0.50134
19	0.40302	0.59698	−0.2443	−0.5159	−0.6619	0.28997	0.52322
20	0.42457	0.57543	−0.1894	−0.5526	−0.5931	0.30984	0.54487
21	0.44612	0.55388	−0.1349	−0.5908	−0.5263	0.32993	0.56629
22	0.46767	0.53233	−0.0808	−0.6305	−0.4612	0.35025	0.58749
23	0.48922	0.51078	−0.0269	−0.6718	−0.3978	0.37078	0.60846
24	0.51078	0.48922	0.0269	−0.7149	−0.3356	0.39154	0.62922
25	0.53233	0.46767	0.0809	−0.7600	−0.2745	0.41251	0.64975
26	0.55388	0.44612	0.1352	−0.8072	−0.2142	0.43371	0.67007
27	0.57543	0.42457	0.1899	−0.8567	−0.1547	0.45513	0.69016

Table 2 *cont.*

i	M'(i	1 − M'(i)	Xn(i)	Xe(i)	Xw(i)	p0.05	p0.95
28	0.59698	0.40302	0.2451	−0.9088	−0.0957	0.47678	0.71002
29	0.61853	0.38147	0.3012	−0.9637	−0.0369	0.49866	0.72966
30	0.64009	0.35991	0.3582	−1.0219	0.0217	0.52078	0.74905
31	0.66164	0.33836	0.4165	−1.0836	0.0803	0.54315	0.76819
32	0.68319	0.31681	0.4762	−1.1495	0.1393	0.56578	0.78708
33	0.70474	0.29526	0.5377	−1.2199	0.1988	0.58868	0.80569
34	0.72629	0.27371	0.6013	−1.2957	0.2590	0.61187	0.82402
35	0.74784	0.25216	0.6674	−1.3777	0.3204	0.63537	0.84204
36	0.76940	0.23060	0.7366	−1.4671	0.3833	0.65921	0.85972
37	0.79095	0.20905	0.8095	−1.5652	0.4480	0.68341	0.87704
38	0.81250	0.18750	0.8870	−1.6740	0.5152	0.70805	0.89395
39	0.83405	0.16595	0.9702	−1.7961	0.5856	0.73309	0.91041
40	0.85560	0.14440	1.0608	−1.9352	0.6602	0.75870	0.92633
41	0.87716	0.12284	1.1610	−2.0968	0.7404	0.78494	0.94164
42	0.89871	0.10129	1.2744	−2.2897	0.8284	0.81196	0.95618
43	0.92026	0.07974	1.4071	−2.5290	0.9278	0.83998	0.96975
44	0.94181	0.05819	1.5705	−2.8440	1.0452	0.86939	0.98199
45	0.96336	0.03664	1.7915	−3.3067	1.1959	0.90098	0.99222
46	0.98491	0.01509	2.1683	−4.1940	1.4336	0.93695	0.99889
			S3: 0.6522	−44.9749	−25.7552		
			S4: 40.3638	82.9450	81.1492		

Total number of results: 47

i	M'(i	1 − M'(i)	Xn(i)	Xe(i)	Xw(i)	p0.05	p0.95
1	0.01477	0.98523	−1.9462	−0.0149	−4.2079	0.00109	0.06175
2	0.03586	0.96414	−1.6805	−0.0365	−3.3098	0.00761	0.09700
3	0.05696	0.94304	−1.5027	−0.0586	−2.8362	0.01762	0.12796
4	0.07806	0.92194	−1.3630	−0.0813	−2.5099	0.02959	0.15679
5	0.09916	0.90084	−1.2454	−0.1044	−2.2593	0.04286	0.18427
6	0.12025	0.87975	−1.1424	−0.1281	−2.0548	0.05708	0.21076
7	0.14135	0.85865	−1.0499	−0.1524	−1.8813	0.07205	0.23650
8	0.16245	0.83755	−0.9652	−0.1773	−1.7301	0.08762	0.26162
9	0.18354	0.81646	−0.8866	−0.2028	−1.5956	0.10370	0.28622
10	0.20464	0.79536	−0.8129	−0.2290	−1.4742	0.12023	0.31037
11	0.22574	0.77426	−0.7431	−0.2558	−1.3632	0.13715	0.33413
12	0.24684	0.75316	−0.6765	−0.2835	−1.2606	0.15443	0.35753
13	0.26793	0.73207	−0.6127	−0.3119	−1.1651	0.17203	0.38060
14	0.28903	0.71097	−0.5511	−0.3411	−1.0755	0.18993	0.40338
15	0.31013	0.68987	−0.4914	−0.3712	−0.9909	0.20810	0.42587
16	0.33122	0.66878	−0.4334	−0.4023	−0.9105	0.22654	0.44810
17	0.35232	0.64768	−0.3766	−0.4344	−0.8339	0.24523	0.47009
18	0.37342	0.62658	−0.3209	−0.4675	−0.7604	0.26416	0.49183
19	0.39451	0.60549	−0.2662	−0.5017	−0.6897	0.28331	0.51334
20	0.41561	0.58439	−0.2121	−0.5372	−0.6214	0.30269	0.53463
21	0.43671	0.56329	−0.1587	−0.5740	−0.5552	0.32229	0.55570
22	0.45781	0.54219	−0.1056	−0.6121	−0.4908	0.34210	0.57556
23	0.47890	0.52110	−0.0527	−0.6518	−0.4280	0.36212	0.59721
24	0.50000	0.50000	0.0000	−0.6931	−0.3665	0.38235	0.61765
25	0.52110	0.47890	0.0528	−0.7363	−0.3062	0.40279	0.63787
26	0.54219	0.45781	0.1057	−0.7813	−0.2468	0.42344	0.65790

Table 2 *cont.*

i	M'(i	1 − M'(i)	Xn(i)	Xe(i)	Xw(i)	p0.05	p0.95
27	0.56329	0.43671	0.1590	−0.8285	−0.1882	0.44430	0.67771
28	0.58439	0.41561	0.2127	−0.8780	−0.1301	0.46537	0.69730
29	0.60549	0.39451	0.2671	−0.9301	−0.0725	0.48666	0.71668
30	0.62658	0.37342	0.3224	−0.9851	−0.0151	0.50817	0.73584
31	0.64768	0.35232	0.3786	−1.0432	0.0423	0.52991	0.75477
32	0.66878	0.33122	0.4361	−1.1050	0.0998	0.55190	0.77346
33	0.68987	0.31013	0.4951	−1.1708	0.1577	0.57413	0.79190
34	0.71097	0.28903	0.5558	−1.2412	0.2161	0.59662	0.81007
35	0.73207	0.26793	0.6187	−1.3170	0.2754	0.61940	0.82797
36	0.75316	0.24684	0.6842	−1.3990	0.3358	0.64247	0.84557
37	0.77426	0.22574	0.7527	−1.4884	0.3977	0.66587	0.86285
38	0.79536	0.20464	0.8250	−1.5865	0.4615	0.68963	0.87977
39	0.81646	0.18354	0.9018	−1.6953	0.5279	0.71378	0.89630
40	0.83755	0.16245	0.9844	−1.8174	0.5974	0.73838	0.91238
41	0.85865	0.14135	1.0743	−1.9565	0.6712	0.76350	0.92795
42	0.87975	0.12025	1.1738	−2.1182	0.7505	0.78924	0.94291
43	0.90084	0.09916	1.2866	−2.3111	0.8377	0.81573	0.95714
44	0.92194	0.07806	1.4185	−2.5503	0.9362	0.84321	0.97041
45	0.94304	0.05696	1.5811	−2.8654	1.0527	0.87204	0.98238
46	0.96414	0.03586	1.8012	−3.3280	1.2024	0.90300	0.99239
47	0.98523	0.01477	2.1767	−4.2153	1.4387	0.93825	0.99891
			S3: 0.6686	−45.9707	−26.3287		
			S4: 41.2923	84.8842	83.0642		

Total number of results: 48

i	M'(i	1 − M'(i)	Xn(i)	Xe(i)	Xw(i)	p0.05	p0.95
1	0.01446	0.98554	−1.9514	−0.0146	−4.2289	0.00107	0.06050
2	0.03512	0.96488	−1.6879	−0.0358	−3.3311	0.00745	0.09506
3	0.05579	0.94421	−1.5114	−0.0574	−2.8577	0.01725	0.12541
4	0.07645	0.92355	−1.3728	−0.0795	−2.5317	0.02897	0.15369
5	0.09711	0.90289	−1.2561	−0.1022	−2.2813	0.04195	0.18064
6	0.11777	0.88223	−1.1539	−0.1253	−2.0770	0.05586	0.20663
7	0.13843	0.86157	−1.0622	−0.1490	−1.9038	0.07050	0:23188
8	0.15909	0.84091	−0.9782	−0.1733	−1.7529	0.08573	0.25623
9	0.17975	0.82025	−0.9004	−0.1981	−1.6187	0.10146	0.28068
10	0.20041	0.79959	−0.8273	−0.2237	−1.4976	0.11762	0.30439
11	0.22107	0.77893	−0.7582	−0.2498	−1.3869	0.13416	0.32772
12	0.24174	0.75826	−0.6924	−0.2767	−1.2847	0.15105	0.35069
13	0.26240	0.73760	−0.6292	−0.3043	−1.1896	0.16825	0.37336
14	0.28306	0.71694	−0.5684	−0.3328	−1.1003	0.18574	0.39573
15	0.30372	0.69628	−0.5094	−0.3620	−1.0161	0.20350	0.41783
16	0.32438	0.67562	−0.4520	−0.3921	−0.9362	0.22151	0.43968
17	0.34504	0.65496	−0.3960	−0.4232	−0.8600	0.23977	0.46129
18	0.36570	0.63430	−0.3412	−0.4552	−0.7869	0.25825	0.48266
19	0.38636	0.61364	−0.2872	−0.4884	−0.7167	0.27696	0.50382
20	0.40702	0.59298	−0.2341	−0.5226	−0.6489	0.29588	0.52476
21	0.42769	0.57231	−0.1815	−0.5581	−0.5833	0.31500	0.54549
22	0.44835	0.55165	−0.1293	−0.5948	−0.5195	0.33434	0.56602
23	0.46901	0.53099	−0.0775	−0.6330	−0.4573	0.35387	0.58634
24	0.48967	0.51033	−0.0258	−0.6727	−0.3965	0.37360	0.60647

Table 2 *cont.*

i	M'(i	1 − M'(i)	Xn(i)	Xe(i)	Xw(i)	p0.05	p0.95
25	0.51033	0.48967	0.0258	−0.7140	−0.3368	0.39353	0.62640
26	0.53099	0.46901	0.0776	−0.7571	−0.2782	0.41366	0.64613
27	0.55165	0.44835	0.1296	−0.8022	−0.2204	0.43398	0.66566
28	0.57231	0.42769	0.1819	−0.8494	−0.1633	0.45451	0.68500
29	0.59298	0.40702	0.2348	−0.8989	−0.1066	0.47524	0.70412
30	0.61364	0.38636	0.2884	−0.9510	−0.0503	0.49618	0.72304
31	0.63430	0.36570	0.3428	−1.0059	0.0059	0.51734	0.74175
32	0.65496	0.34504	0.3983	−1.0641	0.0621	0.53871	0.76023
33	0.67562	0.32438	0.4551	−1.1258	0.1185	0.56032	0.77848
34	0.69628	0.30372	0.5133	−1.1917	0.1753	0.58217	0.79650
35	0.71694	0.28306	0.5734	−1.2621	0.2328	0.60427	0.81426
36	0.73760	0.26240	0.6356	−1.3379	0.2911	0.62664	0.83175
37	0.75826	0.24174	0.7005	−1.4199	0.3506	0.64931	0.84895
38	0.77893	0.22107	0.7683	−1.5093	0.4116	0.67228	0.86584
39	0.79959	0.20041	0.8400	−1.6074	0.4746	0.69561	0.88238
40	0.82025	0.17975	0.9162	−1.7162	0.5401	0.71932	0.89854
41	0.84091	0.15909	0.9982	−1.8383	0.6088	0.74347	0.91427
42	0.86157	0.13843	1.0874	−1.9774	0.6818	0.76812	0.92950
43	0.88223	0.11777	1.1863	−2.1390	0.7604	0.79337	0.94414
44	0.90289	0.09711	1.2984	−2.3319	0.8467	0.81926	0.95805
45	0.92355	0.07645	1.4297	−2.5712	0.9444	0.84631	0.97103
46	0.94421	0.05579	1.5915	−2.8862	1.0600	0.87459	0.98275
47	0.96488	0.03512	1.8107	−3.3489	1.2086	0.90494	0.99255
48	0.98554	0.01446	2.1849	−4.2362	1.4437	0.93950	0.99893
			S3: 0.6850	−46.9665	−26.9023		
			S4: 42.2211	86.8244	84.9802		

Total number of results: 49

i	M'(i	1 − M'(i)	Xn(i)	Xe(i)	Xw(i)	p0.05	p0.95
1	0.01417	0.98583	−1.9565	−0.0143	−4.2495	0.00105	0.05931
2	0.03441	0.96559	−1.6950	−0.0350	−3.3519	0.00730	0.09319
3	0.05466	0.94534	−1.5198	−0.0562	−2.8787	0.01689	0.12297
4	0.07490	0.92510	−1.3823	−0.0779	−2.5529	0.02836	0.15071
5	0.09514	0.90486	−1.2665	−0.1000	−2.3028	0.04108	0.17715
6	0.11538	0.88462	−1.1651	−0.1226	−2.0988	0.05469	0.20266
7	0.13563	0.86437	−1.0741	−0.1458	−1.9259	0.06902	0.22744
8	0.15587	0.84413	−0.9909	−0.1694	−1.7752	0.08392	0.25164
9	0.17611	0.82389	−0.9137	−0.1937	−1.6413	0.09931	0.27535
10	0.19636	0.80364	−0.8413	−0.2186	−1.5205	0.11512	0.29864
11	0.21660	0.78340	−0.7729	−0.2441	−1.4101	0.13130	0.32154
12	0.23684	0.76316	−0.7077	−0.2703	−1.3083	0.14782	0.34411
13	0.25709	0.74292	−0.6452	−0.2972	−1.2134	0.16464	0.36698
14	0.27733	0.72267	−0.5850	−0.3248	−1.1245	0.18174	0.38836
15	0.29757	0.70243	−0.5268	−0.3532	−1.0407	0.19910	0.41008
16	0.31781	0.68219	−0.4701	−0.3825	−0.9612	0.21671	0.43156
17	0.33806	0.66194	−0.4148	−0.4126	−0.8853	0.23455	0.45280
18	0.35830	0.64170	−0.3607	−0.4436	−0.8128	0.25261	0.47382
19	0.37854	0.62146	−0.3076	−0.4757	−0.7430	0.27088	0.49463
20	0.39879	0.60121	−0.2552	−0.5088	−0.6757	0.28936	0.51523
21	0.41903	0.58097	−0.2034	−0.5431	−0.6105	0.30804	0.53563

Table 2 *cont.*

i	$M'(i$	$1 - M'(i)$	$Xn(i)$	$Xe(i)$	$Xw(i)$	$p0.05$	$p0.95$
22	0.43927	0.56073	−0.1522	−0.5785	−0.5473	0.32692	0.55584
23	0.45951	0.54049	−0.1013	−0.6153	−0.4857	0.34599	0.57585
24	0.47976	0.52024	−0.0506	−0.6535	−0.4255	0.36524	0.59568
25	0.50000	0.50000	0.0000	−0.6931	−0.3665	0.38469	0.61531
26	0.52024	0.47976	0.0506	−0.7345	−0.3086	0.40432	0.63476
27	0.54049	0.45951	0.1014	−0.7776	−0.2516	0.42415	0.65401
28	0.56073	0.43927	0.1525	−0.8226	−0.1952	0.44416	0.67308
29	0.58097	0.41903	0.2040	−0.8698	−0.1395	0.46436	0.69196
30	0.60121	0.39879	0.2561	−0.9193	−0.0841	0.48477	0.71064
31	0.62146	0.37854	0.3089	−0.9714	−0.0290	0.50537	0.72912
32	0.64170	0.35830	0.3626	−1.0264	0.0260	0.52617	0.74739
33	0.66194	0.33806	0.4173	−1.0845	0.0812	0.54720	0.76545
34	0.68219	0.31781	0.4734	−1.1463	0.1365	0.56844	0.78329
35	0.70243	0.29757	0.5310	−1.2121	0.1924	0.58991	0.80090
36	0.72267	0.27733	0.5904	−1.2826	0.2489	0.61164	0.81826
37	0.74292	0.25709	0.6520	−1.3583	0.3063	0.63362	0.83536
38	0.76316	0.23684	0.7162	−1.4404	0.3649	0.65589	0.85218
39	0.78340	0.21660	0.7835	−1.5297	0.4251	0.67846	0.86870
40	0.80364	0.19636	0.8546	−1.6278	0.4872	0.70136	0.88488
41	0.82389	0.17611	0.9302	−1.7366	0.5519	0.72465	0.90069
42	0.84413	0.15587	1.0116	−1.8587	0.6199	0.74836	0.91608
43	0.86437	0.13563	1.1002	−1.9978	0.6921	0.77256	0.93098
44	0.88462	0.11538	1.1985	−2.1595	0.7699	0.79734	0.94531
45	0.90486	0.09514	1.3099	−2.3524	0.8554	0.82285	0.95892
46	0.92510	0.07490	1.4405	−2.5916	0.9523	0.84929	0.97163
47	0.94534	0.05466	1.6016	−2.9067	1.0670	0.87703	0.98311
48	0.96559	0.03441	1.8200	−3.3693	1.2147	0.90681	0.99270
49	0.98583	0.01417	2.1930	−4.2566	1.4485	0.94069	0.99895
			S3: 0.7013	−47.9624	−27.4759		
			S4: 43.1501	87.7656	86.8974		

Total number of results: 50

i	$M'(i$	$1 - M'(i)$	$Xn(i)$	$Xe(i)$	$Xw(i)$	$p0.05$	$p0.95$
1	0.01389	0.98611	−1.9614	−0.0140	−4.2697	0.00102	0.05816
2	0.03373	0.96627	−1.7019	−0.0343	−3.3723	0.00715	0.09140
3	0.05357	0.94643	−1.5281	−0.0551	−2.8993	0.01655	0.12061
4	0.07341	0.92659	−1.3915	−0.0762	−2.5738	0.02779	0.14784
5	0.09325	0.90675	−1.2766	−0.0979	−2.3239	0.04024	0.17379
6	0.11310	0.88690	−1.1760	−0.1200	−2.1201	0.05357	0.19883
7	0.13294	0.86706	−1.0857	−0.1426	−1.9474	0.06760	0.22317
8	0.15278	0.84722	−1.0032	−0.1658	−1.7970	0.08218	0.24694
9	0.17262	0.82738	−0.9266	−0.1895	−1.6634	0.09725	0.27022
10	0.19246	0.80754	−0.8549	−0.2138	−1.5429	0.11272	0.29309
11	0.21230	0.78770	−0.7871	−0.2386	−1.4328	0.12856	0.31560
12	0.23214	0.76786	−0.7226	−0.2642	−1.3312	0.14472	0.33778
13	0.25198	0.74802	−0.6607	−0.2903	−1.2367	0.16117	0.35933
14	0.27183	0.72817	−0.6012	−0.3172	−1.1482	0.17790	0.38126
15	0.29167	0.70833	−0.5436	−0.3448	−1.0647	0.19488	0.40262
16	0.31151	0.68849	−0.4876	−0.3733	−0.9855	0.21210	0.42373
17	0.33135	0.66865	−0.4330	−0.4025	−0.9101	0.22955	0.44462

Table 2 *cont.*

i	M'(i	1 − M'(i)	Xn(i)	Xe(i)	Xw(i)	p0.05	p0.95
18	0.35119	0.64881	−0.3796	−0.4326	−0.8379	0.24721	0.46530
19	0.37103	0.62897	−0.3272	−0.4637	−0.7686	0.26507	0.48577
20	0.39087	0.60913	−0.2756	−0.4957	−0.7017	0.28313	0.50604
21	0.41071	0.58929	−0.2246	−0.5288	−0.6371	0.30138	0.52612
22	0.43056	0.56944	−0.1742	−0.5631	−0.5743	0.31980	0.54801
23	0.45040	0.54960	−0.1242	−0.5986	−0.5132	0.33845	0.56572
24	0.47024	0.52976	−0.0744	−0.6353	−0.4536	0.35726	0.58524
25	0.49008	0.50992	−0.0248	−0.6735	−0.3953	0.37625	0.60459
26	0.50992	0.49008	0.0248	−0.7132	−0.3380	0.39541	0.62375
27	0.52976	0.47024	0.0745	−0.7545	−0.2817	0.41476	0.64274
28	0.54960	0.45040	0.1244	−0.7976	−0.2261	0.43428	0.66155
29	0.56944	0.43056	0.1746	−0.8427	−0.1712	0.45399	0.68017
30	0.58929	0.41071	0.2253	−0.8899	−0.1167	0.47388	0.69862
31	0.60913	0.39087	0.2766	−0.9394	−0.0625	0.49396	0.71687
32	0.62897	0.37103	0.3287	−0.9915	−0.0086	0.51423	0.73493
33	0.64881	0.35119	0.3817	−1.0464	0.0454	0.53470	0.75279
34	0.66865	0.33135	0.4358	−1.1046	0.0995	0.55538	0.77045
35	0.68849	0.31151	0.4912	−1.1663	0.1539	0.57627	0.78790
36	0.70833	0.29167	0.5481	−1.2321	0.2088	0.59738	0.80511
37	0.72817	0.27183	0.6070	−1.3026	0.2644	0.61874	0.82210
38	0.74802	0.25198	0.6680	−1.3784	0.3209	0.64034	0.83882
39	0.76786	0.23214	0.7316	−1.4604	0.3787	0.66222	0.85528
40	0.78770	0.21230	0.7983	−1.5497	0.4381	0.68440	0.87144
41	0.80754	0.19246	0.8687	−1.6479	0.4995	0.70691	0.88728
42	0.82738	0.17262	0.9438	−1.7567	0.5634	0.72978	0.90275
43	0.84722	0.15278	1.0246	−1.8788	0.6306	0.75306	0.91781
44	0.86706	0.13294	1.1127	−2.0179	0.7020	0.77683	0.93240
45	0.88690	0.11310	1.2104	−2.1795	0.7791	0.80117	0.94643
46	0.90675	0.09325	1.3212	−2.3724	0.8639	0.82621	0.95976
47	0.92659	0.07341	1.4511	−2.6117	0.9600	0.85216	0.97221
48	0.94643	0.05357	1.6115	−2.9267	1.0739	0.87939	0.99345
49	0.96627	0.03373	1.8290	−3.3894	1.2206	0.90860	0.99285
50	0.98611	0.01389	2.2009	−4.2767	1.4532	0.94184	0.99897
			S3: 0.7178	−48.9584	−28.0496		
			S4: 44.0792	90.7079	88.8156		

Cumulative probability and corresponding linearized values

$M(.)$	X_n	X_c	X_w
0.001	−2.3522	−0.0010	−6.9072
0.002	−2.2881	−0.0020	−6.2136
0.003	−2.2402	−0.0030	−5.8076
0.004	−2.2007	−0.0040	−5.5194
0.005	−2.1664	−0.0050	−5.2958
0.006	−2.1359	−0.0060	−5.1129
0.007	−2.1082	−0.0070	−4.9583
0.008	−2.0827	−0.0080	−4.8243
0.009	−2.0591	−0.0090	−4.7060
0.010	−2.0369	−0.0101	−4.6001
0.011	−2.0160	−0.0111	−4.5043
0.012	−1.9962	−0.0121	−4.4168
0.013	−1.9774	−0.0131	−4.3362
0.014	−1.9595	−0.0141	−4.2616
0.015	−1.9423	−0.0151	−4.1921
0.016	−1.9258	−0.0161	−4.1271
0.017	−1.9099	−0.0171	−4.0659
0.018	−1.8945	−0.0182	−4.0083
0.019	−1.8797	−0.0192	−3.9537
0.020	−1.8653	−0.0202	−3.9019
0.021	−1.8514	−0.0212	−3.8526
0.022	−1.8379	−0.0222	−3.8056
0.023	−1.8248	−0.0233	−3.7606
0.024	−1.8120	−0.0243	−3.7175
0.025	−1.7995	−0.0253	−3.6762
0.026	−1.7873	−0.0263	−3.6365
0.027	−1.7755	−0.0274	−3.5982
0.028	−1.7639	−0.0284	−3.5613
0.029	−1.7525	−0.0294	−3.5257
0.030	−1.7414	−0.0305	−3.4913
0.031	−1.7305	−0.0315	−3.4580
0.032	−1.7199	−0.0325	−3.4258
0.033	−1.7094	−0.0336	−3.3945
0.034	−1.6991	−0.0346	−3.3641
0.035	−1.6891	−0.0356	−3.3346
0.036	−1.6792	−0.0367	−3.3059
0.037	−1.6695	−0.0377	−3.2780
0.038	−1.6599	−0.0387	−3.2508
0.039	−1.6505	−0.0398	−3.2243
0.040	−1.6413	−0.0408	−3.1985
0.041	−1.6322	−0.0419	−3.1733
0.042	−1.6232	−0.0429	−3.1487
0.043	−1.6144	−0.0440	−3.1246
0.044	−1.6057	−0.0450	−3.1011
0.045	−1.5971	−0.0460	−3.0781
0.046	−1.5886	−0.0471	−3.0556
0.047	−1.5803	−0.0481	−3.0336
0.048	−1.5720	−0.0492	−3.0120
0.049	−1.5639	−0.0502	−2.9909

Table 3 *cont.*

M(.)	X_n	X_c	X_w
0.050	−1.5559	−0.0513	−2.9701
0.051	−1.5480	−0.0523	−2.9498
0.052	−1.5402	−0.0534	−2.9299
0.053	−1.5324	−0.0545	−2.9103
0.054	−1.5248	−0.0555	−2.8911
0.055	−1.5172	−0.0566	−2.8722
0.056	−1.5098	−0.0576	−2.8537
0.057	−1.5024	−0.0587	−2.8355
0.058	−1.4951	−0.0598	−2.8175
0.059	−1.4879	−0.0608	−2.7999
0.060	−1.4807	−0.0619	−2.7826
0.061	−1.4737	−0.0629	−2.7655
0.062	−1.4667	−0.0640	−2.7487
0.063	−1.4598	−0.0651	−2.7322
0.064	−1.4529	−0.0661	−2.7159
0.065	−1.4461	−0.0672	−2.6999
0.066	−1.4394	−0.0683	−2.6841
0.067	−1.4327	−0.0694	−2.6685
0.068	−1.4261	−0.0704	−2.6532
0.069	−1.4196	−0.0715	−2.6381
0.070	−1.4131	−0.0726	−2.6231
0.071	−1.4067	−0.0737	−2.6084
0.072	−1.4004	−0.0747	−2.5939
0.073	−1.3941	−0.0758	−2.5796
0.074	−1.3878	−0.0769	−2.5654
0.075	−1.3816	−0.0780	−2.5515
0.076	−1.3755	−0.0790	−2.5377
0.077	−1.3694	−0.0801	−2.5241
0.078	−1.3633	−0.0812	−2.5107
0.079	−1.3573	−0.0823	−2.4974
0.080	−1.3514	−0.0834	−2.4843
0.081	−1.3455	−0.0845	−2.4713
0.082	−1.3396	−0.0856	−2.4585
0.083	−1.3338	−0.0867	−2.4459
0.084	−1.3280	−0.0877	−2.4333
0.085	−1.3223	−0.0888	−2.4210
0.086	−1.3166	−0.0899	−2.4087
0.087	−1.3110	−0.0910	−2.3966
0.088	−1.3054	−0.0921	−2.3847
0.089	−1.2998	−0.0932	−2.3728
0.090	−1.2943	−0.0943	−2.3611
0.091	−1.2888	−0.0954	−2.3495
0.092	−1.2833	−0.0965	−2.3381
0.093	−1.2779	−0.0976	−2.3267
0.094	−1.2725	−0.0987	−2.3155
0.095	−1.2672	−0.0998	−2.3043
0.096	−1.2619	−0.1009	−2.2933
0.097	−1.2566	−0.1020	−2.2824
0.098	−1.2514	−0.1031	−2.2716

Table 3 *cont.*

$M(.)$	X_n	X_c	X_w
0.099	−1.2462	−0.1043	−2.2609
0.100	−1.2410	−0.1054	−2.2503
0.101	−1.2359	−0.1065	−2.2398
0.102	−1.2307	−0.1076	−2.2294
0.103	−1.2257	−0.1087	−2.2191
0.104	−1.2206	−0.1098	−2.2089
0.105	−1.2156	−0.1109	−2.1988
0.106	−1.2106	−0.1121	−2.1888
0.107	−1.2056	−0.1132	−2.1788
0.108	−1.2007	−0.1143	−2.1690
0.109	−1.1958	−0.1154	−2.1592
0.110	−1.1909	−0.1165	−2.1495
0.111	−1.1861	−0.1177	−2.1399
0.112	−1.1812	−0.1188	−2.1304
0.113	−1.1764	−0.1199	−2.1210
0.114	−1.1717	−0.1210	−2.1116
0.115	−1.1669	−0.1222	−2.1023
0.116	−1.1622	−0.1233	−2.0931
0.117	−1.1575	−0.1244	−2.0840
0.118	−1.1528	−0.1256	−2.0749
0.119	−1.1482	−0.1267	−2.0659
0.120	−1.1436	−0.1278	−2.0570
0.121	−1.1390	−0.1290	−2.0481
0.122	−1.1344	−0.1301	−2.0393
0.123	−1.1298	−0.1313	−2.0306
0.124	−1.1253	−0.1324	−2.0220
0.125	−1.1208	−0.1335	−2.0134
0.126	−1.1163	−0.1347	−2.0048
0.127	−1.1118	−0.1358	−1.9964
0.128	−1.1074	−0.1370	−1.9880
0.129	−1.1029	−0.1381	−1.9796
0.130	−1.0985	−0.1393	−1.9713
0.131	−1.0941	−0.1404	−1.9631
0.132	−1.0898	−0.1416	−1.9550
0.133	−1.0854	−0.1427	−1.9468
0.134	−1.0811	−0.1439	−1.9388
0.135	−1.0768	−0.1450	−1.9308
0.136	−1.0725	−0.1462	−1.9228
0.137	−1.0682	−0.1473	−1.9150
0.138	−1.0640	−0.1485	−1.9071
0.139	−1.0597	−0.1497	−1.8993
0.140	−1.0555	−0.1508	−1.8916
0.141	−1.0513	−0.1520	−1.8839
0.142	−1.0471	−0.1532	−1.8763
0.143	−1.0430	−0.1543	−1.8687
0.144	−1.0388	−0.1555	−1.8612
0.145	−1.0347	−0.1567	−1.8537
0.146	−1.0306	−0.1578	−1.8462
0.147	−1.0265	−0.1590	−1.8388

Table 3 *cont.*

M(.)	X_n	X_c	X_w
0.148	−1.0224	−0.1602	−1.8315
0.149	−1.0184	−0.1614	−1.8242
0.150	−1.0143	−0.1625	−1.8169
0.151	−1.0103	−0.1637	−1.8097
0.152	−1.0063	−0.1649	−1.8025
0.153	−1.0023	−0.1661	−1.7954
0.154	−0.9983	−0.1672	−1.7883
0.155	−0.9943	−0.1684	−1.7813
0.156	−0.9903	−0.1696	−1.7742
0.157	−0.9864	−0.1708	−1.7673
0.158	−0.9825	−0.1720	−1.7604
0.159	−0.9786	−0.1732	−1.7535
0.160	−0.9747	−0.1744	−1.7466
0.161	−0.9708	−0.1756	−1.7398
0.162	−0.9669	−0.1767	−1.7330
0.163	−0.9630	−0.1779	−1.7263
0.164	−0.9592	−0.1791	−1.7196
0.165	−0.9554	−0.1803	−1.7130
0.166	−0.9516	−0.1815	−1.7063
0.167	−0.9477	−0.1827	−1.6997
0.168	−0.9440	−0.1839	−1.6932
0.169	−0.9402	−0.1851	−1.6867
0.170	−0.9364	−0.1863	−1.6802
0.171	−0.9327	−0.1875	−1.6737
0.172	−0.9289	−0.1888	−1.6673
0.173	−0.9252	−0.1900	−1.6609
0.174	−0.9215	−0.1912	−1.6546
0.175	−0.9178	−0.1924	−1.6483
0.176	−0.9141	−0.1936	−1.6420
0.177	−0.9104	−0.1948	−1.6357
0.178	−0.9067	−0.1960	−1.6295
0.179	−0.9031	−0.1972	−1.6233
0.180	−0.8994	−0.1985	−1.6172
0.181	−0.8958	−0.1997	−1.6110
0.182	−0.8922	−0.2009	−1.6049
0.183	−0.8885	−0.2021	−1.5989
0.184	−0.8849	−0.2034	−1.5928
0.185	−0.8814	−0.2046	−1.5868
0.186	−0.8778	−0.2058	−1.5808
0.187	−0.8742	−0.2070	−1.5749
0.188	−0.8706	−0.2083	−1.5689
0.189	−0.8671	−0.2095	−1.5630
0.190	−0.8636	−0.2107	−1.5572
0.191	−0.8600	−0.2120	−1.5513
0.192	−0.8565	−0.2132	−1.5455
0.193	−0.8530	−0.2144	−1.5397
0.194	−0.8495	−0.2157	−1.5339
0.195	−0.8460	−0.2169	−1.5282
0.196	−0.8425	−0.2182	−1.5225

Table 3 *cont.*

M(.)	X_n	X_c	X_w
0.197	−0.8391	−0.2194	−1.5168
0.198	−0.8356	−0.2207	−1.5111
0.199	−0.8321	−0.2219	−1.5055
0.200	−0.8287	−0.2232	−1.4999
0.201	−0.8253	−0.2244	−1.4943
0.202	−0.8218	−0.2257	−1.4887
0.203	−0.8184	−0.2269	−1.4832
0.204	−0.8150	−0.2282	−1.4777
0.205	−0.8116	−0.2294	−1.4722
0.206	−0.8082	−0.2307	−1.4667
0.207	−0.8049	−0.2319	−1.4613
0.208	−0.8015	−0.2332	−1.4558
0.209	−0.7981	−0.2345	−1.4504
0.210	−0.7948	−0.2357	−1.4450
0.211	−0.7914	−0.2370	−1.4397
0.212	−0.7881	−0.2383	−1.4344
0.213	−0.7848	−0.2395	−1.4290
0.214	−0.7814	−0.2408	−1.4237
0.215	−0.7781	−0.2421	−1.4185
0.216	−0.7748	−0.2434	−1.4132
0.217	−0.7715	−0.2446	−1.4080
0.218	−0.7682	−0.2459	−1.4028
0.219	−0.7650	−0.2472	−1.3976
0.220	−0.7617	−0.2485	−1.3924
0.221	−0.7584	−0.2498	−1.3873
0.222	−0.7552	−0.2510	−1.3821
0.223	−0.7519	−0.2523	−1.3770
0.224	−0.7487	−0.2536	−1.3719
0.225	−0.7454	−0.2549	−1.3669
0.226	−0.7422	−0.2562	−1.3618
0.227	−0.7390	−0.2575	−1.3568
0.228	−0.7358	−0.2588	−1.3518
0.229	−0.7326	−0.2601	−1.3468
0.230	−0.7294	−0.2614	−1.3418
0.231	−0.7262	−0.2627	−1.3368
0.232	−0.7230	−0.2640	−1.3319
0.233	−0.7198	−0.2653	−1.3270
0.234	−0.7166	−0.2666	−1.3221
0.235	−0.7135	−0.2679	−1.3172
0.236	−0.7103	−0.2692	−1.3123
0.237	−0.7072	−0.2705	−1.3074
0.238	−0.7040	−0.2718	−1.3026
0.239	−0.7009	−0.2731	−1.2978
0.240	−0.6978	−0.2745	−1.2930
0.241	−0.6946	−0.2758	−1.2882
0.242	−0.6915	−0.2771	−1.2834
0.243	−0.6884	−0.2784	−1.2787
0.244	−0.6853	−0.2797	−1.2739
0.245	−0.6822	−0.2811	−1.2692

Table 3 *cont.*

M(.)	X_n	X_e	X_w
0.246	−0.6791	−0.2824	−1.2645
0.247	−0.6760	−0.2837	−1.2598
0.248	−0.6729	−0.2850	−1.2551
0.249	−0.6698	−0.2864	−1.2505
0.250	−0.6668	−0.2877	−1.2458
0.251	−0.6637	−0.2890	−1.2412
0.252	−0.6607	−0.2904	−1.2366
0.253	−0.6576	−0.2917	−1.2320
0.254	−0.6546	−0.2930	−1.2274
0.255	−0.6515	−0.2944	−1.2229
0.256	−0.6485	−0.2957	−1.2183
0.257	−0.6454	−0.2971	−1.2138
0.258	−0.6424	−0.2984	−1.2092
0.259	−0.6394	−0.2998	−1.2047
0.260	−0.6364	−0.3011	−1.2002
0.261	−0.6334	−0.3025	−1.1958
0.262	−0.6304	−0.3038	−1.1913
0.263	−0.6274	−0.3052	−1.1868
0.264	−0.6244	−0.3065	−1.1824
0.265	−0.6214	−0.3079	−1.1780
0.266	−0.6184	−0.3093	−1.1736
0.267	−0.6154	−0.3106	−1.1692
0.268	−0.6125	−0.3120	−1.1648
0.269	−0.6095	−0.3134	−1.1604
0.270	−0.6065	−0.3147	−1.1560
0.271	−0.6036	−0.3161	−1.1517
0.272	−0.6006	−0.3175	−1.1474
0.273	−0.5977	−0.3188	−1.1430
0.274	−0.5947	−0.3202	−1.1387
0.275	−0.5918	−0.3216	−1.1344
0.276	−0.5889	−0.3230	−1.1302
0.277	−0.5860	−0.3244	−1.1259
0.278	−0.5830	−0.3257	−1.1216
0.279	−0.5801	−0.3271	−1.1174
0.280	−0.5772	−0.3285	−1.1132
0.281	−0.5743	−0.3299	−1.1089
0.282	−0.5714	−0.3313	−1.1047
0.283	−0.5685	−0.3327	−1.1005
0.284	−0.5656	−0.3341	−1.0963
0.285	−0.5627	−0.3355	−1.0922
0.286	−0.5598	−0.3369	−1.0880
0.287	−0.5569	−0.3383	−1.0838
0.288	−0.5540	−0.3397	−1.0797
0.289	−0.5512	−0.3411	−1.0756
0.290	−0.5483	−0.3425	−1.0715
0.291	−0.5454	−0.3439	−1.0673
0.292	−0.5426	−0.3453	−1.0633
0.293	−0.5397	−0.3467	−1.0592
0.294	−0.5369	−0.3482	−1.0551

Table 3 *cont.*

$M(.)$	X_n	X_c	X_w
0.295	−0.5340	−0.3496	−1.0510
0.296	−0.5312	−0.3510	−1.0470
0.297	−0.5283	−0.3524	−1.0429
0.298	−0.5255	−0.3538	−1.0389
0.299	−0.5227	−0.3553	−1.0349
0.300	−0.5198	−0.3567	−1.0309
0.301	−0.5170	−0.3581	−1.0269
0.302	−0.5142	−0.3596	−1.0229
0.303	−0.5114	−0.3610	−1.0189
0.304	−0.5086	−0.3624	−1.0149
0.305	−0.5057	−0.3639	−1.0110
0.306	−0.5029	−0.3653	−1.0070
0.307	−0.5001	−0.3667	−1.0031
0.308	−0.4973	−0.3682	−0.9992
0.309	−0.4945	−0.3696	−0.9952
0.310	−0.4918	−0.3711	−0.9913
0.311	−0.4890	−0.3725	−0.9874
0.312	−0.4862	−0.3740	−0.9835
0.313	−0.4834	−0.3754	−0.9797
0.314	−0.4806	−0.3769	−0.9758
0.315	−0.4779	−0.3784	−0.9719
0.316	−0.4751	−0.3798	−0.9681
0.317	−0.4723	−0.3813	−0.9642
0.318	−0.4696	−0.3827	−0.9604
0.319	−0.4668	−0.3842	−0.9566
0.320	−0.4640	−0.3857	−0.9527
0.321	−0.4613	−0.3872	−0.9489
0.322	−0.4585	−0.3886	−0.9451
0.323	−0.4558	−0.3901	−0.9413
0.324	−0.4530	−0.3916	−0.9376
0.325	−0.4503	−0.3931	−0.9338
0.326	−0.4476	−0.3945	−0.9300
0.327	−0.4448	−0.3960	−0.9263
0.328	−0.4421	−0.3975	−0.9225
0.329	−0.4394	−0.3990	−0.9188
0.330	−0.4366	−0.4005	−0.9150
0.331	−0.4339	−0.4020	−0.9113
0.332	−0.4312	−0.4035	−0.9076
0.333	−0.4285	−0.4050	−0.9039
0.334	−0.4258	−0.4065	−0.9002
0.335	−0.4231	−0.4080	−0.8965
0.336	−0.4203	−0.4095	−0.8928
0.337	−0.4176	−0.4110	−0.8892
0.338	−0.4149	−0.4125	−0.8855
0.339	−0.4122	−0.4140	−0.8818
0.340	−0.4095	−0.4155	−0.8782
0.341	−0.4069	−0.4171	−0.8745
0.342	−0.4042	−0.4186	−0.8709
0.343	−0.4015	−0.4201	−0.8673

Table 3 *cont.*

M(.)	X_n	X_e	X_w
0.344	−0.3988	−0.4216	−0.8637
0.345	−0.3961	−0.4231	−0.8600
0.346	−0.3934	−0.4247	−0.8564
0.347	−0.3907	−0.4262	−0.8528
0.348	−0.3881	−0.4277	−0.8492
0.349	−0.3854	−0.4293	−0.8457
0.350	−0.3827	−0.4308	−0.8421
0.351	−0.3801	−0.4323	−0.8385
0.352	−0.3774	−0.4339	−0.8350
0.353	−0.3747	−0.4354	−0.8314
0.354	−0.3721	−0.4370	−0.8279
0.355	−0.3694	−0.4385	−0.8243
0.356	−0.3668	−0.4401	−0.8208
0.357	−0.3641	−0.4416	−0.8173
0.358	−0.3615	−0.4432	−0.8137
0.359	−0.3588	−0.4448	−0.8102
0.360	−0.3562	−0.4463	−0.8067
0.361	−0.3535	−0.4479	−0.8032
0.362	−0.3509	−0.4494	−0.7997
0.363	−0.3482	−0.4510	−0.7963
0.364	−0.3456	−0.4526	−0.7928
0.365	−0.3430	−0.4542	−0.7893
0.366	−0.3403	−0.4557	−0.7858
0.367	−0.3377	−0.4573	−0.7824
0.368	−0.3351	−0.4589	−0.7789
0.369	−0.3325	−0.4605	−0.7755
0.370	−0.3298	−0.4621	−0.7721
0.371	−0.3272	−0.4637	−0.7686
0.372	−0.3246	−0.4652	−0.7652
0.373	−0.3220	−0.4668	−0.7618
0.374	−0.3194	−0.4684	−0.7584
0.375	−0.3167	−0.4700	−0.7550
0.376	−0.3141	−0.4716	−0.7516
0.377	−0.3115	−0.4732	−0.7482
0.378	−0.3089	−0.4748	−0.7448
0.379	−0.3063	−0.4765	−0.7414
0.380	−0.3037	−0.4781	−0.7380
0.381	−0.3011	−0.4797	−0.7346
0.382	−0.2985	−0.4813	−0.7313
0.383	−0.2959	−0.4829	−0.7279
0.384	−0.2933	−0.4845	−0.7246
0.385	−0.2907	−0.4862	−0.7212
0.386	−0.2881	−0.4878	−0.7179
0.387	−0.2855	−0.4894	−0.7145
0.388	−0.2829	−0.4911	−0.7112
0.389	−0.2804	−0.4927	−0.7079
0.390	−0.2778	−0.4943	−0.7046
0.391	−0.2752	−0.4960	−0.7012
0.392	−0.2726	−0.4976	−0.6979

Table 3 *cont.*

$M(.)$	X_n	X_c	X_w
0.393	−0.2700	−0.4993	−0.6946
0.394	−0.2675	−0.5009	−0.6913
0.395	−0.2649	−0.5026	−0.6880
0.396	−0.2623	−0.5042	−0.6848
0.397	−0.2597	−0.5059	−0.6815
0.398	−0.2572	−0.5075	−0.6782
0.399	−0.2546	−0.5092	−0.6749
0.400	−0.2520	−0.5109	−0.6717
0.401	−0.2494	−0.5125	−0.6684
0.402	−0.2469	−0.5142	−0.6652
0.403	−0.2443	−0.5159	−0.6619
0.404	−0.2417	−0.5175	−0.6587
0.405	−0.2392	−0.5192	−0.6554
0.406	−0.2366	−0.5209	−0.6522
0.407	−0.2341	−0.5226	−0.6490
0.408	−0.2315	−0.5243	−0.6457
0.409	−0.2290	−0.5260	−0.6425
0.410	−0.2264	−0.5277	−0.6393
0.411	−0.2238	−0.5294	−0.6361
0.412	−0.2213	−0.5311	−0.6329
0.413	−0.2187	−0.5328	−0.6297
0.414	−0.2162	−0.5345	−0.6265
0.415	−0.2136	−0.5362	−0.6233
0.416	−0.2111	−0.5379	−0.6201
0.417	−0.2086	−0.5396	−0.6169
0.418	−0.2060	−0.5413	−0.6137
0.419	−0.2035	−0.5430	−0.6106
0.420	−0.2009	−0.5448	−0.6074
0.421	−0.1984	−0.5465	−0.6042
0.422	−0.1958	−0.5482	−0.6011
0.423	−0.1933	−0.5499	−0.5979
0.424	−0.1908	−0.5517	−0.5948
0.425	−0.1882	−0.5534	−0.5916
0.426	−0.1857	−0.5552	−0.5885
0.427	−0.1832	−0.5569	−0.5854
0.428	−0.1806	−0.5587	−0.5822
0.429	−0.1781	−0.5604	−0.5791
0.430	−0.1756	−0.5622	−0.5760
0.431	−0.1730	−0.5639	−0.5729
0.432	−0.1705	−0.5657	−0.5697
0.433	−0.1680	−0.5674	−0.5666
0.434	−0.1654	−0.5692	−0.5635
0.435	−0.1629	−0.5710	−0.5604
0.436	−0.1604	−0.5727	−0.5573
0.437	−0.1579	−0.5745	−0.5542
0.438	−0.1553	−0.5763	−0.5511
0.439	−0.1528	−0.5781	−0.5481
0.440	−0.1503	−0.5799	−0.5450
0.441	−0.1478	−0.5816	−0.5419

Table 3 *cont.*

M(.)	X_n	X_c	X_w
0.442	−0.1453	−0.5834	−0.5388
0.443	−0.1427	−0.5852	−0.5358
0.444	−0.1402	−0.5870	−0.5327
0.445	−0.1377	−0.5888	−0.5296
0.446	−0.1352	−0.5906	−0.5266
0.447	−0.1327	−0.5924	−0.5235
0.448	−0.1301	−0.5942	−0.5205
0.449	−0.1276	−0.5961	−0.5174
0.450	−0.1251	−0.5979	−0.5144
0.451	−0.1226	−0.5997	−0.5113
0.452	−0.1201	−0.6015	−0.5083
0.453	−0.1176	−0.6033	−0.5053
0.454	−0.1151	−0.6052	−0.5022
0.455	−0.1126	−0.6070	−0.4992
0.456	−0.1100	−0.6088	−0.4962
0.457	−0.1075	−0.6107	−0.4932
0.458	−0.1050	−0.6125	−0.4902
0.459	−0.1025	−0.6144	−0.4871
0.460	−0.1000	−0.6162	−0.4841
0.461	−0.0975	−0.6181	−0.4811
0.462	−0.0950	−0.6199	−0.4781
0.463	−0.0925	−0.6218	−0.4751
0.464	−0.0900	−0.6237	−0.4721
0.465	−0.0875	−0.6255	−0.4692
0.466	−0.0850	−0.6274	−0.4662
0.467	−0.0825	−0.6293	−0.4632
0.468	−0.0800	−0.6312	−0.4602
0.469	−0.0775	−0.6330	−0.4572
0.470	−0.0750	−0.6349	−0.4543
0.471	−0.0724	−0.6368	−0.4513
0.472	−0.0699	−0.6387	−0.4483
0.473	−0.0674	−0.6406	−0.4454
0.474	−0.0649	−0.6425	−0.4424
0.475	−0.0624	−0.6444	−0.4394
0.476	−0.0599	−0.6463	−0.4365
0.477	−0.0574	−0.6482	−0.4335
0.478	−0.0549	−0.6501	−0.4306
0.479	−0.0524	−0.6520	−0.4276
0.480	−0.0499	−0.6540	−0.4247
0.481	−0.0474	−0.6559	−0.4218
0.482	−0.0449	−0.6578	−0.4188
0.483	−0.0424	−0.6598	−0.4159
0.484	−0.0399	−0.6617	−0.4130
0.485	−0.0374	−0.6636	−0.4100
0.486	−0.0349	−0.6656	−0.4071
0.487	−0.0324	−0.6675	−0.4042
0.488	−0.0299	−0.6695	−0.4013
0.489	−0.0274	−0.6714	−0.3983
0.490	−0.0249	−0.6734	−0.3954

Table 3 *cont.*

M(.)	X_n	X_c	X_w
0.491	−0.0224	−0.6754	−0.3925
0.492	−0.0199	−0.6773	−0.3896
0.493	−0.0174	−0.6793	−0.3867
0.494	−0.0149	−0.6813	−0.3838
0.495	−0.0124	−0.6832	−0.3809
0.496	−0.0099	−0.6852	−0.3780
0.497	−0.0074	−0.6872	−0.3751
0.498	−0.0049	−0.6892	−0.3722
0.499	−0.0024	−0.6912	−0.3693
0.500	0.0001	−0.6932	−0.3664
0.501	0.0026	−0.6952	−0.3636
0.502	0.0051	−0.6972	−0.3607
0.503	0.0076	−0.6992	−0.3578
0.504	0.0101	−0.7012	−0.3549
0.505	0.0126	−0.7032	−0.3520
0.506	0.0151	−0.7053	−0.3492
0.507	0.0176	−0.7073	−0.3463
0.508	0.0201	−0.7093	−0.3434
0.509	0.0226	−0.7114	−0.3406
0.510	0.0251	−0.7134	−0.3377
0.511	0.0276	−0.7154	−0.3349
0.512	0.0301	−0.7175	−0.3320
0.513	0.0326	−0.7195	−0.3291
0.514	0.0351	−0.7216	−0.3263
0.515	0.0376	−0.7237	−0.3234
0.516	0.0401	−0.7257	−0.3206
0.517	0.0426	−0.7278	−0.3177
0.518	0.0451	−0.7299	−0.3149
0.519	0.0476	−0.7319	−0.3121
0.520	0.0501	−0.7340	−0.3092
0.521	0.0526	−0.7361	−0.3064
0.522	0.0551	−0.7382	−0.3035
0.523	0.0576	−0.7403	−0.3007
0.524	0.0601	−0.7424	−0.2979
0.525	0.0626	−0.7445	−0.2951
0.526	0.0651	−0.7466	−0.2922
0.527	0.0676	−0.7487	−0.2894
0.528	0.0701	−0.7508	−0.2866
0.529	0.0726	−0.7529	−0.2838
0.530	0.0751	−0.7551	−0.2809
0.531	0.0777	−0.7572	−0.2781
0.532	0.0802	−0.7593	−0.2753
0.533	0.0827	−0.7615	−0.2725
0.534	0.0852	−0.7636	−0.2697
0.535	0.0877	−0.7658	−0.2669
0.536	0.0902	−0.7679	−0.2641
0.537	0.0927	−0.7701	−0.2613
0.538	0.0952	−0.7722	−0.2585
0.539	0.0977	−0.7744	−0.2557

Table 3 *cont.*

M(.)	X_n	X_e	X_w
0.540	0.1003	−0.7766	−0.2529
0.541	0.1028	−0.7788	−0.2501
0.542	0.1053	−0.7809	−0.2473
0.543	0.1078	−0.7831	−0.2445
0.544	0.1103	−0.7853	−0.2417
0.545	0.1128	−0.7875	−0.2389
0.546	0.1154	−0.7897	−0.2361
0.547	0.1179	−0.7919	−0.2333
0.548	0.1204	−0.7941	−0.2305
0.549	0.1229	−0.7963	−0.2277
0.550	0.1254	−0.7986	−0.2249
0.551	0.1280	−0.8008	−0.2222
0.552	0.1305	−0.8030	−0.2194
0.553	0.1330	−0.8053	−0.2166
0.554	0.1355	−0.8075	−0.2138
0.555	0.1381	−0.8097	−0.2110
0.556	0.1406	−0.8120	−0.2083
0.557	0.1431	−0.8142	−0.2055
0.558	0.1457	−0.8165	−0.2027
0.559	0.1482	−0.8188	−0.2000
0.560	0.1507	−0.8210	−0.1972
0.561	0.1533	−0.8233	−0.1944
0.562	0.1558	−0.8256	−0.1916
0.563	0.1583	−0.8279	−0.1889
0.564	0.1609	−0.8302	−0.1861
0.565	0.1634	−0.8325	−0.1834
0.566	0.1659	−0.8348	−0.1806
0.567	0.1685	−0.8371	−0.1778
0.568	0.1710	−0.8394	−0.1751
0.569	0.1736	−0.8417	−0.1723
0.570	0.1761	−0.8440	−0.1696
0.571	0.1786	−0.8464	−0.1668
0.572	0.1812	−0.8487	−0.1641
0.573	0.1837	−0.8510	−0.1613
0.574	0.1863	−0.8534	−0.1586
0.575	0.1888	−0.8557	−0.1558
0.576	0.1914	−0.8581	−0.1531
0.577	0.1939	−0.8604	−0.1503
0.578	0.1965	−0.8628	−0.1476
0.579	0.1990	−0.8652	−0.1448
0.580	0.2016	−0.8676	−0.1421
0.581	0.2041	−0.8699	−0.1393
0.582	0.2067	−0.8723	−0.1366
0.583	0.2093	−0.8747	−0.1338
0.584	0.2118	−0.8771	−0.1311
0.585	0.2144	−0.8795	−0.1284
0.586	0.2169	−0.8820	−0.1256
0.587	0.2195	−0.8844	−0.1229
0.588	0.2221	−0.8868	−0.1201

Table 3 *cont.*

$M(.)$	X_n	X_c	X_w
0.589	0.2246	−0.8892	−0.1174
0.590	0.2272	−0.8917	−0.1147
0.591	0.2298	−0.8941	−0.1119
0.592	0.2324	−0.8966	−0.1092
0.593	0.2349	−0.8990	−0.1065
0.594	0.2375	−0.9015	−0.1037
0.595	0.2401	−0.9039	−0.1010
0.596	0.2427	−0.9064	−0.0983
0.597	0.2453	−0.9089	−0.0955
0.598	0.2478	−0.9114	−0.0928
0.599	0.2504	−0.9139	−0.0901
0.600	0.2530	−0.9164	−0.0873
0.601	0.2556	−0.9189	−0.0846
0.602	0.2582	−0.9214	−0.0819
0.603	0.2608	−0.9239	−0.0792
0.604	0.2634	−0.9264	−0.0764
0.605	0.2660	−0.9289	−0.0737
0.606	0.2686	−0.9315	−0.0710
0.607	0.2712	−0.9340	−0.0683
0.608	0.2738	−0.9366	−0.0655
0.609	0.2764	−0.9391	−0.0628
0.610	0.2790	−0.9417	−0.0601
0.611	0.2816	−0.9442	−0.0574
0.612	0.2842	−0.9468	−0.0546
0.613	0.2868	−0.9494	−0.0519
0.614	0.2894	−0.9520	−0.0492
0.615	0.2920	−0.9546	−0.0465
0.616	0.2946	−0.9572	−0.0438
0.617	0.2973	−0.9598	−0.0410
0.618	0.2999	−0.9624	−0.0383
0.619	0.3025	−0.9650	−0.0356
0.620	0.3051	−0.9677	−0.0329
0.621	0.3077	−0.9703	−0.0302
0.622	0.3104	−0.9729	−0.0274
0.623	0.3130	−0.9756	−0.0247
0.624	0.3156	−0.9782	−0.0220
0.625	0.3183	−0.9809	−0.0193
0.626	0.3209	−0.9836	−0.0166
0.627	0.3236	−0.9863	−0.0138
0.628	0.3262	−0.9889	−0.0111
0.629	0.3288	−0.9916	−0.0084
0.630	0.3315	−0.9943	−0.0057
0.631	0.3341	−0.9970	−0.0030
0.632	0.3368	−0.9998	−0.0002
0.633	0.3394	−1.0025	0.0025
0.634	0.3421	−1.0052	0.0052
0.635	0.3448	−1.0079	0.0079
0.636	0.3474	−1.0107	0.0106
0.637	0.3501	−1.0134	0.0133

Table 3 *cont.*

M(.)	X_n	X_e	X_w
0.638	0.3528	−1.0162	0.0161
0.639	0.3554	−1.0190	0.0188
0.640	0.3581	−1.0217	0.0215
0.641	0.3608	−1.0245	0.0242
0.642	0.3634	−1.0273	0.0269
0.643	0.3661	−1.0301	0.0297
0.644	0.3688	−1.0329	0.0324
0.645˙	0.3715	−1.0357	0.0351
0.646	0.3742	−1.0385	0.0378
0.647	0.3769	−1.0414	0.0405
0.648	0.3796	−1.0442	0.0433
0.649	0.3823	−1.0471	0.0460
0.650	0.3850	−1.0499	0.0487
0.651	0.3877	−1.0528	0.0514
0.652	0.3904	−1.0556	0.0541
0.653	0.3931	−1.0585	0.0569
0.654	0.3958	−1.0614	0.0596
0.655	0.3985	−1.0643	0.0623
0.656	0.4012	−1.0672	0.0650
0.657	0.4039	−1.0701	0.0678
0.658	0.4067	−1.0730	0.0705
0.659	0.4094	−1.0760	0.0732
0.660	0.4121	−1.0789	0.0759
0.661	0.4148	−1.0818	0.0787
0.662	0.4176	−1.0848	0.0814
0.663	0.4203	−1.0878	0.0841
0.664	0.4231	−1.0907	0.0869
0.665	0.4258	−1.0937	0.0896
0.666	0.4285	−1.0967	0.0923
0.667	0.4313	−1.0997	0.0950
0.668	0.4340	−1.1027	0.0978
0.669	0.4368	−1.1057	0.1005
0.670	0.4396	−1.1088	0.1032
0.671	0.4423	−1.1118	0.1060
0.672	0.4451	−1.1148	0.1087
0.673	0.4479	−1.1179	0.1114
0.674	0.4506	−1.1210	0.1142
0.675	0.4534	−1.1240	0.1169
0.676	0.4562	−1.1271	0.1197
0.677	0.4590	−1.1302	0.1224
0.678	0.4618	−1.1333	0.1251
0.679	0.4646	−1.1364	0.1279
0.680	0.4674	−1.1395	0.1306
0.681	0.4702	−1.1427	0.1334
0.682	0.4730	−1.1458	0.1361
0.683	0.4758	−1.1490	0.1389
0.684	0.4786	−1.1521	0.1416
0.685	0.4814	−1.1553	0.1443
0.686	0.4842	−1.1585	0.1471

Table 3 *cont.*

$M(.)$	X_n	X_e	X_w
0.687	0.4870	−1.1617	0.1498
0.688	0.4899	−1.1649	0.1526
0.689	0.4927	−1.1681	0.1553
0.690	0.4955	−1.1713	0.1581
0.691	0.4984	−1.1745	0.1609
0.692	0.5012	−1.1778	0.1636
0.693	0.5041	−1.1810	0.1664
0.694	0.5069	−1.1843	0.1691
0.695	0.5098	−1.1876	0.1719
0.696	0.5126	−1.1908	0.1747
0.697	0.5155	−1.1941	0.1774
0.698	0.5183	−1.1974	0.1802
0.699	0.5212	−1.2008	0.1829
0.700	0.5241	−1.2041	0.1857
0.701	0.5270	−1.2074	0.1885
0.702	0.5299	−1.2108	0.1913
0.703	0.5327	−1.2141	0.1940
0.704	0.5356	−1.2175	0.1968
0.705	0.5385	−1.2209	0.1996
0.706	0.5414	−1.2243	0.2024
0.707	0.5443	−1.2277	0.2051
0.708	0.5473	−1.2311	0.2079
0.709	0.5502	−1.2345	0.2107
0.710	0.5531	−1.2380	0.2135
0.711	0.5560	−1.2414	0.2163
0.712	0.5590	−1.2449	0.2191
0.713	0.5619	−1.2484	0.2219
0.714	0.5648	−1.2519	0.2246
0.715	0.5678	−1.2554	0.2274
0.716	0.5707	−1.2589	0.2302
0.717	0.5737	−1.2624	0.2330
0.718	0.5766	−1.2660	0.2358
0.719	0.5796	−1.2695	0.2386
0.720	0.5826	−1.2731	0.2414
0.721	0.5856	−1.2767	0.2443
0.722	0.5885	−1.2803	0.2471
0.723	0.5915	−1.2839	0.2499
0.724	0.5945	−1.2875	0.2527
0.725	0.5975	−1.2911	0.2555
0.726	0.6005	−1.2948	0.2583
0.727	0.6035	−1.2984	0.2611
0.728	0.6065	−1.3021	0.2640
0.729	0.6095	−1.3058	0.2668
0.730	0.6126	−1.3095	0.2696
0.731	0.6156	−1.3132	0.2724
0.732	0.6186	−1.3169	0.2753
0.733	0.6217	−1.3206	0.2781
0.734	0.6247	−1.3244	0.2810
0.735	0.6278	−1.3282	0.2838

Table 3 *cont.*

$M(.)$	X_n	X_c	X_w
0.736	0.6308	−1.3319	0.2866
0.737	0.6339	−1.3357	0.2895
0.738	0.6370	−1.3395	0.2923
0.739	0.6400	−1.3434	0.2952
0.740	0.6431	−1.3472	0.2980
0.741	0.6462	−1.3511	0.3009
0.742	0.6493	−1.3549	0.3037
0.743	0.6524	−1.3588	0.3066
0.744	0.6555	−1.3627	0.3095
0.745	0.6586	−1.3666	0.3123
0.746	0.6618	−1.3706	0.3152
0.747	0.6649	−1.3745	0.3181
0.748	0.6680	−1.3785	0.3210
0.749	0.6712	−1.3824	0.3239
0.750	0.6743	−1.3864	0.3267
0.751	0.6775	−1.3904	0.3296
0.752	0.6806	−1.3945	0.3325
0.753	0.6838	−1.3985	0.3354
0.754	0.6870	−1.4026	0.3383
0.755	0.6901	−1.4066	0.3412
0.756	0.6933	−1.4107	0.3441
0.757	0.6965	−1.4148	0.3470
0.758	0.6997	−1.4190	0.3499
0.759	0.7029	−1.4231	0.3528
0.760	0.7061	−1.4273	0.3558
0.761	0.7094	−1.4314	0.3587
0.762	0.7126	−1.4356	0.3616
0.763	0.7158	−1.4398	0.3645
0.764	0.7191	−1.4441	0.3675
0.765	0.7223	−1.4483	0.3704
0.766	0.7256	−1.4526	0.3733
0.767	0.7289	−1.4569	0.3763
0.768	0.7321	−1.4612	0.3792
0.769	0.7354	−1.4655	0.3822
0.770	0.7387	−1.4698	0.3851
0.771	0.7420	−1.4742	0.3881
0.772	0.7453	−1.4786	0.3911
0.773	0.7486	−1.4830	0.3940
0.774	0.7520	−1.4874	0.3970
0.775	0.7553	−1.4918	0.4000
0.776	0.7586	−1.4963	0.4030
0.777	0.7620	−1.5007	0.4060
0.778	0.7653	−1.5052	0.4090
0.779	0.7687	−1.5098	0.4119
0.780	0.7721	−1.5143	0.4149
0.781	0.7755	−1.5188	0.4180
0.782	0.7789	−1.5234	0.4210
0.783	0.7823	−1.5280	0.4240
0.784	0.7857	−1.5326	0.4270

Table 3 *cont.*

$M(.)$	X_n	X_c	X_w
0.785	0.7891	−1.5373	0.4300
0.786	0.7925	−1.5419	0.4330
0.787	0.7960	−1.5466	0.4361
0.788	0.7994	−1.5513	0.4391
0.789	0.8029	−1.5561	0.4422
0.790	0.8064	−1.5608	0.4452
0.791	0.8098	−1.5656	0.4483
0.792	0.8133	−1.5704	0.4513
0.793	0.8168	−1.5752	0.4544
0.794	0.8203	−1.5801	0.4575
0.795	0.8238	−1.5849	0.4605
0.796	0.8274	−1.5898	0.4636
0.797	0.8309	−1.5947	0.4667
0.798	0.8345	−1.5997	0.4698
0.799	0.8380	−1.6046	0.4729
0.800	0.8416	−1.6096	0.4760
0.801	0.8452	−1.6146	0.4791
0.802	0.8488	−1.6197	0.4822
0.803	0.8524	−1.6247	0.4853
0.804	0.8560	−1.6298	0.4885
0.805	0.8596	−1.6349	0.4916
0.806	0.8632	−1.6401	0.4947
0.807	0.8669	−1.6453	0.4979
0.808	0.8705	−1.6504	0.5010
0.809	0.8742	−1.6557	0.5042
0.810	0.8779	−1.6609	0.5074
0.811	0.8816	−1.6662	0.5105
0.812	0.8853	−1.6715	0.5137
0.813	0.8890	−1.6768	0.5169
0.814	0.8927	−1.6822	0.5201
0.815	0.8965	−1.6876	0.5233
0.816	0.9002	−1.6930	0.5265
0.817	0.9040	−1.6985	0.5297
0.818	0.9078	−1.7039	0.5329
0.819	0.9116	−1.7095	0.5362
0.820	0.9154	−1.7150	0.5394
0.821	0.9192	−1.7206	0.5427
0.822	0.9231	−1.7262	0.5459
0.823	0.9269	−1.7318	0.5492
0.824	0.9308	−1.7375	0.5524
0.825	0.9346	−1.7432	0.5557
0.826	0.9385	−1.7489	0.5590
0.827	0.9424	−1.7547	0.5623
0.828	0.9464	−1.7605	0.5656
0.829	0.9503	−1.7663	0.5689
0.830	0.9542	−1.7722	0.5722
0.831	0.9582	−1.7781	0.5755
0.832	0.9622	−1.7840	0.5789
0.833	0.9662	−1.7900	0.5822

Table 3 *cont.*

M(.)	X_n	X_e	X_w
0.834	0.9702	−1.7960	0.5856
0.835	0.9742	−1.8020	0.5889
0.836	0.9782	−1.8081	0.5923
0.837	0.9823	−1.8142	0.5957
0.838	0.9864	−1.8204	0.5990
0.839	0.9905	−1.8266	0.6024
0.840	0.9946	−1.8328	0.6059
0.841	0.9987	−1.8391	0.6093
0.842	1.0028	−1.8454	0.6127
0.843	1.0070	−1.8517	0.6161
0.844	1.0112	−1.8581	0.6196
0.845	1.0154	−1.8646	0.6230
0.846	1.0196	−1.8710	0.6265
0.847	1.0238	−1.8776	0.6300
0.848	1.0280	−1.8841	0.6335
0.849	1.0323	−1.8907	0.6370
0.850	1.0366	−1.8974	0.6405
0.851	1.0409	−1.9041	0.6440
0.852	1.0452	−1.9108	0.6475
0.853	1.0496	−1.9176	0.6511
0.854	1.0539	−1.9244	0.6546
0.855	1.0583	−1.9313	0.6582
0.856	1.0627	−1.9382	0.6618
0.857	1.0671	−1.9452	0.6653
0.858	1.0716	−1.9522	0.6690
0.859	1.0760	−1.9593	0.6726
0.860	1.0805	−1.9664	0.6762
0.861	1.0850	−1.9735	0.6798
0.862	1.0896	−1.9808	0.6835
0.863	1.0941	−1.9880	0.6872
0.864	1.0987	−1.9954	0.6908
0.865	1.1033	−2.0028	0.6945
0.866	1.1079	−2.0102	0.6982
0.867	1.1126	−2.0177	0.7020
0.868	1.1172	−2.0252	0.7057
0.869	1.1219	−2.0328	0.7094
0.870	1.1266	−2.0405	0.7132
0.871	1.1314	−2.0482	0.7170
0.872	1.1361	−2.0560	0.7208
0.873	1.1409	−2.0639	0.7246
0.874	1.1458	−2.0718	0.7284
0.875	1.1506	−2.0797	0.7322
0.876	1.1555	−2.0878	0.7361
0.877	1.1604	−2.0959	0.7400
0.878	1.1653	−2.1040	0.7439
0.879	1.1703	−2.1123	0.7478
0.880	1.1753	−2.1206	0.7517
0.881	1.1803	−2.1289	0.7556
0.882	1.1853	−2.1374	0.7596

Table 3 *cont.*

M(.)	X_n	X_c	X_w
0.883	1.1904	−2.1459	0.7636
0.884	1.1955	−2.1545	0.7676
0.885	1.2007	−2.1631	0.7716
0.886	1.2058	−2.1719	0.7756
0.887	1.2110	−2.1807	0.7796
0.888	1.2163	−2.1896	0.7837
0.889	1.2216	−2.1986	0.7878
0.890	1.2269	−2.2076	0.7919
0.891	1.2322	−2.2167	0.7960
0.892	1.2376	−2.2260	0.8002
0.893	1.2430	−2.2353	0.8044
0.894	1.2484	−2.2447	0.8086
0.895	1.2539	−2.2541	0.8128
0.896	1.2595	−2.2637	0.8170
0.897	1.2650	−2.2734	0.8213
0.898	1.2706	−2.2831	0.8256
0.899	1.2763	−2.2930	0.8299
0.900	1.2819	−2.3030	0.8342
0.901	1.2877	−2.3130	0.8386
0.902	1.2934	−2.3232	0.8429
0.903	1.2992	−2.3334	0.8473
0.904	1.3051	−2.3438	0.8518
0.905	1.3110	−2.3543	0.8562
0.906	1.3169	−2.3649	0.8607
0.907	1.3229	−2.3756	0.8652
0.908	1.3290	−2.3864	0.8698
0.909	1.3351	−2.3973	0.8743
0.910	1.3412	−2.4084	0.8789
0.911	1.3474	−2.4195	0.8836
0.912	1.3536	−2.4308	0.8882
0.913	1.3599	−2.4423	0.8929
0.914	1.3663	−2.4538	0.8977
0.915	1.3727	−2.4655	0.9024
0.916	1.3791	−2.4774	0.9072
0.917	1.3857	−2.4894	0.9120
0.918	1.3922	−2.5015	0.9169
0.919	1.3989	−2.5138	0.9218
0.920	1.4056	−2.5262	0.9267
0.921	1.4123	−2.5388	0.9317
0.922	1.4192	−2.5515	0.9367
0.923	1.4261	−2.5644	0.9417
0.924	1.4330	−2.5775	0.9468
0.925	1.4401	−2.5908	0.9520
0.926	1.4472	−2.6042	0.9571
0.927	1.4544	−2.6178	0.9623
0.928	1.4616	−2.6316	0.9676
0.929	1.4689	−2.6456	0.9729
0.930	1.4764	−2.6598	0.9782
0.931	1.4839	−2.6742	0.9836

Table 3 *cont.*

M(.)	X_n	X_e	X_w
0.932	1.4914	−2.6888	0.9891
0.933	1.4991	−2.7036	0.9946
0.934	1.5069	−2.7187	1.0001
0.935	1.5147	−2.7339	1.0057
0.936	1.5226	−2.7495	1.0114
0.937	1.5307	−2.7652	1.0171
0.938	1.5388	−2.7812	1.0229
0.939	1.5471	−2.7975	1.0287
0.940	1.5554	−2.8140	1.0346
0.941	1.5639	−2.8308	1.0406
0.942	1.5724	−2.8480	1.0466
0.943	1.5811	−2.8654	1.0527
0.944	1.5899	−2.8831	1.0589
0.945	1.5989	−2.9011	1.0651
0.946	1.6079	−2.9195	1.0714
0.947	1.6171	−2.9382	1.0778
0.948	1.6265	−2.9572	1.0843
0.949	1.6359	−2.9767	1.0908
0.950	1.6456	−2.9965	1.0974
0.951	1.6554	−3.0167	1.1042
0.952	1.6653	−3.0373	1.1110
0.953	1.6754	−3.0584	1.1179
0.954	1.6857	−3.0799	1.1249
0.955	1.6962	−3.1019	1.1320
0.956	1.7068	−3.1244	1.1392
0.957	1.7177	−3.1474	1.1466
0.958	1.7287	−3.1710	1.1540
0.959	1.7400	−3.1951	1.1616
0.960	1.7515	−3.2198	1.1693
0.961	1.7632	−3.2452	1.1772
0.962	1.7752	−3.2712	1.1851
0.963	1.7875	−3.2978	1.1933
0.964	1.8000	−3.3253	1.2016
0.965	1.8128	−3.3535	1.2100
0.966	1.8259	−3.3825	1.2186
0.967	1.8393	−3.4124	1.2274
0.968	1.8531	−3.4432	1.2364
0.969	1.8672	−3.4750	1.2456
0.970	1.8818	−3.5078	1.2550
0.971	1.8967	−3.5417	1.2646
0.972	1.9120	−3.5769	1.2745
0.973	1.9279	−3.6133	1.2846
0.974	1.9442	−3.6511	1.2950
0.975	1.9610	−3.6904	1.3057
0.976	1.9785	−3.7313	1.3167
0.977	1.9965	−3.7739	1.3281
0.978	2.0152	−3.8184	1.3398
0.979	2.0347	−3.8650	1.3520
0.980	2.0550	−3.9139	1.3645

Table 3 *cont.*

$M(.)$	X_n	X_e	X_w
0.981	2.0761	−3.9653	1.3776
0.982	2.0982	−4.0195	1.3911
0.983	2.1214	−4.0767	1.4053
0.984	2.1458	−4.1375	1.4201
0.985	2.1715	−4.2022	1.4356
0.986	2.1988	−4.2714	1.4519
0.987	2.2278	−4.3457	1.4692
0.988	2.2588	−4.4260	1.4875
0.989	2.2921	−4.5133	1.5070
0.990	2.3282	−4.6089	1.5280
0.991	2.3676	−4.7147	1.5507
0.992	2.4111	−4.8330	1.5755
0.993	2.4596	−4.9672	1.6029
0.994	2.5148	−5.1222	1.6336
0.995	2.5788	−5.3058	1.6688
0.996	2.6556	−5.5308	1.7103
0.997	2.7523	−5.8217	1.7616
0.998	2.8844	−6.2335	1.8299

Gamma function

B	E(X)/A	V(X)/A²	B	E(X)/A	V(X)/A²
0.5	2.000	20.000	5.4	0.922	0.039
0.6	1.505	6.997	5.5	0.923	0.038
0.7	1.266	3.427	5.6	0.924	0.037
0.8	1.133	2.040	5.7	0.925	0.036
0.9	1.052	1.372	5.8	0.926	0.035
1.0	1.000	1.000	5.9	0.927	0.034
1.1	0.965	0.771	6.0	0.927	0.033
1.2	0.941	0.620	6.1	0.928	0.032
1.3	0.924	0.513	6.2	0.929	0.031
1.4	0.911	0.435	6.3	0.930	0.030
1.5	0.903	0.376	6.4	0.931	0.029
1.6	0.897	0.329	6.5	0.932	0.028
1.7	0.892	0.292	6.6	0.932	0.028
1.8	0.889	0.261	6.7	0.933	0.027
1.9	0.887	0.236	6.8	0.934	0.026
2.0	0.886	0.215	6.9	0.934	0.026
2.1	0.886	0.196	7.0	0.935	0.025
2.2	0.886	0.181	7.1	0.936	0.024
2.3	0.886	0.167	7.2	0.937	0.024
2.4	0.886	0.155	7.3	0.937	0.023
2.5	0.887	0.144	7.4	0.938	0.023
2.6	0.888	0.135	7.5	0.938	0.022
2.7	0.889	0.126	7.6	0.939	0.022
2.8	0.890	0.119	7.7	0.940	0.021
2.9	0.892	0.112	7.8	0.940	0.021
3.0	0.893	0.106	7.9	0.941	0.020
3.1	0.894	0.100	8.0	0.941	0.020
3.2	0.896	0.095	8.1	0.942	0.019
3.3	0.897	0.090	8.2	0.943	0.019
3.4	0.898	0.085	8.3	0.943	0.019
3.5	0.900	0.081	8.4	0.944	0.018
3.6	0.901	0.078	8.5	0.944	0.018
3.7	0.902	0.074	8.6	0.945	0.018
3.8	0.904	0.071	8.7	0.945	0.017
3.9	0.905	0.068	8.8	0.946	0.017
4.0	0.906	0.065	8.9	0.946	0.017
4.1	0.908	0.062	9.0	0.947	0.016
4.2	0.909	0.060	9.1	0.947	0.016
4.3	0.910	0.057	9.2	0.948	0.016
4.4	0.911	0.055	9.3	0.948	0.015
4.5	0.912	0.053	9.4	0.949	0.015
4.6	0.914	0.051	9.5	0.949	0.015
4.7	0.915	0.049	9.6	0.949	0.014
4.8	0.916	0.048	9.7	0.950	0.014
4.9	0.917	0.046	9.8	0.950	0.014
5.0	0.918	0.045	9.9	0.951	0.014
5.1	0.919	0.043	10.0	0.951	0.013
5.2	0.920	0.042	10.1	0.951	0.013
5.3	0.921	0.040	10.2	0.952	0.013

Table 4 *cont.*

B	E(X)/A	V(X)/A²	B	E(X)/A	V(X)/A²
10.3	0.952	0.013	15.2	0.966	0.007
10.4	0.953	0.013	15.3	0.966	0.006
10.5	0.953	0.012	15.4	0.966	0.006
10.6	0.953	0.012	15.5	0.966	0.006
10.7	0.954	0.012	15.6	0.966	0.006
10.8	0.954	0.012	15.7	0.967	0.006
10.9	0.954	0.012	15.8	0.967	0.006
11.0	0.955	0.011	15.9	0.967	0.006
11.1	0.955	0.011	16.0	0.967	0.006
11.2	0.955	0.011	16.1	0.967	0.006
11.3	0.956	0.011	16.2	0.968	0.006
11.4	0.956	0.011	16.3	0.968	0.006
11.5	0.956	0.011	16.4	0.968	0.006
11.6	0.957	0.010	16.5	0.968	0.006
11.7	0.957	0.010	16.6	0.968	0.006
11.8	0.957	0.010	16.7	0.968	0.006
11.9	0.958	0.010	16.8	0.969	0.005
12.0	0.958	0.010	16.9	0.969	0.005
12.1	0.958	0.010	17.0	0.969	0.005
12.2	0.959	0.010	17.1	0.969	0.005
12.3	0.959	0.009	17.2	0.969	0.005
12.4	0.959	0.009	17.3	0.969	0.005
12.5	0.959	0.009	17.4	0.970	0.005
12.6	0.960	0.009	17.5	0.970	0.005
12.7	0.960	0.009	17.6	0.970	0.005
12.8	0.960	0.009	17.7	0.970	0.005
12.9	0.960	0.009	17.8	0.970	0.005
13.0	0.961	0.009	17.9	0.970	0.005
13.1	0.961	0.008	18.0	0.970	0.005
13.2	0.961	0.008	18.1	0.971	0.005
13.3	0.961	0.008	18.2	0.971	0.005
13.4	0.962	0.008	18.3	0.971	0.005
13.5	0.962	0.008	18.4	0.971	0.005
13.6	0.962	0.008	18.5	0.971	0.005
13.7	0.962	0.008	18.6	0.971	0.005
13.8	0.963	0.008	18.7	0.971	0.005
13.9	0.963	0.008	18.8	0.972	0.005
14.0	0.963	0.008	18.9	0.972	0.004
14.1	0.963	0.007	19.0	0.972	0.004
14.2	0.964	0.007	19.1	0.972	0.004
14.3	0.964	0.007	19.2	0.972	0.004
14.4	0.964	0.007	19.3	0.972	0.004
14.5	0.964	0.007	19.4	0.972	0.004
14.6	0.964	0.007	19.5	0.972	0.004
14.7	0.965	0.007	19.6	0.973	0.004
14.8	0.965	0.007	19.7	0.973	0.004
14.9	0.965	0.007	19.8	0.973	0.004
15.0	0.965	0.007	19.9	0.973	0.004
15.1	0.966	0.007	20.0	0.973	0.004

Kolmogorov–Smirnov test

n	dp			n	dp		
	$p=0.90$	$p=0.95$	$p=0.99$		$p=0.90$	$p=0.95$	$p=0.99$
1	0.9500	0.9750	0.9950	25	0.2377	0.2640	0.3166
2	0.7764	0.8419	0.9293	26	0.2332	0.2591	0.3106
3	0.6360	0.7076	0.8290	27	0.2290	0.2544	0.3050
4	0.5652	0.6239	0.7342	28	0.2250	0.2499	0.2997
5	0.5094	0.5633	0.6685	29	0.2212	0.2457	0.2947
6	0.4680	0.5193	0.6166	30	0.2176	0.2417	0.2899
7	0.4361	0.4834	0.5758	31	0.2141	0.2379	0.2853
8	0.4096	0.4543	0.5418	32	0.2108	0.2342	0.2809
9	0.3875	0.4300	0.5133	33	0.2077	0.2308	0.2768
10	0.3687	0.4092	0.4889	34	0.2047	0.2274	0.2728
11	0.3524	0.3912	0.4677	35	0.2018	0.2242	0.2690
12	0.3382	0.3754	0.4490	36	0.1991	0.2212	0.2653
13	0.3255	0.3614	0.4325	37	0.1965	0.2183	0.2618
14	0.3142	0.3489	0.4176	38	0.1939	0.2154	0.2584
15	0.3040	0.3376	0.4042	39	0.1915	0.2127	0.2552
16	0.2947	0.3273	0.3920	40	0.1891	0.2101	0.2521
17	0.2863	0.3180	0.3809	45	0.1786	0.1984	0.2380
18	0.2785	0.3094	0.3706	50	0.1696	0.1884	0.2260
19	0.2714	0.3014	0.3612	60	0.1551	0.1723	0.2067
20	0.2647	0.2941	0.3524	70	0.1438	0.1587	0.2521
21	0.2586	0.2872	0.3443	80	0.1347	0.1496	0.1795
22	0.2528	0.2809	0.3367	90	0.1271	0.1412	0.1694
23	0.2475	0.2749	0.3295	100	0.1207	0.1340	0.1608
24	0.2424	0.2693	0.3229				

References

Abramovitz, M. and Stegan, A.I. (1964) *Handbook of Mathematical Functions*, Dover, New York.

Anderson, R.T. and Neri, L. (1990) *Reliability-Centred Maintenance: Management and Engineering Methods*, Elsevier Science Publishers, London.

Blanchard, B.S. (1976) *Engineering Organization and Management*, Prentice-Hall, New Jersey.

Blanchard, B.S. (1986) *Logistics Engineering and Management*, 3rd edition, Prentice-Hall, New Jersey.

Blanchard, B.S. (1991) *Systems Engineering Management*, John Wiley & Sons, New York.

Blanchard, B.S., Verma, D. and Peterson, E.L. (1995) *Maintainability*, John Wiley & Sons, New York.

CADAM-ADAM (1985) *Computer Aided Design and Manufacturing – Anthropometric Design Aid Manikins*, Cadam Inc., Burbank, California.

CARD (1987) *Center for Anthropometric Research Data*, Armstrong Aerospace Medical Research Laboratory (AAMRI)/Human Engineering Group (HEG), Wright–Patterson AFB, Ohio.

Cox, D.R. (1962) *Renewal Theory*, Methuen, London.

Friend, C.H. (1992) *Aircraft Maintenance Management*, Longman Scientific & Technical, Harlow.

Gleick, J. (1993) *Genius, The Life and Science of Richard Feynman*, Pantheon Books, New York.

Hunt, S. (1992) Maintainability allocation – An alternative technique, in *Proceedings International Symposium, Advances in Logistics* (ed. J. Knezevic), Research Centre MIRCE, Exeter, UK, pp. 141–51.

Hunt, S. (1993) Time dependent maintainability analysis, in *Proceedings International Symposium, Advances in Logistics* (ed. J. Knezevic), Research Centre MIRCE, Exeter, UK, pp. 60–7.

International Atomic Energy Association (1993) *IAEA Training Courses Series, Handbook on Safety Related Maintenance* (draft), IAEA, Vienna.

Johnson, M. (1993) Boeing 777 airplane information management system–philosophy and display. *Royal Aeronautical Conference Proceedings*, Advanced Avionics on the Airbus A330/A340 and the Boeing 777 Aircraft, 17 November 1993.

Knezevic, J. (1985) Methodology 'EXETER' for selecting an optimal part-replacement policy. *Maintenance Management International*, **5**, 209–18.

Knezevic, J. (1990) Reliability and maintenance: fundamentals. Lecture Notes, School of Engineering, University of Exeter, UK.

Knezevic, J. (1993) *Reliability, Maintainability and Supportability Engineering – A Probabilistic Approach*, McGraw-Hill, London, 292 pp. plus software PROBCHAR.

Knezevic, J. (1994) Maintainability prediction at the design stage. *Journal of Communications in Reliability, Maintainability and Supportability*, **1**(1), 24–9.

Knezevic, J. (1994) Life cycle engineering. Lecture Notes, School of Engineering, University of Exeter, UK.

Knezevic, J. (1995a) Effective analysis of existing maintainability data. *Journal of Communications in Reliability, Maintainability and Supportability*, **2**(1), 18–22.

Knezevic, J. (1995b) Operation and maintenance. Lecture Notes, School of Engineering, University of Exeter, UK.

MIL-H-46855B (1981) *Human Engineering Requirements for Military Systems*, Data Item Description DI-H-7057 Human Engineering Design Approach Document-Maintainer, Department of Defense, Washington, DC.

MIL-HDBK-759A (1987) *Human Engineering Design for Army Material*, Department of Defense, Washington, DC.

MIL-STD-721C (1966) *Military Standard, Definitions of Effectiveness Terms for Reliability, Maintainability, Human Factors and Safety*, Department of Defense, Washington, DC.

MIL-STD-1472C (1979) (D to be released), *Human Engineering Design Criteria for Military Systems, Equipment and Facilities*, Department of Defense, Washington, DC.

Mobley, R.K. (1990) *An Introduction to Predictive Maintenance*, Van Nostrand Reinhold, New York.

Niczyporuk, Z.T. (1994) Role of technical diagnostics in improvements of safety in coal mines, in *Condition Monitoring '94, Proceedings International Conference on Condition Monitoring*, Swansea, UK, pp. 34–50.

Patton, J.D. (1983) *Preventive Maintenance*, Instrument Society of America, Research Triangle Park, North Carolina.

Patton, J.D. (1988) *Maintainability and Maintenance Management*, 2nd edition, Instrument Society of America, Research Triangle Park, North Carolina.

Further reading

Alexander, M., Garrett, J.W. and Flanner, M.P. (1969) *Anthropometric Dimensions of Air Force Personnel for Workspace and Design Criteria*, Wright–Patterson AFB, Ohio, Aerospace Medical Research Laboratory (AMRL-TM-69-6).

Al-Najjar, B. (1991) On the selection of condition-based maintenance for mechanical systems, in *Operational Reliability and Systematic Maintenance*, (eds K. Holmberg and A. Folkeson), Elsevier Science Publisher, UK, pp. 153–73.

Army Regulation 602–2 (1979) *Manpower and Personnel Integration (MANPRINT) in Material Acquisition Process*, MRSA, Lexington, USA.

Barlow, R.E. and Proschan, F. (1975) *Statistical Theory of Reliability and Life Testing*, Holt, Rhinehart & Winston, New York.

Biferno, M. (1991) *Computerised Design Using Human Model Simulation: A Need for Standards*, SAE, Chicago.

Billinton, A. (1983) *Reliability Evaluation of Engineering Systems Concept and Techniques*, Pitman Books, London.

Blanchard, B.S. and Lowery, E.E. (1969) *Maintainability Principles and Practices*, McGraw-Hill, New York.

Blanchard, B.S. and Fabrycky, W.J. (1981) *Systems Engineering and Analysis*, Prentice-Hall, New Jersey.

Bland, R. and Knezevic, J. (1987) *A practical application of a new method for condition-based maintenance. Maintenance Management International*, **7**(1), 31–7.

Bogdanoff, J.L. and Kozin, F. (1985) *Probabilistic Models of Cumulative Damage*, John Wiley & Sons, New York.

British Standard BS 3811 (1983) *Glossary of Maintenance Management Terms in Terotechnology*, British Standards Institution, London.

British Standard BS 4778 (1991) *Section 3.2 Glossary of International Terms*, British Standards Institution, London.

British Standard BS 5760 (1981) *Reliability of Systems, Equipment and Components*, British Standards Institution, London.

British Standard BS 6548 part 2 (1992) *Guide to Maintainability Studies During the Design Phase*. British Standards Institutes, London.

Carter, A.D.S. (1972) *Mechanical Reliability*, Macmillan, London.

Carter, G.M. Henshall, J.L. and Wakeman, R.J. (1991) Influence of surfactants on the mechanical properties and comminution of wet-milled calcite. *Powder Technology*, **65**, 403–410.

Collacott, R.A. (1977) *Mechanical Fault Diagnosis and Condition Monitoring*, Chapman & Hall, London.

Collins, J.A. (1981) *Failure of Materials in Mechanical Design; Analysis, Prediction, Prevention*, John Wiley & Sons, New York.

Cox, D.R. and Miller, H.D. (1965) *The Theory of Stochastic Processes*, Methuen, London.

CREW CHIEF (1983) *A Computer Graphics Model of a Maintenance Technician*,

Armstrong Aerospace Medical Research Laboratory (AAMRL)/Human Engineering Group (HEG), Wright–Patterson AFB, Ohio.

Cunningham, C.E. and Cox, W. (1972) *Applied Maintainability Engineering*, John Wiley & Sons, New York.

DEF STAN 00-25 (1989) Human Factors for Designers of Equipment, MoD, London.

DEF STAN 00-41 *MoD Practices and Procedures for Reliability and Maintainability*, Issue 3, MoD, London.

Dekker, P. (1994) *Applications of Maintenance Optimization Models: A Review and Analysis*, Report 9228/A, Econometric Institute, Erasmus University Rotterdam, The Netherlands.

Dhillon, B.S. and Chanan Singh (1981) *Engineering Reliability, New Techniques and Applications*, John Wiley & Sons, New York.

DOD Directive 5000.40 (1991) *Reliability and Maintainability*, Department of Defense, Washington, DC.

DOD-HDBK-743 (1981) Anthropometry of U.S. Military Personnel, Department of Defense, Washington, DC.

Dreyfuss, H. (1967) *The Measure of Man*, Whitney Library of Design, New York.

Einstein, A. (1991) *The World As I See It*, translated by Alan Harris, Citadel Press Books, New York.

El-Haram, M. and Knezevic, J. (1994) The new developments in condition-based approach to reliability, in *Proceedings 10th International Logistics Congress*, Exeter, UK, pp. 163–9.

El-Haram, M. and Knezevic, J. (1994) Practical application of the condition-based approach to monitoring particle size distribution, in *Proceedings Condition Monitoring and Diagnostic Engineering Management*, COMADEM '94, New Delhi, pp. 162–8.

Gaertner, J.P. (1989) *Demonstration of Reliability-Centred Maintenance*, Electric Power Research Institute, Palo Alto, California.

Garrett, J.W. (1971) *A Collation of Anthropometry*. Wright–Patterson AFB, Ohio, Wright Air Development Center (AMRL-TR-68-1, 2 vols).

Gertsbakh, I.B. and Kordonskiy, Kh.B. (1969) *Models of Failure*, Springer-Verlag, New York.

Gits, C.W. (1986) On the maintenance concept for a technical system: II. Literature review. *Maintenance Management International*, **6**, 181–96.

Goldman, A. and Slattery, T. (1967) *Maintainability – A Major Element of System Effectiveness*, John Wiley & Sons, New York.

Hahn, G.J. and Shapiro, S. (1967) *Statistical Models in Engineering*, John Wiley & Sons, New York.

HUMANSCALE 1/2/3 (1974) Henry Dreyfuss and Associates, Massachusetts Institute of Technology, MIT Press, Cambridge, Mass.

Hurd, W.L. (1962) Full design definition is essential to missile maintainability, in *Proceedings Eighth Annual Maintainability Quality Control National Symposium*, Washington, D.C.

Irving, C. (1993) *Wide-Body, The Making of the 747*, Hodder & Stoughton, London.

Jardine, A.K.S. (1973) *Maintenance, Replacement, and Reliability*, A Halsted Press Book, John Wiley & Sons, New York.

Johnson, F.J. (1984) *Anthropometry of the Clothes U.S. Army Ground Troop and Combat Vehicle Crewman*, Natick MS: US Army Natick Research & Development Center (TR-84/034).

Jones, P.F. (1992) *CAD/CAM: Features, Applications and Management*. Macmillan, Hong Kong.

Kapur, K.C. and Lamberson, L.R. (1977) *Reliability in Engineering Design*, John Wiley & Sons, New York.

Knezevic, J. (1987) Condition parameter based approach to calculation of reliability characteristics. *Reliability Engineering*, **19**(1), 29–39.

Knezevic, J. (1987) On the application of a condition parameter based reliability approach to pipeline item maintenance, in *Proceedings 2nd International Conference on Pipes, Pipeline and Pipe Line Items*, Utrecht, The Netherlands, June.

Knezevic, J. and Henshall, L. (1988) Estimation of the design life for notched components subjected to creep cracking using a condition parameter based reliability approach, in *Proceedings Conference 'Materials and Engineering Design'*, London, 9–13 May 1988, pp. 193–201.

Kolmogorov, A.N. (1950) *Foundation of the Theory of Probability*, Chelsea Publishing Company, New York.

Lansdown, A.R. and Price, A.L. (1986) *Materials to Resist Wear: A Guide to their Selection and Use*, Pergamon Press, New York.

Lubelsky, B.L. (1962) Maintainability techniques in production. Paper presented at *Seventh Military–Industry Missile and Space Maintainability Symposium*, San Diego, California.

Lyderson, S. (1988) Reliability testing based on deterioration measurements, PhD. Thesis, Trondheim, Norway.

Malik, M.A.K. (1979) Reliable preventive maintenance scheduling. *AIIE Transactions*, **11**, 221–8.

Mann, L. (1976) *Maintenance Management*, Lexington Books, D.C. Heath & Co., Lexington, Massachusetts.

MIL-HDBK-472 (1969) *Military Handbook, Maintainability Prediction*, Department of Defense, Washington, DC.

MIL-STD-470B (1983) *Military Standard, Maintainability Program for Systems and Equipment*, Department of Defense, Washington, DC.

MIL-STD-471A (1969) *Military Standard, Maintainability Verification, Demonstration, Evaluation*, Department of Defense, Washington, DC.

MIL-STD-2084 (1983) *Military Standard, General Requirements for Maintainability*, Department of Defense, Washington, DC.

MIL-STD-2165 (1976) *Military Standard, Testability Program for Electronic Systems and Equipments*, Department of Defense, Washington, DC.

Missile Systems Division Maintainability Engineering Manual (1963) Lockheed Missiles and Space Co. Rept. 801178, Aug. 1.

Moss, M.A. (1985) *Designing for Minimal Maintenance Expense: The Practical Application of Reliability and Maintainability*, Marcel Dekker, New York.

Nakajima, S. (1988) *Total Productive Maintenance (TPM)*, Productivity Press, Cambridge, Massachusetts.

Nakajima, S. (ed.) (1989) *TPM Development Program: Implementing Total Productive Maintenance*, Productivity Press, Cambridge, Massachusetts.

Neale, M.J. (1987) Trends in maintenance and condition monitoring, in *Condition Monitoring '87, Proceedings International Conference on Condition Monitoring*, Swansea, UK, pp. 2–12.

Newborough, E.T. (1967) *Effective Maintenance Management*, McGraw-Hill, New York.

Nicholas, C. (1994) Life cycle cost. Course notice for the 4th International Industrial Summer School, Research Centre MIRCE, University of Exeter, UK.

Niebel, B.W. (1985) *Engineering Maintenance Management*, Marcel Dekker, New York.

Onsoyen, E. (1990) Accelerated testing and mechanical evaluation of components exposed to wear, PhD. Thesis, University of Trondheim, The Norwegian Institute of Technology, Norway.

Parzen, E. (1962) *Stochastic Processes*, Holden–Day, New York.

Paulsen, J.L. and Lauridsen, K. (1991) Information flow in a decision support system for maintenance planning, in *Operational Reliability and Systematic Maintenance* (eds K. Holmberg and A. Folkeson), Elsevier Science Publishers, UK, pp. 261–70.

Petroski, H. (1985) *To Engineer is Human*, Macmillan, London.

Roebuck, J.A., Kroemer, K.H.E. and Thomson, E.G. (1975) *Engineering Anthropometry Methods*, John Wiley & Sons, New York.

Sanders, M.S. (1983) *U.S. Truck Driver Anthropometric Truck Workspace Data Survey*. Final Report, Society of Automotive Engineers, Warrendale, Pennsylvania.

Sanders, M.S. and McCormick, E.J. (1993) *Human Factors in Engineering and Design*, 7th edition, McGraw-Hill, New York.

Sanders, M.S. and Shaw, B.E. (1984) *Female U.S. Truck Driver Anthropometric Truck Workspace Data Survey*, SAE, Chicago.

Scully, J.C. (1990) *The Fundamentals of Corrosion*, Pergamon Press, Oxford.

Sherif, Y.S. and Smith, M.L. (1981) Optimal maintenance models for systems subject to failure – a review. *Naval Research Logistics Quarterly*, **28**, 47–74.

Smiley, R.W. (1963) Design review and field feedback trigger system improvement. *Transactions Joint AIAA, SAE, ASME Aerospace Maintainability Conference*, Washington, D.C.

Standtorv, H. and Rausand, M. (1991) RCM – Closing the loop between design reliability and operational reliability. *Maintenance Journal*, **6**(1), 13–21.

Sternstein, E. and Gold, T. (1991) *From Takeoff To Landing*, Pocket Books, Simon & Schuster, New York.

Willmott, P. (1989) Maintenance engineering in Europe – the scope for collaborative technology transfer and joint venture. *Maintenance Journal*, **4**(4), 10–13.

Yan, X.P. (1994) Oil monitoring based condition maintenance management, in *Proceedings Condition Monitoring and Diagnostic Engineering Management, COMADEM '94*, New Delhi, pp. 154–61.